Role of Nanomaterials in Sustainable Agriculture

Nanomaterials have rapidly emerged as a revolutionary tool across diverse fields including modern agriculture. The properties of nanomaterials, such as size in nanometers and high surface area-to-volume ratio, offer innovative solutions to enable nutrient uptake and targeted and efficient release to tissues to aid in enhancing nutrient assimilation and metabolism. These formulations enhance crop productivity, resilience to climatic change, disease management, and stress tolerance. This book provides a comprehensive and up-to-date elucidation of nanomaterials in sustainable agriculture with a strong emphasis on green synthesis approaches and their applications. The distinguishing feature of this book lies in its integrative action, plant–nanomaterial interactions, uptake and translocation, and plant responses at cellular, physiological, and biochemical levels. By combining mechanistic insights with applied perspectives, the book bridges fundamental research and field-level applications, offering a holistic understanding of advancements in the delivery system of nano-fertilizers and the significance of nanomaterials in plant growth, development, and stress tolerance.

Key Features:

- Current status of nanomaterials in the agricultural system
- Green synthesis routes and strategies for characterization of nanomaterials
- Uptake, translocation, and interaction with plant tissues
- Impact of nanomaterials on plant growth, development, and physiology
- Significance of nanomaterials in mitigating biotic and abiotic stresses
- Recent advances and applications in nano-enabled delivery systems
- Phytotoxicity, biosafety, and risk assessment of agricultural nanomaterials

This book is anticipated by postgraduate students, academicians, researchers, and professionals in fields of agriculture, plant sciences, plant stress management, and environmental biotechnology. It will also serve as a valuable reference for industry partners seeking sustainable and safe nanotechnology-based agricultural solutions.

Role of Nanomaterials in Sustainable Agriculture

Edited by
Priyanka Sharma, Renu Bhardwaj,
and Anil Kumar

CRC Press
Taylor & Francis Group
Boca Raton London New York

CRC Press is an imprint of the
Taylor & Francis Group, an **informa** business

Designed cover image: Getty

First edition published 2027
by CRC Press
2385 NW Executive Center Drive, Suite 320, Boca Raton FL 33431

and by CRC Press
4 Park Square, Milton Park, Abingdon, Oxon, OX14 4RN

CRC Press is an imprint of Taylor & Francis Group, LLC

Library of Congress CataloginginPublication Data
Names: Sharma, Priyanka, (Professor of bioengineering) editor | Bhardwaj,
Renu editor | Kumar, Anil, (expert in vaccine biotechnology) editor
Title: Role of nanomaterials in sustainable agriculture / edited by
Priyanka Sharma, Renu Bhardwaj, Anil Kumar.
Description: First edition. | Boca Raton, FL : CRC Press, 2026. |
Includes bibliographical references and index.
Identifiers: LCCN 2026014537 (print) | LCCN 2026014538 (ebook) |
ISBN 9781032659404 hardback | ISBN 9781032659411 paperback |
ISBN 9781032659435 ebook
Subjects: LCSH: Nanostructured materials | Agricultural innovations
Classification: LCC TA418.9.N35 R65 2026 (print) | LCC TA418.9.N35 (ebook)
LC record available at https://lccn.loc.gov/2026014537
LC ebook record available at https://lccn.loc.gov/2026014538

ISBN: 9781032659404 (hbk)
ISBN: 9781032659411 (pbk)
ISBN: 9781032659435 (ebk)

DOI: 10.1201/9781032659435

Typeset in Times
by codeMantra

Contents

Preface

Worldwide climate change driven by intensifying anthropogenic burdens on natural resources has accentuated challenges in modern agriculture. Overreliance on chemical fertilizers, herbicides, and pesticides has degraded soil and ecosystem health, imposing sustainable and climate-resilient technological solutions.

Nanotechnology has emerged as a promising, transformative interdisciplinary field with significant potential in agricultural systems. The inherent physical and chemical properties of nanomaterials, like high surface area, reactivity, and controlled release, offer new prospects to improve nutrient-uptake efficiency, crop protection, and better yield through precise modulation of agrochemicals. Nano-enabled fertilizers, pesticides, and delivery systems are advancing precision agriculture by enabling targeted, site-specific, and efficient application of active ingredients.

This book, *Role of Nanomaterials in Sustainable Agriculture*, provides a comprehensive overview of the current status and emerging applications of nanomaterials in agriculture, with a strong emphasis on sustainability and climate resilience. This book covers fundamental aspects including green synthesis and characterization of nanoparticles, their uptake, translocation within plant systems, and their role in mitigating abiotic and biotic stresses. Special attention is given to smart nanotechnology interdisciplinary approaches to enhance nanoparticle-mediated delivery and utilization of nutrients to crops and ensure food safety through nano-sensors. Each chapter includes high-quality illustrations to enhance understanding of the topics.

This book is intended for plant scientists, agricultural biotechnologists, environmental researchers, nano-biotechnologists, research scholars, and postgraduate students. We gratefully acknowledge the contributions of all authors and thank the Taylor & Francis team for their support throughout the publication process. We hope this volume will stimulate interdisciplinary research and contribute meaningfully to the advancement of nano-enabled sustainable agriculture.

About the Editors

Priyanka Sharma is an associate professor at Bioengineering Sciences & Research, MIT-Art, Design and Technology University, Pune. She has over 15 years of research and teaching experience in phytoremediation, plant stress physiology, heavy metal detoxification, advanced molecular biology, cell biology, and microbial bioengineering for sustainable agriculture. She obtained her Ph.D. in botany from Guru Nanak Dev University, Amritsar, and has completed prestigious postdoctoral fellowships under the UGC–Dr. D.S. Kothari and DBT-BioCARe (Govt. of India) at Savitribai Phule Pune University. She is also mentoring several M.Sc, B.Tech, M.Tech, and Ph.D. students.

Her research focuses on plant host–microbial interactions, molecular mechanisms of heavy metal stress tolerance, secondary metabolites, nano-biofertilizer formulations, and metabolomic profiling of stress responses. She has authored plentiful high-impact research papers, book chapters, and textbooks, with over 1,500 citations (h-index: 15). She serves as an editorial member for *Frontiers in Microbiomes*, *Frontiers in Plant Science*, and *Current Indian Science* (Bentham) and as a reviewer in many leading international journals like *Acta Physiologiae Plantarum* (Springer), *Ecotoxicology and Environmental Safety* (Elsevier), *International Journal of Green Pharmacy*, *Journal of Hazardous Materials* (Elsevier), *Journal of Plant Growth Regulation*, *BMC Plant Biology*, and *Frontier Plant Science*.
She is the recipient of the Best Women Young Scientist Award (2020). She is a fellow of the Young Academy of India (2021). Her recent academic endeavors integrate climate-resilient and microbial roles in sustainable approaches in agribiotech, reflecting her commitment to advancing sustainable applied agribiotechnology.

Renu Bhardwaj is a distinguished plant scientist and academic leader with over four decades of experience in teaching, research, and higher education administration. She holds a Ph.D. in plant physiology from Guru Nanak Dev University (GNDU), Amritsar, where she has served in several key leadership roles, including Controller of Examinations, Director of Research, Dean of Life Sciences, and Director of the Internal Quality Assurance Cell (IQAC). Her academic career reflects an exceptional commitment to research excellence, institutional development, and quality assurance in higher education.

Her research expertise spans plant stress physiology, plant growth regulators, biochemistry, molecular biology, ecophysiology, phytoremediation, and environmental sciences. She has authored over 500 scholarly publications with over 28,855 citations (h-index: 82), including high-impact research articles and book chapters, and she has edited nine internationally published books with leading publishers such as Springer, Elsevier, Wiley, Nova Science, and CRC Press. She is also an accomplished inventor with three patents and four designs in the area of thermal insulating materials and soundproofing system. She has also guided 48 Ph.D. scholars and successfully completed more than 20 major funded research projects supported by national agencies including DST, DBT, CSIR, UGC, and MoEF.

An internationally recognized scholar, Bhardwaj is a Clarivate Highly Cited Researcher, is listed among the Top 2% of Scientists worldwide by Stanford–Elsevier, and ranks prominently in Scilit, Scholar GPS, AD Scientific Index, and Research.com. Her contributions have been acknowledged through numerous honors, including Lifetime Achievement Award, Scientist of the Year, and fellowships of the National Environmental Science Academy and the Linnean Society of London. She has served extensively as a NAAC Peer Team Coordinator and Member, contributing significantly to accreditation and quality enhancement across Indian higher education institutions. As an editor, she brings deep scientific insight, editorial rigor, and a strong interdisciplinary vision to advancing contemporary research in plant and environmental sciences.

Anil Kumar holds a Ph.D. in biotechnology from Savitribai Phule University in Pune, India. His research topic focused on studying *Bordetella pertussis* antigens and their nasal delivery, where he developed proteomics and immunoproteomics approaches to assess the quality of whole-cell pertussis vaccine and encapsulated aP antigens using PLGA microparticles to develop a nasal pertussis vaccine.

He is a highly accomplished professional in the field of biotechnology with a specialization in new product development and analytical method design for biologicals and immunobiologicals. He has over 23 years of industry experience in the field of vaccines in Pune, India. He has expertise in managing quality control laboratories and ensuring compliance with strict regulatory requirements. He has a proven track record in designing, developing, and validating quality control assays for various vaccines including acellular pertussis antigens and combination vaccines, inactivated polio vaccine (IPV), oral polio vaccine (OPV), and whole-cell pertussis-based combination vaccines.

In addition to his technical expertise, Kumar has extensive experience in collaborating with international regulatory agencies and ensuring compliance with standards set by organizations such as the World Health Organization (WHO), U.S. Food and Drug Administration (FDA), Medicines and Healthcare products Regulatory Agency (MHRA), and Therapeutic Goods Administration (TGA).

Contributors

Mohd Ali
Department of Zoology
Guru Nanak Dev University
Amritsar, India

Upma Arora
Department of Botany
Lyallpur Khalsa College
Jalandhar, India

Aagam Bamb
MIT School of Bioengineering Sciences & Research
MIT-Art, Design and Technology University
Pune, India

Renu Bhardwaj
University Centre for Research & Development
Chandigarh University
Mohali, India

Savita Bhardwaj
Department of Botany
MCM DAV College
Kangra, India

Tamanna Bhardwaj
Department of Botanical and Environmental Sciences
Guru Nanak Dev University
Amritsar, India

Harish Changotra
Department of Molecular Biology and Biochemistry
Guru Nanak Dev University
Amritsar, India

Pooja Deshmukh
CSIR-National Chemical Laboratory
Pune, India
and
Academy of Scientific and Innovative Research (AcSIR)
Ghaziabad, India

Satish U. Deshmukh
Department of Chemistry
Deogiri College
Chhatrapati Sambhajinagar, India

Ajit K. Dhas
Department of Chemistry
Deogiri College
Chhatrapati Sambhajinagar, India

Apurva Gosavi
MIT School of Bioengineering Sciences & Research
MIT-Art, Design and Technology University
Pune, India

Sarvesh Jadhav
MIT School of Bioengineering Sciences & Research
MIT-Art, Design and Technology University
Pune, India

Jahanvi
Science Department
Government College of Education, Jalandhar
Guru Nanak Dev University
Amristar, India

Jetina
Science Department
Government College of Education, Jalandhar
Guru Nanak Dev University
Amristar, India

Narendra Kadoo
CSIR-National Chemical Laboratory
Pune, India
and
Academy of Scientific and Innovative Research (AcSIR)
Ghaziabad, India

Amol Kahandal
MIT School of Bioengineering Sciences & Research
MIT-Art, Design and Technology Universitycopy
Pune, India

Anant B. Kanagare
Department of Chemistry
Deogiri College
Chhatrapati Sambhajinagar, India

Dhriti Kapoor
School of Biological and Environmental Sciences
Shoolini University
Solan, India

Gurneet Kaur
Department of Zoology
Government College of Education
Jalandhar, India

Gurvarinder Kaur
Department of Botany
Punjabi University
Patiala, India

Harsimran Kaur
Post Graduate Department of Botany
Khalsa College Amritsar
Amritsar, India

Manjot Kaur
Science Department
Government College of Education, Jalandhar
Guru Nanak Dev University
Amritsar, India

Simranpreet Kaur
Science Department
Government College of Education, Jalandhar
Guru Nanak Dev University
Amritsar, India

Talwinder Kaur
Department of Microbiology
Guru Nanak Dev University
Amritsar, India

Tarandeep Kaur
Science Department
Government College of Education, Jalandhar
Guru Nanak Dev University
Amristar, India

Laxmi G. Kathawate
Rayat Shikshan Sanstha's
Radhabai Kale Mahila Mahavidyalaya
Ahmednagar, India

Sukhmeen Kaur Kohli
Department of Earth and Climate Science
Indian Institute of Science, Education and Research (IISER)
Tirupati, India

Sandeep Kour
Department of Zoology
Guru Nanak Dev University
Amritsar, India

Anil Kumar
Serum Institute of India Ltd
Hadapsar, Pune, India

Deepak Kumar
Department of Zoology
Guru Nanak Dev University
Amritsar, India

Gitika Kumawat
School of Bioengineering Science & Research
MIT-Art, Design and Technology University
Pune, India

Isha Madaan
Department of Botany
Punjabi University
Patiala, India
and
Science Department
Government College of Education
Jalandhar, India

Varsha Arun Mahadik
CSIR-National Chemical Laboratory
Pune, India
and
Academy of Scientific and Innovative Research (AcSIR)
Ghaziabad, India

Rajesh Kumari Manhas
Department of Microbiology
Guru Nanak Dev University
Amritsar, India

Puja Ohri
Department of Zoology
Guru Nanak Dev University
Amritsar, India

Parkirti
Department of Zoology
Guru Nanak Dev University
Amritsar, India

Varsha Rattan
Faculty of Agricultural Sciences
DAV University
Jalandhar, India

Akanksha Sharma
Government College of Education, Jalandhar
Guru Nanak Dev University
Amritsar, India

Ashish Sharma
Department of Botany and Environment Studies
DAV University
Jalandhar, India

Ashutosh Sharma
Faculty of Agricultural Sciences
DAV University
Jalandhar, India

Dhriti Sharma
Siddarth Govt. College
Nadaun, Hamirpur, India

Indu Sharma
Department of Life Sciences, University Institute of Sciences
Sant Baba Bhag Singh University
Jalandhar, India

Nivedita Sharma
Science Department
Government College of Education, Jalandhar
Guru Nanak Dev University
Amristar, India

Pratibha Sharma
Faculty of Agricultural Sciences
DAV University
Jalandhar, India

Priyanka Sharma
MIT School of Bioengineering Sciences & Research
MIT-Art, Design and Technology University
Pune, India

Roohi Sharma
Department of Zoology
Guru Nanak Dev University
Amritsar, India

Sonika Sharma
Faculty of Agricultural Sciences
DAV University
Jalandhar, India

Vikas Sharma
Department of Molecular Biology and Genetic Engineering
Lovely Professional University
Jalandhar, India

Shamsher Singh
Department Agriculture
Government Brijindra College
Punjabi University
Patiala, India

Palampreet Singh
Department of Botany
Punjabi University
Patiala, India

Geetika Sirhindi
Department of Botany
Punjabi University
Patiala, India

Chandrakant K. Tagad
Department of Biochemistry
S.B.E.S. College of Science
Chhatrapati Sambhajinagar, India

Aparna S. Taware
Department of Botany
Deogiri College
Chhatrapati Sambhajinagar, India

Jeenia Thalyari
Faculty of Agricultural Sciences
DAV University
Jalandhar, India

Usha Thakur
Department of Botany
MCM DAV College
Kangra, India

Vikram
Department of Zoology
Guru Nanak Dev University
Amritsar, India

1 Current Status of Nanomaterials in the Agricultural System

Varsha Arun Mahadik, Pooja Deshmukh, Narendra Kadoo, Anil Kumar, and Priyanka Sharma

1.1 INTRODUCTION

Nanotechnology is a material science field emerging as a pioneering way with vast potential applications across various industries, including agriculture. Over the last decade, there has been a significant surge in nanotechnology research, which has profoundly shaped various aspects of modern life and driven transformative societal impacts (Surendranath and Mohanan 2021). Its applications span a multifaceted array of industries (pharmaceuticals, food technology, animal nutrition, cosmetics, electronics) and medical and agricultural systems (Tripathi et al. 2023). Sustainability has emerged as a central concern across the medical, industrial, and agricultural domains. In agriculture, nanomaterials have found diverse applications, including as nanofertilizers and nanopesticides. The adoption of sustainable practices is vital for the future of agriculture, with scientists focusing on the development of numerous nanoparticles (NPs), comprising silver (AgNPs), gold (AuNPs), copper (CuNPs), iron oxide (Fe_3O_4NPs), zinc oxide (ZnONPs), etc. These nanomaterials significantly contribute to improving agricultural efficiency and productivity. Additionally, advanced nanomaterials, namely quantum dots, polymeric NPs, silica NPs, carbon nanotubes, and liposome-derived, are pivotal in driving sustainable agricultural innovations (Balusamy et al. 2023; Prasad et al. 2017).

The integration of NPs in the field of agriculture offers numerous advantages. Their application promotes seed germination, growth of the plant, and tolerance mechanisms against biotic and abiotic stresses, thereby facilitating the cultivation of stress-tolerant crops and higher yields. Furthermore, the utilization of NPs facilitates the enhancement of soil quality, promoting a sustainable ecosystem. Nanoscale delivery systems can also augment the effectiveness of fertilizers, herbicides, and pesticides by targeting delivery to plants, thereby reducing waste and environmental pollution while maximizing nutrient uptake and disease management (Tripathi et al. 2023). Nanomaterials, such as nanoclays, nanocomposites, and nanofertilizers, increase soil fertility and water retention, thereby improving soil health, preventing degradation, and supporting sustainable agriculture. Nanosensors enable the maintenance of soil quality, plant health, and ecological conditions, delivering

DOI: 10.1201/9781032659435-1

data specifically on nutrients, moisture, and diseases to facilitate precision farming and optimize resource utilization (Fogel and Limson 2016; Miguel-Rojas and Pérez-de-Luque 2023).

NPs and nanomaterials enhance crop resilience to environmental stress by regulating physiological, biochemical, and molecular processes, while facilitating nutrient biofortification through improved uptake of essential micronutrients, such as iron, zinc, and selenium. However, while nanotechnology offers eco-friendly solutions and improved crop nutrition, it raises concerns about safety, regulatory challenges, and ethical implications in food production and consumption (Hofmann et al. 2020). In summary, nanomaterial science holds the prospect to revolutionize agriculture by enabling sustainable approaches to enhance agricultural yield and food safety while minimizing environmental impact and addressing resource constraints.

1.2 MOLECULAR INSIGHTS INTO NANOPARTICLES FOR PLANT GROWTH AND DEVELOPMENT

In agriculture, the utilization of NPs began to gain attention and research interest in the early 2000s. Initially, the focus of NP research in agriculture was centered on the green synthesis of NPs. Early studies emphasized more on physicochemical attributes of NPs, like charge, size, surface area, and shape as well as those that influence their interaction with plants, soil, and microorganisms. As the understanding of these properties expanded, the use of NPs, viz gold, copper, silver, zinc, and iron oxide, for enhancing agricultural efficiency increased rapidly (Wang et al. 2023). The impact and significance of these nanomaterials in regulating physiological, cellular, biochemical, and molecular processes of crops, even under adverse environmental conditions, laid the foundation for their application in crop protection and growth promotion.

NP-based delivery systems enabled precise, targeted, and controlled delivery of essential nutrients, fertilizers, and pesticides to the plant system, augmenting uptake of nutrients and their efficiency. Their formulations serve as carriers for the targeted and controlled release of herbicides, pesticides, fungicides, etc., aiding effective pest and disease management while reducing chemical usage. They also exhibit unique adsorption, catalytic, and ion exchange properties that can be exploited for soil remediation, nutrient retention, and improving soil structure, soil fertility, water retention, and overall crop productivity (Qurat ul Ain et al. 2023). NPs are composed of organic materials and metals that range in size from 1 to 100 nm gaining importance in agriculture owing to their nanoscale dimensions, greater reactivity, high surface-to-volume ratio, and efficient diffusion into plant tissues. When applied to plants, NPs facilitate improved nutrient uptake, stimulate metabolic processes, and modulate stress responses, thereby significantly impacting physiological and biochemical pathways (Ullah and Khan 2022).

They serve as nanofertilizers by gradually releasing nutrients, with ZnO, CuO, and Fe_2O_3 NPs being particularly suitable due to their controlled-release profiles. Additionally, NPs like TiO_2, carbon nanotubes, and graphene oxide enhance plant growth, combat pathogens at low concentrations, and improve seed germination while reducing contamination (Safdar et al. 2022). NPs offer a distinct advantage due to their small size, enabling uptake through stomata and root hairs and subsequent transport via the xylem and phloem. Their application in plants can be achieved through methods like foliar and root application and seed soaking treatment, reliant on the intended use. Various properties of NPs, viz. size, surface, shape, pH, and other ions or compounds conjugated significantly, influence their uptake and transport. Studies reveal that rod-shaped NPs, compared to spherical ones, are more easily absorbed and transported in plants like *Arabidopsis*, highlighting the critical role of NP shape and charge in their effective delivery to plant cells (Klaine et al. 2008; Wang et al. 2023; Zhang et al. 2022).

Abdel-Aziz et al. (2016) studied the outcome of foliar spraying of chitosan-based nanofertilizers coated with phosphorus (P), nitrogen (N), and potassium (K) on wheat plants, confirming penetration and translocation of NPs via transmission electron microscopy. Treated plants exhibited enhanced growth and yield. Copper NPs, widely used for crop protection, interact with organelles of pathogens, generating reactive oxygen species (ROS) that trigger lipid and protein degradation, DNA damage, and microbial death. In tomato, CuO, Cu_2O, and Cu/Cu_2 nanocomposites showed superior fungicidal activity against *Phytophthora infestans* at low concentrations compared to commercial fungicides. ZnONPs tested on coffee (*Coffea arabica* L.) increased zinc accumulation, resulting in greater fresh and dry weight of different plant parts. The dsRNA nanosheets coated with layered double hydroxide (LDH) demonstrated efficient delivery and protection against plant viruses, with stability and efficacy persisting for up to 20 days after a single spray (Giannousi et al. 2013; Mitter et al. 2017; Rossi et al. 2019). Several researchers have shown that these NPs also modulate hormonal pathways. For instance, carbon-based NPs, such as graphene oxide, influence auxin transport and signaling, which are essential for root elongation and lateral root development. Moreover, cerium oxide NPs regulate abscisic acid (ABA) signaling, thereby enhancing drought tolerance through improved stomatal regulation and water uptake efficiency (Li et al. 2024; Vankova et al. 2017).

1.3 DIVERSE NANOPARTICLES USED FOR THE PLANT SYSTEM

The activity of NPs has been illustrated in a wide range of plants, highlighting their role in germination, increasing biomass, and enhancing photosynthesis capacity. NPs have been used extensively to address plant diseases and to increase growth and crop productivity (Kaningini et al. 2022). NPs can be synthesized from different metal salts and are known to have applications in plant growth and protection (Ajilogba et al. 2021; Ovais et al. 2018). Zinc NPs (ZnNPs) are found to be less toxic (Kinraide et al. 2011) and are important for plant defense against many pathogens (Cabot et al. 2019). Silicon NPs (SiNPs) are biocompatible, affordable, and non-toxic and possess high adsorption power. Recently, SiNPs were demonstrated to advance plant growth, crop yield, and resistance against diverse diseases (Rastogi et al. 2019). Antimicrobial activity of CuNPs has been tested for growth inhibition of plant pathogenic bacteria (*Pseudomonas aeruginosa*, *Staphylococcus aureus*, *Propionibacterium acnes*, *Enterococcus faecalis*, *Salmonella typhimurium*, *Proteus vulgaris*, and *Shigella flexneri*) as well as fungi (*Phytophthora infestans* and *Fusarium spp.*) (Kasana et al. 2016). Similarly, AgNPs have been shown to enhance germination rate and plant growth efficiency. To protect plants against pathogens, unlike other NPs, AgNPs are used as nanofertilizers, nanopesticides, and pest management of crops in agriculture, while in the food industry, they have been used for food packaging and food protection due to their various benefits as antibacterial, antifungal, and antiviral agents (Khan et al. 2023). Based on the elements predominant in the NP, they are categorized, and their application in agriculture is summarized in Table 1.1.

TABLE 1.1
Nanoparticle Class and Their Role in Different Functions of Plants

Nanoparticle Class	Plants	Function	References
SiO_2	*Larix olgensis*	Seedling growth	Bao-shan et al. (2004)
	Maize	Seed germination	Rangaraj et al. (2012)
	Tomato	Seed germination	Siddiqui and Al-Whaibi (2014)
	Tomato	Seed germination under abiotic stress	Haghighi and Teixeira da Silva (2014)
nZVI	Peanut	Plant growth and root development	Li et al. (2015)
FeNP	Wheat	Cadmium and drought stress	Adrees et al. (2020)
ZnO	*Brassica nigra*	Seed germination	Sedghi et al. (2013)
	Cyamopsis tetragonoloba	Plant shoot and root growth	Tarafdar et al. (2013)
	Chickpea	Plant shoot and root growth	Burman et al. (2013)
	Onion	Plant growth promotion	Lawre and Raskar (2014)
	Wheat and soybean	Plant growth promotion	Ramesh et al. (2014)
	Capsicum	Secondary metabolite production	García-López et al. (2018)
	Soybean	Increase seed yield	Yusefi-Tanha et al. (2020)
TiO_2	*Mentha piperita* L.	Protect the chloroplast from aging	Ahmad et al. (2018)
	Dracocephalum moldavica	Plant growth in stressed conditions	Gohari et al. (2020)
CuNP	Bean	Elevation of chlorophyll a content	Pradhan et al. (2015)
	Watermelon	Suppress fungal growth in the root	Borgatta et al. (2018)
AgNP	Wheat	Antifungal activity against *Bipolaris sorokiniana*	Mishra et al. (2014)
Carbon-based NP	Wheat and other cereals	Strong antifungal activity against *Fusarium graminearum*	Wang et al. (2014)
MgONP	Tomato	Resistance of tomato to *Ralstonia solanacearum*	Imada et al. (2016)
CeNP	Tomato	Fusarium wilt suppression	Adisa et al. (2018)

1.4 FACETS OF NANOTECHNOLOGY FOR SUSTAINABLE AGRICULTURE

The global food demand is rising on account of the human population. Hence, there is a need to find novel approaches that optimize resource utilization and mitigate environmental stress. Among them, nanotechnology is a cutting-edge field in providing solutions for challenges such as efficient nutrient uptake, stress tolerance in combating disease-causing pathogens, targeted delivery of antimicrobial agents, and helping plants enhance resilience and achieve sustainable crop productivity. Nanotechnology is a multifaceted technology that encompasses nanofertilizers, nanopesticides, nanocarriers, nanosensors, nanoenzymes, and extensive applications in agricultural farming. This section explores three primary facets of nanotechnology as sustainable agricultural practices: nanofertilizers, nanopesticides, and nanocarriers.

1.4.1 Nanofertilizers

Nowadays, farmers apply large quantities of fertilizers to enhance crop productivity; however, uncontrolled application of synthetic fertilizers has destructive effects on the soil as well as on the environment, leading to soil organic matter depletion, acidification, and shifts in soil microbial diversity due to changes in nutrient content (Wu et al. 2020). Thus, nanofertilizers may be considered as an eco-friendly and promising alternative to chemical fertilizers. They are composed of nanostructured formulations containing macro- and micronutrients that can be applied to plants in very small amounts, possessing properties of slow nutrient release and efficient plant uptake. Their distinctive physical and chemical characteristics provide clear benefits over conventional chemical fertilizers. The nanoscale size ensures nanofertilizers penetrate plant tissues effectively through both foliar application and soil incorporation. Many reports have proved that nanofertilizers can boost crop production by inducing germination rate, biosynthetic pathway of protein, carbohydrate, photosynthesis, and stress tolerance (Nongbet et al. 2022).

Nanofertilizers are categorized into three types of formulations: (i) nanoscale fertilizers, a chemical fertilizer condensed to NP sizes; (ii) nano-based additive fertilizers, supplemented with nanomaterials; and (iii) nanoscale coating fertilizers as nanofilms, which consist of encapsulated nutrients (Mastronardi et al. 2015; Mejias et al. 2021). Nanofertilizers are mainly created by enclosing essential nutrients within nanomaterial structures. First, these nanomaterials are synthesized through physical top-down methods or chemical bottom-up techniques; then, specific nutrients get loaded into nanoporous carriers, covered with ultra-thin polymer layers, or stabilized in nanoscale emulsions, directly for cations like NH_4^+, K^+, Ca^{2+}, and Mg^{2+}, or following surface adjustments for anions such as NO_3^-, PO_4^{3-}, and SO_4^{2-} (Shang et al. 2019). Researchers have evaluated diverse inorganic, organic, and hybrid nanomaterials across multiple crop species to determine their effects on growth, physiological processes, and yield outcomes (Raliya et al. 2018). Field experiments comparing encapsulated nanofertilizers against conventional chemical fertilizers in farming systems demonstrated high yield for many crops such as wheat (5.35%), sesame (24.24%), and mustard (8.4%) on nitrogen–zinc-based nanofertilizers coupled through organic practices (Kumar et al. 2022). Moreover, the widespread adoption of nanofertilizers could reduce the necessity for massive industrial interference, as crops require notably smaller amounts than traditional synthetic options (Verma et al. 2022).

1.4.2 Nanopesticides

Nanopesticides, which utilize nanomaterials for targeted and slow release of active agrochemical pesticides, represent an emerging approach in modern agriculture. The nanoformulation of pesticides has garnered substantial consideration owing to its potential to enhance efficacy, reduce pesticide dosages, and minimize environmental and human health risks (Saini et al. 2015). A variety of nanoemulsion and micro-emulsion formulations are available to control pests and diseases in different farming systems. The components of these nanoemulsion formulations can help plant systems fight against pathogens and regulate their growth. Key characteristics of NPs for nanopesticide applications encompass distinct attributes such as chemical makeup (organic, inorganic, metallic, polymeric, or carbon-based), morphology (rods, spheres, irregular forms, or tubes), surface-to-volume ratio, dimensions, crystalline structure (amorphous or crystalline), and toxicity profile (Deka et al. 2021).

Major groups of nanopesticides can be created based on the target organism (herbicides, insecticides, acaricides, molluscicides, rodenticides, bactericides, and fungicides) and based on the delivery system (nanosuspensions, nanoemulsions, and nanocapsules) (Baliyarsingh and Pradhan 2023). AgNPs stabilized using polyvinyl pyrrolidone were sprayed onto *Ricinus communis* foliage, which exhibited strong insect-repellent activity against *Achaea janata* L. and *Spodoptera litura* (F) (Athanassiou et al. 2018).

1.4.3 Nanocarriers

Nanocarriers are advanced delivery vehicles with unique structural and functional properties that enhance the efficacy of agrochemicals and reduce their environmental impact. Liposomes and dendrimers are two excellent examples of nanocarriers, characterized by high encapsulation efficiency, physicochemical stability, minimal environmental impact, and robust stability under various environmental conditions. Pereira et al. (2019) assessed the effects of chitosan/tripolyphosphate (nanoCS/TPP) and alginate/chitosan (nanoALG/CS), along with GA_3 (gibberellic acid), by the seed priming method for growth and yield of tomato plants in the field. The study demonstrated that nanocarrier systems, such as nanoALG/CS-GA_3, could enhance fruit production by nearly fourfold, highlighting their potential, alongside plant growth regulators, to improve both the quality and quantity of crops.

Kwak et al. (2019) demonstrated the feasibility of chloroplast genetic modification via chitosan-bound single-walled carbon nanotubes as delivery vehicles. This NP-based approach for transgene introduction into chloroplasts offers significant practical benefits compared to existing methods, serving as a promising technique for transforming fully developed plants to advance bioengineering and fundamental plant research. Nanoencapsulation further shields agrochemicals, nutrients, and bioactive compounds from degradation by oxidation, light exposure, and environmental stressors (Martins et al. 2024; Machado et al. 2022). Applications extend to nanofertilizers, nanoherbicides, nanopesticides, and nanoinsecticides, all of which leverage nanocarriers to promote eco-friendly farming methods.

1.5 NANOPARTICLES IN PLANT DISEASE CONTROL

Along with climate change, plant diseases are also one of the important problems for global agriculture, impacting crop yields, food security, and ecosystem health. Plant pathogens (fungi, bacteria, viruses, nematodes), insects, and weeds cause devastating crop losses. Plant pathogens and pests contribute to 40% of the global crop yield losses of maize, potato, rice, soybean, and wheat worldwide (He et al. 2020). Addressing these major issues requires innovative solutions, such as enhanced disease monitoring systems, advanced pathogen detection tools (including specialized sensors), predictive analytics models, and robust data processing methods, to avert potential future epidemics (Ristaino et al. 2021).

NPs serve as multifaceted tools in plant disease management, addressing crop health through protection, nutrition, and diagnostics. The research in plant pathology is increasingly integrating a diverse array of materials, including metalloids, metallic oxides, nonmetals, and carbon nanomaterials, such as graphene oxides and nanotubes, along with advanced structures like liposomes, dendrimers, and quantum dots (Elmer et al. 2018; Elmer and White 2018). These materials function through several key mechanisms to ensure plant vitality. Direct protection is achieved through antimicrobial activity, where NPs, such as silver and copper, disrupt the structures of pathogens and utilize them as elicitors to stimulate the plant's own defense systems. Further, nanotechnology enhances chemical efficiency by utilizing nanoscale carriers, which improve the delivery of antifungal agents and controlled-release systems that optimize the application of pesticides and fertilizers. This is complemented by nanofertilizers that improve nutrient uptake and nanosensors that provide early, sustainable disease detection. Collectively, these applications represent a highly precise and integrated approach to modern agricultural challenges.

Metalloid NPs, such as silicon NPs (SiNPs), have shown potential in enhancing plant resistance against various pathogens. The porous structure of SiNPs, characterized by extensive surface area, positions them as excellent vehicles for delivering pesticides, herbicides, and fertilizers, thereby enabling innovations such as nanopesticides, nanoherbicides, and nanofertilizers that could substantially enhance crop yields across various species (Yuvaraj et al. 2023). In the past decade, nanomaterials, predominantly metal oxide NPs (MONPs), have acquired prominence in agriculture for enhancing crop performance, thanks to their extensive surface area, superior chemical and thermal stability, and customizable physicochemical properties (Maity et al. 2022). To date, carbon nanotubes (single- and multi-wall), graphene oxide (in its reduced and oxidized forms), and fullerenes have shown a significant impact on plant health, highlighting their promise as innovative tools in plant disease management (Elmer et al. 2018).

- **Early Detection of Disease Using Nanosensors**: Advancements in chemical and biological sensing technologies have driven a range of practical applications, including the evaluation of the quality of medicinal and industrial products. Lately, numerous NP-enhanced chemo- and biosensors have emerged and reached commercial stages for crop health diagnostics (Li et al. 2020). Some NPs, including AuNP-ssDNA, PtNP-IgG, and AgNR, function as nanosensors and can be utilized for the early detection of plant pathogens. Early detection of disease enables timely intervention, which, in turn, prevents the spread of disease and minimizes crop losses. Singh et al. (2010) utilized immunosensors employing AuNPs to enable the detection of Karnal bunt (*Tilletia indica*) in wheat through surface plasmon resonance technology.

- **Targeted Delivery of Antimicrobial Agents Using Nanomaterials**: Nanomaterials, such as liposomes, polymeric NPs, and dendrimers, can be used to encapsulate fungicides, ensuring their gradual and controlled release. This approach minimizes environmental impact and reduces the quantity of chemical inputs while maximizing efficacy against plant pathogens. AgNPs, due to their antimicrobial properties, are mainly used for disinfection (Baker et al. 2005). CuNPs and AgNPs effectively inhibit the development of the fungal pathogens *Alternaria alternata* and *Botrytis cinerea*. ZnONPs and magnesium oxide MgONP NPs demonstrated antifungal effects against *Alternaria alternata*, *Rhizopus stolonifer*, *Fusarium oxysporum*, and *Mucor plumbeus*. A hybrid MWCNT-g-PCA material that encapsulates the traditional pesticides zineb and mancozeb proved more effective as a fungicide targeting *A. alternata* (Mittal et al. 2020).
- **Nanoparticles as Elicitors**: NPs can act as signaling molecules, which induce plant defense pathways/mechanisms to counter pathogens. NPs interact with plant tissues to activate multiple defense mechanisms, including the upregulation of protective genes, ROS signaling, and hormone-mediated biochemical processes. These reactions equip plants to identify and resist pathogen attacks, establishing a natural barrier against infections (Noman et al. 2023). AgNPs applied at 50 µg/mL trigger salicylic acid-dependent defense mechanisms in wheat, providing effective resistance against the yellow rust pathogen *Puccinia striiformis* f.sp. *tritici* (Bano and Ummat-ul-Habib 2020; Noman et al. 2023). Nanopesticides (e.g., CuNPs), nanoinsecticides (e.g., nanostructured alumina, i.e., NSA), antimicrobial agent NPs (e.g., TiO_2NPs, ZnONPs), nanoemulsions, nanoherbicides (polyepsilon–caprolactone, i.e., PCL nanocapsule carrier for atrazine), and nanofungicides help the plant system in mitigating plant diseases. The primary objective of exploring nanomaterials in agriculture is to achieve cost savings and operational efficiency, specifically, reducing agrochemical use, mitigating nutrient runoff, and enhancing crop yields through practical and time-efficient approaches (Singh et al. 2021). The application of NPs for managing various plant diseases facilitates sustainable and effective crop protection.

1.6 NANOTECHNOLOGY AND PLANT STRESS MANAGEMENT

Plants face biotic and abiotic stresses. Biotic stress can be caused by various disease-causing plant pathogens (viruses, bacteria, fungi, nematodes), insects, and herbivores. Environmental adverse conditions, such as heat, cold, salinity, drought, nutrient-deficient soil, light conditions, humidity, oxidative stress, precipitation, heavy metals, radiation, and altitude, are abiotic stress factors that can have an adverse impact on plant growth, development, and productivity. The physical and chemical attributes of NPs, including their crystalline structure, catalytic potential, porous nature, and aggregation tendencies, enable them to cross cellular barriers and exert effects on biological systems. Plants under abiotic stress experience excessive buildup of ROS, which harms cell membranes, DNA, proteins, and other cellular structures, ultimately stunting growth. NPs such as C60-NPs, CeO_2-NPs, and Fe_2O_3-NPs have the ability to scavenge ROS (Zhao et al. 2020) and mitigate plant stress. When beneficial NPs reach various plant tissues, they enhance photosynthesis rates, biomass accumulation, chlorophyll levels, sugar concentrations, osmolyte production, and antioxidant defenses. Additionally, these NPs optimize nitrogen metabolism, elevate chlorophyll and protein levels, and activate genes associated with tolerance to both abiotic and biotic stresses (A. Singh et al. 2021).

"Nanozymes" refer to NPs that mimic enzyme functions, providing multiple benefits compared to natural enzymes. Key advantages encompass straightforward synthesis processes, affordability, exceptional stability, strong catalytic efficiency, and easy surface functionalization. These materials are actively investigated for diverse uses in biosensors, immunoassays, medical diagnostics and treatments, theranostics, tissue regeneration, oxidative stress mitigation, and pollutant degradation (Singh 2019). Boghossian et al. (2013) demonstrated ROS production in isolated chloroplasts from *Spinacia oleracea* when directly exposed to NP antioxidants like CeO_2NPs, including dextran-coated nanoceria (dNC), which had previously proven effective as a ROS quencher. In various crops such as cereals, legumes, oilseeds, and vegetables, NPs can support plant nutrition during abiotic stress, helping to mitigate its effects and improve overall plant performance (Tighe-Neira et al. 2023).

Engineered nanomaterials (ENMs) offer diverse applications, serving as seed-coating agents and nanobiosensors, nanofertilizers to enhance soil microbiomes and health, nanopesticides for targeted pathogen management, and nanocarriers to support crop genetic modification. ENMs also enable targeted nutrient delivery for better uptake, support economical nutrient recycling, transform plants into stress-forecasting sensors, foster resilient varieties via genomics, optimize crop handling to cut losses, and enrich soil microbiomes for improved fertility and plant vigor (Ijaz et al. 2023).

1.7 NANOTECHNOLOGY FOR CROP IMPROVEMENT

Nanotechnology has the potential to solve agricultural problems with affordable and straightforward solutions. Several industries like Tropical AgroSystem, Alert Biotech, and Nano Agro Science Co-operative Society have launched nanofertilizers TAG NANO ZINC, TAG NANO CAL, TAG NANO POTASH, Nano Bor 20%, and NASCO Escort-P, which are useful to farmers in increasing crop productivity (Mittal et al. 2020). The field application of these nanomaterials is crucial for understanding their impact on plants under field conditions. In Table 1.2, we provide examples of field experiments involving NPs on plants.

TABLE 1.2
Use of Nanoparticles for Crop Improvement in Field Conditions

Plants	Type of Nanoparticle	Effect on Crop	Field Application	References
Brassica juncea	AuNPs	Seed yield increased by 19%	Foliar spraying (10 ppm)	Kumar et al. (2019)
Oat and berseem	ZnO, TiO_2, CuO, and AgNPs	Enhance seed germination and plant growth	Seed priming (750 mg)	Maity et al. (2018)
Foxtail millet	ZnONPs	Improved nitrogen content and millet grain oil	Foliar spraying of ZnO (2.6 mg/L twice)	Kolenčík et al. (2019)
Maize	Zn-encapsulated chitosan NPs	Rise in grain yield	Seed priming and foliar application (0.01%–0.16%)	Choudhary et al. (2019)
Fenugreek	AgNPs	Increased yield	Foliar application (40 mg/L)	Sadak (2019)
Maize	CuNPs	Elevation in seed number and grain yield in draught stress conditions	Seed priming (50–90μM)	Van Nguyen et al. (2022)

In *Brassica juncea*, AuNPs at concentrations of 0, 10, 25, 50, and 100 ppm were tested under field conditions through foliar spray. The uptake of NPs was confirmed by atomic absorption spectrometry analysis of leaf tissue. Upon treatment, it was observed that the number of branches, stem diameter, seed yield, and pod number, among other factors, were positively affected (Kumar et al. 2019). In a seed priming study, the combination of ZnO, TiO_2, CuO, and AgNPs was examined for its effect on oat and berseem. Seeds of these two plants were treated with NPs at three concentrations: 750 mg, 1,000 mg, and 1,250 mg/kg of seed. After treatment, the highest seed yield was given by plants treated with TiO_2 at a concentration of 1,250 mg/kg seeds (Maity et al. 2018). Apart from the plant effect, the group also characterized the effect of NPs on soil microbiota, where no significant change in their population was observed. Additionally, another group performed seed priming, as well as foliar application of Zn–chitosan NPs, on maize plants. In the field of maize, the treatment resulted in a yield increase of up to 40% and could alleviate Curvularia leaf spot (CLS) disease by enhancing plant innate immunity (Kolenčík et al. 2019).

1.8 CONCLUSIONS AND FUTURE PERSPECTIVES

The assimilation of nanomaterials in agricultural systems paves the way for addressing crucial challenges in modern agriculture. Nanomaterials, including NPs, nanocomposites, and carbon-based nanostructures, have evidenced potential in improving crop productivity, soil health, and mitigating wide stresses. These nanomaterials have emerged as effective in several applications as nanofertilizers, nanopesticides, and nanocarriers for the targeted and controlled delivery of agrochemicals, thereby enlightening their efficacy by lowering environmental contamination. Furthermore, the role of nano-enabled technology in advancing precision agriculture by use of nanosensors to monitor real-time crop health, soil properties, and pest outbreaks is revolutionary. The emergence of nanomaterial-enabled innovative delivery systems for biocontrol agents and nutrients exemplifies monitoring the synergy between nanotechnology and sustainable agriculture practices. Therefore, in the future, prioritizing green synthesis methods to produce eco-friendly nanomaterials is key to reducing ecological toxicity and promoting biocompatibility within agricultural ecosystems. Nevertheless, gaining a thorough understanding of the long-term behavior and ecological consequences of nanomaterials remains a significant challenge in agronomy and environmental research. Joint initiatives across academic groups investigating varied nanomaterial uses will prove essential for crafting high-performing, versatile, durable, and economical innovations.

REFERENCES

Abdel-Aziz, H.M.M., Hasaneen, M.N.A., Omer, A.M. 2016. Nano chitosan-NPK fertilizer enhances the growth and productivity of wheat plants grown in sandy soil. *Spanish Journal of Agricultural Research* 14, e0902–e0902. https://doi.org/10.5424/sjar/2016141-8205

Adisa, I.O., Reddy Pullagurala, V.L., Rawat, S., Hernandez-Viezcas, J.A., Dimkpa, C.O., Elmer, W.H., White, J.C., Peralta-Videa, J.R., Gardea-Torresdey, J.L. 2018. Role of cerium compounds in fusarium wilt suppression and growth enhancement in tomato (*Solanum lycopersicum*). *Journal of Agricultural and Food Chemistry* 66, 5959–5970. https://doi.org/10.1021/acs.jafc.8b01345

Adrees, M., Khan, Z.S., Ali, S., Hafeez, M., Khalid, S., ur Rehman, M.Z., Hussain, A., Hussain, K., Shahid Chatha, S.A., Rizwan, M. 2020. Simultaneous mitigation of cadmium and drought stress in wheat by soil application of iron nanoparticles. *Chemosphere* 238, 124681. https://doi.org/10.1016/j.chemosphere.2019.124681

Ahmad, B., Shabbir, A., Jaleel, H., Khan, M.M.A., Sadiq, Y. 2018. Efficacy of titanium dioxide nanoparticles in modulating photosynthesis, peltate glandular trichomes and essential oil production and quality in *Mentha piperita* L. *Current Plant Biology* 13, 6–15. https://doi.org/10.1016/j.cpb.2018.04.002

Ajilogba, C., Babalola, O., Nikoro, D. 2021. Nanotechnology as vehicle for biocontrol of plant diseases in crop production. In: Babalola, O.O. (eds), *Food Security and Safety*, pp. 709–724. Springer, Cham. https://doi.org/10.1007/978-3-030-50672-8_35

Athanassiou, C.G., Kavallieratos, N.G., Benelli, G., Losic, D., Usha Rani, P., Desneux, N. 2018. Nanoparticles for pest control: current status and future perspectives. *Journal of Pest Science* 91, 1–15. https://doi.org/10.1007/s10340-017-0898-0

Baker, C., Pradhan, A., Pakstis, L., Pochan, Darrin J., Shah, S.I. 2005. Synthesis and antibacterial properties of silver nanoparticles. *Journal of Nanoscience and Nanotechnology* 5, 244–249. https://doi.org/10.1166/jnn.2005.034

Baliyarsingh, B., Pradhan, C.K. 2023. Prospects of plant-derived metallic nanopesticides against storage pests - a review. *Journal of Agriculture and Food Research* 14, 100687. https://doi.org/10.1016/j.jafr.2023.100687

Balusamy, S.R., Joshi, A.S., Perumalsamy, H., Mijakovic, I., Singh, P. 2023. Advancing sustainable agriculture: a critical review of smart and eco-friendly nanomaterial applications. *Journal of Nanobiotechnology* 21, 372. https://doi.org/10.1186/s12951-023-02135-3

Bano, A., Ummat-ul-Habib, U. 2020. Interactive effects of Ag-nanoparticles, salicylic acid, and plant growth promoting rhizobacteria on the physiology of wheat infected with yellow rust. *Journal of Plant Pathology* 102, 1215–1225. https://doi.org/10.1007/s42161-020-00626-y

Bao-shan, L., Shao-qi, D., Chun-hui, L., Li-jun, F., Shu-chun, Q., Min, Y. 2004. Effect of TMS (nanostructured silicon dioxide) on growth of *Changbai larch* seedlings. *Journal of Forestry Research* 15, 138–140. https://doi.org/10.1007/BF02856749

Boghossian, A.A., Sen, F., Gibbons, B.M., Sen, S., Faltermeier, S.M., Giraldo, J.P. et al. 2013. Application of nanoparticle antioxidants to enable hyperstable chloroplasts for solar energy harvesting. *Advanced Energy Materials* 3, 881–893. https://doi.org/10.1002/aenm.201201014

Borgatta, J., Ma, C., Hudson-Smith, N., Elmer, W., Plaza Pérez, C.D., De La Torre-Roche, R. et al. 2018. Copper based nanomaterials suppress root fungal disease in watermelon (*Citrullus lanatus*): role of particle morphology, composition and dissolution behavior. *ACS Sustainable Chemistry & Engineering* 6, 14847–14856. https://doi.org/10.1021/acssuschemeng.8b03379

Burman, U., Saini, M., Kumar, P. 2013. Effect of zinc oxide nanoparticles on growth and antioxidant system of chickpea seedlings. *Toxicological & Environmental Chemistry* 95, 605–612. https://doi.org/10.1080/02772248.2013.803796

Cabot, C., Martos, S., Llugany, M., Gallego, B., Tolrà, R., Poschenrieder, C. 2019. A role for zinc in plant defense against pathogens and herbivores. *Frontiers in Plant Science* 10, 1171.

Choudhary, R.C., Kumaraswamy, R.V., Kumari, S., Sharma, S.S., Pal, A., Raliya, R. et al. 2019. Zinc encapsulated chitosan nanoparticle to promote maize crop yield. *International Journal of Biological Macromolecules* 127, 126–135. https://doi.org/10.1016/j.ijbiomac.2018.12.274

Deka, B., Babu, A., Baruah, C., Barthakur, M. 2021. Nanopesticides: a systematic review of their prospects with special reference to tea pest management. *Frontiers in Nutrition* 8, 686131. https://doi.org/10.3389/fnut.2021.686131

Elmer, W., Ma, C., White, J. 2018. Nanoparticles for plant disease management. *Current Opinion in Environmental Science & Health* 6, 66–70. https://doi.org/10.1016/j.coesh.2018.08.002

Elmer, W., White, J.C. 2018. The future of nanotechnology in plant pathology. *Annual Review of Phytopathology* 56, 111–133. https://doi.org/10.1146/annurev-phyto-080417-050108

Fogel, R., Limson, J. 2016. Developing biosensors in developing countries: South Africa as a case study. *Biosensors* 6, 5. https://doi.org/10.3390/bios6010005

García-López, J.I., Zavala-García, F., Olivares-Sáenz, E., Lira-Saldívar, R.H., Barriga-Castro, E.D., Ruiz-Torres, N.A. et al. 2018. Zinc oxide nanoparticles boosts phenolic compounds and antioxidant activity of *Capsicum annuum* L. during germination. *Agronomy* 8, 215. https://doi.org/10.3390/agronomy8100215

Giannousi, K., Avramidis, I., Dendrinou-Samara, C. 2013. Synthesis, characterization and evaluation of copper based nanoparticles as agrochemicals against *Phytophthora infestans*. *RSC Advances*. 3, 21743–21752. https://doi.org/10.1039/C3RA42118J

Gohari, G., Mohammadi, A., Akbari, A., Panahirad, S., Dadpour, M.R., Fotopoulos, V., Kimura, S. 2020. Titanium dioxide nanoparticles (TiO_2 NPs) promote growth and ameliorate salinity stress effects on essential oil profile and biochemical attributes of *Dracocephalum moldavica*. *Scientific Reports* 10, 912. https://doi.org/10.1038/s41598-020-57794-1

Haghighi, M., Teixeira da Silva, J.A. 2014. The effect of N-TiO_2 on tomato, onion, and radish seed germination. *Journal of Crop Science and Biotechnology* 17, 221–227. https://doi.org/10.1007/s12892-014-0056-7

He, S., Creasey Krainer, K.M. 2020. Pandemics of people and plants: which is the greater threat to food security? *Molecular Plant* 13, 933–934. https://doi.org/10.1016/j.molp.2020.06.007

Hofmann, T., Lowry, G.V., Ghoshal, S., Tufenkji, N., Brambilla, D., Dutcher, J.R. et al. 2020. Technology readiness and overcoming barriers to sustainably implement nanotechnology-enabled plant agriculture. *Nature Food* 1, 416–425. https://doi.org/10.1038/s43016-020-0110-1

Ijaz, M., Khan, F., Ahmed, T., Noman, M., Zulfiqar, F., Rizwan, M., Chen, J., Siddique, H.M.K., Li, B. 2023. Nanobiotechnology to advance stress resilience in plants: current opportunities and challenges. *Materials Today Bio* 22, 100759. https://doi.org/10.1016/j.mtbio.2023.100759

Imada, K., Sakai, S., Kajihara, H., Tanaka, S., Ito, S. 2016. Magnesium oxide nanoparticles induce systemic resistance in tomato against bacterial wilt disease. *Plant Pathology* 65, 551–560. https://doi.org/10.1111/ppa.12443

Kaningini, A.G., Nelwamondo, A.M., Azizi, S., Maaza, M., Mohale, K.C. 2022. Metal nanoparticles in agriculture: a review of possible use. *Coatings* 12, 1586. https://doi.org/10.3390/coatings12101586

Kasana, R.C., Panwar, N.R., Kaul, R.K., Kumar, P. 2016. Copper nanoparticles in agriculture: biological synthesis and antimicrobial activity. In: Ranjan, S., Dasgupta, N., Lichtfouse, E. (eds), *Nanoscience in Food and Agriculture 3, Sustainable Agriculture Reviews*, pp. 129–143. Springer International Publishing, Cham. https://doi.org/10.1007/978-3-319-48009-1_5

Khan, S., Zahoor, M., Sher Khan, R., Ikram, M., Islam, N.U. 2023. The impact of silver nanoparticles on the growth of plants: the agriculture applications. *Heliyon* 9, e16928. https://doi.org/10.1016/j.heliyon.2023.e16928

Kinraide, T.B., Poschenrieder, C., Kopittke, P.M. 2011. The standard electrode potential ($E\theta$) predicts the prooxidant activity and the acute toxicity of metal ions. *Journal of Inorganic Biochemistry* 105, 1438–1445. https://doi.org/10.1016/j.jinorgbio.2011.08.024

Klaine, S.J., Alvarez, P.J.J., Batley, G.E., Fernandes, T.F., Handy, R.D., Lyon, D.Y. et al. 2008. Nanomaterials in the environment: behavior, fate, bioavailability, and effects. *Environmental Toxicology and Chemistry* 27, 1825–1851. https://doi.org/10.1897/08-090.1

Kolenčík, M., Ernst, D., Komár, M., Urík, M., Šebesta, M., Dobročka, E. et al. 2019. Effect of foliar spray application of zinc oxide nanoparticles on quantitative, nutritional, and physiological parameters of foxtail millet (*Setaria italica* L.) under field conditions. *Nanomaterials* 9, 1559. https://doi.org/10.3390/nano9111559

Kumar, A., Singh, K., Verma, P., Singh, O., Panwar, A., Singh, T., Kumar, Y., Raliya, R. 2022. Effect of nitrogen and zinc nanofertilizer with the organic farming practices on cereal and oil seed crops. *Scientific Reports* 12, 6938. https://doi.org/10.1038/s41598-022-10843-3

Kumar, S., Nehra, M., Dilbaghi, N., Marrazza, G., Hassan, A.A., Kim, K.-H. 2019. Nano-based smart pesticide formulations: emerging opportunities for agriculture. *Journal of Controlled Release* 294, 131–153. https://doi.org/10.1016/j.jconrel.2018.12.012

Kwak, S.-Y., Lew, T.T.S., Sweeney, C.J., Koman, V.B., Wong, M.H., Bohmert-Tatarev, K. et al. 2019. Chloroplast-selective gene delivery and expression in planta using chitosan-complexed single-walled carbon nanotube carriers. *Nature Nanotechnology* 14, 447–455. https://doi.org/10.1038/s41565-019-0375-4

Lawre, S., Raskar, S. 2014. Influence of zinc oxide nanoparticles on growth, flowering and seed productivity in onion. *International Journal of Current Microbiology and Applied Sciences* 3, 874–881.

Li, G., Gao, Q., Nyande, A., Dong, Z., Khan, E.H., Han, Y., Wu, H. 2024. Cerium oxide nanoparticles promoted lateral root formation in Arabidopsis by modulating reactive oxygen species and Ca(2+) level. *Functional Plant Biology* 51, FP24196. https://doi.org/10.1071/FP24196

Li, X., Yang, Y., Gao, B., Zhang, M. 2015. Stimulation of peanut seedling development and growth by zero-valent iron nanoparticles at low concentrations. *PLoS One* 10, e0122884. https://doi.org/10.1371/journal.pone.0122884

Li, Z., Yu, T., Paul, R., Fan, J., Yang, Y., Wei, Q. 2020. Agricultural nanodiagnostics for plant diseases: recent advances and challenges. *Nanoscale Advances* 2, 3083–3094. https://doi.org/10.1039/C9NA00724E

Machado, T.O., Grabow, J., Sayer, C., De Araújo, P.H.H., Ehrenhard, M.L., Wurm, F.R. 2022. Biopolymer-based nanocarriers for sustained release of agrochemicals: a review on materials and social science perspectives for a sustainable future of agri- and horticulture. *Advances in Colloid and Interface Science* 303, 102645. https://doi.org/10.1016/j.cis.2022.102645

Maity, A., Natarajan, N., Vijay, D., Srinivasan, R., Pastor, M., Malaviya, D.R. 2018. Influence of metal nanoparticles (NPs) on germination and yield of oat (*Avena sativa*) and berseem (*Trifolium alexandrinum*). *Proceedings of the National Academy of Sciences, India Section B: Biological Sciences* 88, 595–607. https://doi.org/10.1007/s40011-016-0796-x

Maity, D., Gupta, U., Saha, S. 2022. Biosynthesized metal oxide nanoparticles for sustainable agriculture: next-generation nanotechnology for crop production, protection and management. *Nanoscale* 14, 13950–13989. https://doi.org/10.1039/D2NR03944C

Martins, J.H. da S., Martins, L.H. da S., Silva, J. de M. e, Almada, S.R. de, Oliveira, J.A.R. de, Rai, M. 2024. Nanoencapsulation for delivery of agrochemicals to boost plant health, In: Rai, M., Ingle, K.K. (eds), *Nanotechnology in Plant Health*, pp. 1–11. CRC Press.

Mastronardi, E., Tsae, P., Zhang, X., Monreal, C., DeRosa, M.C. 2015. Strategic role of nanotechnology in fertilizers: potential and limitations, In: Rai, M., Ribeiro, C., Mattoso, L., Duran, N. (eds), *Nanotechnologies in Food and Agriculture*, pp. 25–67. Springer International Publishing, Cham. https://doi.org/10.1007/978-3-319-14024-7_2

Mejias, J.H., Salazar, F., Pérez Amaro, L., Hube, S., Rodriguez, M., Alfaro, M. 2021. Nanofertilizers: a cutting-edge approach to increase nitrogen use efficiency in grasslands. *Frontier Environmental* Science 9, 635114. https://doi.org/10.3389/fenvs.2021.635114

Miguel-Rojas, C., Pérez-de-Luque, A. 2023. Nanobiosensors and nanoformulations in agriculture: new advances and challenges for sustainable agriculture. *Emerging Topics in Life Sciences* 7, 229–238. https://doi.org/10.1042/ETLS20230070

Mishra, V., Sharma, R., Nakuleshawar, D., Jasuja, D., Gupta, K. 2014. A review on green synthesis of nanoparticles and evaluation of antimicrobial activity. *International Journal of Green and Herbal Chemistry* 333, 81–94.

Mittal, D., Kaur, G., Singh, P., Yadav, K., Ali, S.A. 2020. Nanoparticle-based sustainable agriculture and food science: recent advances and future outlook. *Frontier in Nanotechnology* 2, 579954. https://doi.org/10.3389/fnano.2020.579954

Mitter, N., Worrall, E.A., Robinson, K.E., Li, P., Jain, R.G., Taochy, C. et al. 2017. Clay nanosheets for topical delivery of RNAi for sustained protection against plant viruses. *Nature Plants* 3, 16207. https://doi.org/10.1038/nplants.2016.207

Noman, M., Ahmed, T., Wang, J., Ijaz, M., Shahid, M., Islam, M.S. et al. 2023. Nano-enabled crop resilience against pathogens: potential, mechanisms and strategies. *Crop Health* 1, 15. https://doi.org/10.1007/s44297-023-00015-8

Nongbet, A., Mishra, A.K., Mohanta, Y.K., Mahanta, S., Ray, M.K., Khan, M. et al. 2022. Nanofertilizers: a smart and sustainable attribute to modern agriculture. *Plants* 11, 2587. https://doi.org/10.3390/plants11192587

Ovais, M., Khalil, A.T., Islam, N.U., Ahmad, I., Ayaz, M., Saravanan, M. et al. 2018. Role of plant phytochemicals and microbial enzymes in biosynthesis of metallic nanoparticles. *Applied Microbiology and Biotechnology* 102, 6799–6814. https://doi.org/10.1007/s00253-018-9146-7

Pereira, A.D.E.S., Oliveira, H.C., Fraceto, L.F. 2019. Polymeric nanoparticles as an alternative for application of gibberellic acid in sustainable agriculture: a field study. *Scientific Reports* 9, 7135. https://doi.org/10.1038/s41598-019-43494-y

Pradhan, P., Fischer, G., Velthuizen, H. van, Reusser, D.E., Kropp, J.P. 2015. Closing yield gaps: how sustainable can we be? *PLoS One* 10, e0129487. https://doi.org/10.1371/journal.pone.0129487

Prasad, R., Bhattacharyya, A., Nguyen, Q.D. 2017. Nanotechnology in sustainable agriculture: recent developments, challenges, and perspectives. *Frontier in Microbiology* 8, 1014. https://doi.org/10.3389/fmicb.2017.01014

Qurat ul Ain, Hussain, H.A., Zhang, Q., Rasheed, A., Imran, A., Hussain, S., Ahmad, N., Bibi, H., Ali, K.S. 2023. Use of nano-fertilizers to improve the nutrient use efficiencies in plants. In: Aftab, T. and Hakeem, K.R. (eds), *Sustainable Plant Nutrition*, pp. 299–32. Academic Press. https://doi.org/10.1016/B978-0-443-18675-2.00013-4

Raliya, R., Saharan, V., Dimkpa, C., Biswas, P. 2018. Nanofertilizer for precision and sustainable agriculture: current state and future perspectives. *Journal of Agricultural and Food Chemistry* 66, 6487–6503. https://doi.org/10.1021/acs.jafc.7b02178

Ramesh, A., Sharma, S.K., Sharma, M.P., Yadav, N., Joshi, O.P. 2014. Inoculation of zinc solubilizing *Bacillus aryabhattai* strains for improved growth, mobilization and biofortification of zinc in soybean and wheat cultivated in Vertisols of central India. *Applied Soil Ecology* 73, 87–96. https://doi.org/10.1016/j.apsoil.2013.08.009

Rangaraj, S., Gopalu, K., Periasamy, P., Venkatachalam, R., Kannan, N. 2012. Growth and physiological responses of maize (*Zea mays* L.) to porous silica nanoparticles in soil. *Journal of Nanoparticle Research* 14, 6. https://doi.org/10.1007/s11051-012-1294-6

Rastogi, A., Tripathi, D.K., Yadav, S., Chauhan, D.K., Živčák, M., Ghorbanpour, M., El-Sheery, N.I., Brestic, M. 2019. Application of silicon nanoparticles in agriculture. *3 Biotech* 9, 90. https://doi.org/10.1007/s13205-019-1626-7

Ristaino, J.B., Anderson, P.K., Bebber, D.P., Brauman, K.A., Cunniffe, N.J., Fedoroff, N.V. et al. 2021. The persistent threat of emerging plant disease pandemics to global food security. *Proceedings of the National Academy of Sciences of the United States of America* 118, e2022239118. https://doi.org/10.1073/pnas.2022239118

Rossi, L., Fedenia, L.N., Sharifan, H., Ma, X., Lombardini, L. 2019. Effects of foliar application of zinc sulfate and zinc nanoparticles in coffee (*Coffea arabica* L.) plants. *Plant Physiology and Biochemistry* 135, 160–166. https://doi.org/10.1016/j.plaphy.2018.12.005

Sadak, M.S. 2019. Impact of silver nanoparticles on plant growth, some biochemical aspects, and yield of fenugreek plant (*Trigonella foenum-graecum*). *Bulletin of the National Research Centre* 43, 38. https://doi.org/10.1186/s42269-019-0077-y

Safdar, M., Kim, W., Park, S., Gwon, Y., Kim, Y.-O., Kim, J. 2022. Engineering plants with carbon nanotubes: a sustainable agriculture approach. *Journal of Nanobiotechnology* 20, 275. https://doi.org/10.1186/s12951-022-01483-w

Saini, P., Gopal, M., Kumar, R., Gogoi, R. 2015. Residue, dissipation, and safety evaluation of pyridalyl nanoformulation in Okra (*Abelmoschus esculentus* [L.] Moench). *Environmen tal Monitoring and Assessment* 187(3), 123. https://doi.org/10.1007/s10661-015-4361-5

Sedghi, M., Hadi, M., Toluie, S. 2013. Effect of nano zinc oxide on the germination of soybean seeds under drought stress. A*nnals of West University of Timisoara: Series of Biology* 16, 73–78.

Shang, Y., Hasan, Md.K., Ahammed, G.J., Li, M., Yin, H., Zhou, J. 2019. Applications of nanotechnology in plant growth and crop protection: a review. *Molecules* 24, 2558. https://doi.org/10.3390/molecules24142558

Siddiqui, M.H., Al-Whaibi, M.H. 2014. Role of nano-SiO_2 in germination of tomato (*Lycopersicum esculentum* seeds Mill.). *Saudi Journal of Biological Sciences* 21, 13–17. https://doi.org/10.1016/j.sjbs.2013.04.005

Singh, A., Tiwari, S., Pandey, J., Lata, C., Singh, I.K. 2021. Role of nanoparticles in crop improvement and abiotic stress management. *Journal of Biotechnology* 337, 57–70. https://doi.org/10.1016/j.jbiotec.2021.06.022

Singh, P., Singh, R., Verma, P., Bhadouria, R., Kumar, A., Kaushik, M. (eds). 2021. *Plant-Microbes-Engineered Nano-particles (PM-ENPs) Nexus in Agro-Ecosystems: Understanding the Interaction of Plant, Microbes and Engineered Nano-particles (ENPS), Advances in Science, Technology & Innovation.* Springer International Publishing, Cham.

Singh, S. 2019. Nanomaterials exhibiting enzyme-like properties (nanozymes): current advances and future perspectives. *Frontier Chemistry* 7, 46. https://doi.org/10.3389/fchem.2019.00046

Singh, S., Singh, M., Agrawal, V.V., Kumar, A. 2010. An attempt to develop surface plasmon resonance based immunosensor for Karnal bunt (*Tilletia indica*) diagnosis based on the experience of nano-gold based lateral flow immuno-dipstick test. *Thin Solid Films* 519, 1156–1159. https://doi.org/10.1016/j.tsf.2010.08.061

Surendranath, A., Mohanan, P. 2021. Impact of nanoparticles in balancing the ecosystem. *Biointerface Research in Applied Chemistry* 11, 10461–10481. https://doi.org/10.33263/BRIAC113.1046110481

Tarafdar, J., Sharma, S., Raliya, R. 2013. Nanotechnology: interdisciplinary science of applications. *African Journal of Biotechnology* 12, 219–226.

Tighe-Neira, R., Recio, G., Rengel, Z., Inostroza-Blancheteau, C., Reyes-Díaz, M. 2023. Chapter 11 - Role of some nanomaterials in management of nutrients stress in crop plants. In: Husen, A. (ed), *Engineered Nanomaterials for Sustainable Agricultural Production, Soil Improvement and Stress Management, Plant Biology, Sustainability and Climate Change*, pp. 241–255. Academic Press, New York. https://doi.org/10.1016/B978-0-323-91933-3.00004-0

Tripathi, S., Mahra, S.J.V., Tiwari, K., Rana, S., Tripathi, D.K., Sharma, S., Sahi, S. 2023. Recent advances and perspectives of nanomaterials in agricultural management and associated environmental risk: a review. *Nanomaterials* 13, 1604. https://doi.org/10.3390/nano13101604

Ullah, S., Khan, M.N. 2022. Classification, synthetic, and characterization approaches to nanoparticles, and their applications in various fields of nanotechnology: a review. *Catalysts* 12, 386. https://doi.org/10.3390/catal12111386

Van Nguyen, D., Nguyen, H.M., Le, N.T., Nguyen, K.H., Nguyen, H.T., Le, H.M., Nguyen, A.T., Dinh, N.T.T., Hoang, S.A., Van Ha, C. 2022. Copper nanoparticle application enhances plant growth and grain yield in maize under drought stress conditions. *Journal of Plant Growth Regulation* 41, 364–375. https://doi.org/10.1007/s00344-021-10301-w

Vankova, R., Landa, P., Podlipna, R., Dobrev, P.I., Prerostova, S., Langhansova, L., Gaudinova, A., Motkova, K., Knirsch, V., Vanek, T. 2017. ZnO nanoparticle effects on hormonal pools in Arabidopsis thaliana. *Science of the Total Environment* 593–594, 535–542. https://doi.org/10.1016/j.scitotenv.2017.03.160

Verma, K.K., Song, X.-P., Joshi, A., Tian, D.-D., Rajput, V.D., Singh, M., Arora, J., Minkina, T., Li, Y.-R.. 2022. Recent trends in nano-fertilizers for sustainable agriculture under climate change for global food security. *Nanomaterials* 12, 173. https://doi.org/10.3390/nano12010173

Wang, X., Liu, X., Chen, J., Han, H., Yuan, Z. 2014. Evaluation and mechanism of antifungal effects of carbon nanomaterials in controlling plant fungal pathogen. *Carbon* 68, 798–806. https://doi.org/10.1016/j.carbon.2013.11.072

Wang, X., Xie, H., Wang, P., Yin, H. 2023. Nanoparticles in plants: uptake, transport and physiological activity in leaf and root. *Materials* 16, 3097. https://doi.org/10.3390/ma16083097

Wu, L., Jiang, Y., Zhao, F., He, X., Liu, H., Yu, K. 2020. Increased organic fertilizer application and reduced chemical fertilizer application affect the soil properties and bacterial communities of grape rhizosphere soil. *Scientific Reports* 10, 9568. https://doi.org/10.1038/s41598-020-66648-9

Yusefi-Tanha, E., Fallah, S., Rostamnejadi, A., Pokhrel, L.R. 2020. Zinc oxide nanoparticles (ZnONPs) as a novel nanofertilizer: influence on seed yield and antioxidant defense system in soil grown soybean (*Glycine* max cv. Kowsar). *Science of The Total Environment* 738, 140240. https://doi.org/10.1016/j.scitotenv.2020.140240

Yuvaraj, M., Sathya Priya, R., Jagathjothi, N., Saranya, M., Suganthi, N., Sharmila, R., Cyriac, J., Anitha, R., Subramanian, K.S. 2023. Silicon nanoparticles (SiNPs): challenges and perspectives for sustainable agriculture. *Physiological and Molecular Plant Pathology* 128, 102161. https://doi.org/10.1016/j.pmpp.2023.102161

Zhang, H., Goh, N.S., Wang, J.W., Pinals, R.L., González-Grandío, E., Demirer, G.S. et al. 2022. Nanoparticle cellular internalization is not required for RNA delivery to mature plant leaves. *Nature Nanotechnology* 17, 197–205. https://doi.org/10.1038/s41565-021-01018-8

Zhao, L., Lu, L., Wang, A., Zhang, H., Huang, M., Wu, H., Xing, B., Wang, Z., Ji, R. 2020. Nano-biotechnology in agriculture: use of nanomaterials to promote plant growth and stress tolerance. *Journal of Agriculture and Food Chemistry* 68, 1935–1947. https://doi.org/10.1021/acs.jafc.9b06615

2 Green Synthesis and Characterization of Plant-Mediated Metal Nanoparticles

Aparna S. Taware, Chandrakant K. Tagad, Laxmi G. Kathawate, Ajit K. Dhas, Satish U. Deshmukh, and Anant B. Kanagare

2.1 INTRODUCTION

Nanoparticles (NPs) are particles ranging in size from one to one hundred nanometers (nm). Compared to its bulk form, the nanoscale material exhibits unique, extraordinary, and more advantageous physical and chemical properties. This is because the ratio of surface area to volume of the material or particle is significantly increased at the nanoscale. NP research primarily focuses on metal NPs (MNPs) due to their ease of manufacture and the breadth of their potential applications (Ahmed et al. 2016; Gajendiran and Rajendran 2011). There are many different kinds of antimicrobials and antibacterials, as well as detectors, surface coating agents, and catalysts, mostly nano-catalysts for various organic transformations. Metallic NPs, such as silver (Ag), platinum (Pt), gold (Au), and palladium (Pd), have received the most attention from researchers. The investigation of AgNPs is quite interesting, particularly in the fields of medicine and healthcare. Ag is an effective antibacterial agent in addition to being toxic to cells (Rai et al. 2009). It has long been known that Ag can inhibit bacterial growth in both industrial and medical processes. Ointments containing AgNPs and AgNPs are the most common form of application for Ag in the medical business, where it is used to treat burns and open wounds and to prevent infection.

Green NPs have been synthesized using plant extracts, microorganisms, and environmentally and ecologically friendly reducing and capping agents. The use of plants in the production of NPs is a straightforward process that is not only fast and cost-effective but also environmentally beneficial (Anker et al. 2008). Compared to other known methods of synthesis, environmental or plant-mediated NP synthesis is regarded as the method of choice for use in human therapies due to its cost-effectiveness, safety, and environmental friendliness. The employment of biological processes provides straightforward, cost-effective, and friendly to the ecosystem approach for NP synthesis (Yaqoob et al. 2020). The Ag ions were reduced by using plant leaf extracts. The particle size distribution produced by this approach ranged from

 DOI: 10.1201/9781032659435-2

15 to 500 nm, and it was also found that reaction temperature influenced particle size. The rate of reduction in the ions by the plant leaf extract was comparable to that of chemical reduction. A particle with a nanometer size of 1–100 nm is referred to as a NP. Due to a rise in the ratio of surface area to volume of the material/particle, the nanoscale material exhibits novel, exceptional, and superior physical and chemical properties in comparison with its bulk structure (Baig et al. 2021).

This approach has been used to produce MNPs, including Au, Ag, Pt, and copper (Cu), which have shown promising results in applications such as cancer therapy, sensors, and water treatment. In addition, the use of plants in NP synthesis offers a sustainable, eco-friendly approach that can contribute to the development of green technologies and to the protection of the environment (Shankar et al. 2003).

2.2 PHYTO-SYNTHESIS OF METAL NPS

Traditional methods of producing MNPs involve the use of harsh and toxic chemicals, like sodium borohydride, and they have the potential to affect both the environment and human health. In addition, these methods require high temperatures and pressure conditions, which can be energy-intensive and expensive. Chemical reduction is one of the most common methods used for synthesizing MNPs (Guzman et al. 2012). This method involves the application of reducing agents, such as hydrazine or sodium borohydride, to reduce metal ions to NPs in solution. Another traditional method is thermal decomposition, which involves heating metal precursors to high temperatures to produce NPs. However, these methods often yield NPs with broad size distributions and low stability, limiting their potential applications (Hussain et al. 2016). Therefore, there is an increasing need for more sustainable and environmentally friendly methods for producing MNPs. Plant-mediated synthesis has emerged as an attractive alternative method due to its potential to produce NPs with controlled size and shape, high stability, and low toxicity. Using plants as reducing and stabilizing agents, we can make MNPs under ambient conditions, reducing energy consumption and eliminating the need for harsh chemicals. This method also has the potential to be scaled up for commercial production, making it a promising avenue for sustainable nanotechnology.

The production of MNPs by using plants is a rapidly emerging field and is gaining popularity because of its eco-friendliness and cost-effectiveness. This process, also known as green synthesis, is a sustainable and environmentally friendly alternative to conventional methods that use toxic chemicals and solvents. Green synthesis involves the use of plant extracts, which are rich in various biomolecules such as flavonoids, terpenoids, and alkaloids, which can act as stabilizing and reducing agents to produce NPs (Zhou and Tang 2018). The process involves the reduction of metal ions in the presence of these biomolecules, which results in NP production with controlled size and shape. Moreover, green synthesis using plants is highly efficient and can produce NPs in large quantities, making it an attractive alternative for industrial applications. The use of green synthesis not only reduces environmental impact but also offers a safer and eco-friendly approach to NP synthesis. With increasing demand for sustainable production methods, green synthesis using plants is poised to become a major player in nanotechnology.

2.3 A COMMON PROCESS TO SYNTHESIZE METAL NANOPARTICLES USING PLANT-BASED MATERIALS

Synthesis of MNPs using plant extracts involves common steps: collecting plant material, cleaning, drying, extracting with an appropriate solvent, reducing metal ions to form NPs, and stabilizing them (Figure 2.1). However, there are numerous variations with respect to the plant extraction methods and the temperature, pH, and salt concentrations for MNP synthesis in various reports. The standard protocol for the plant-mediated synthesis of MNPs is summarized as follows. Plant materials are collected from various locations on the plant and subjected to washing with detergents, followed by rinsing with double-distilled water for two to three cycles. The plant materials are subsequently allowed to partially dry at room temperature. The plant materials are individually measured and weighed before being boiled in an appropriate quantity of distilled water for 10–15 minutes. The resulting solution is filtered through a nylon mesh cloth to remove any solid residues and then stored at 4°C for subsequent NP synthesis. The collected filtrate is mixed with an aqueous solution of metal precursor salts such as Ag nitrate ($AgNO_3$) and chloroauric acid ($HAuCl_4$) at varying concentrations to form AgNPs and AuNPs. The mixture is kept at room temperature or subjected to varying synthesis conditions such as microwave irradiation, autoclaving, or incubation at 50°C–80°C, depending on the type of plant extract and the precursor salts (Kahandal et al. 2023). The formation of Ag/Au NPs is indicated by rapid changes in the color of the reaction mixture, resulting from the reduction of metal ions to their NP form. The ultraviolet–visible (UV–vis) spectrum of the NP solution is recorded. The characteristic surface plasmon resonance (SPR) peak in the visible spectrum further confirms the formation of MNPs.

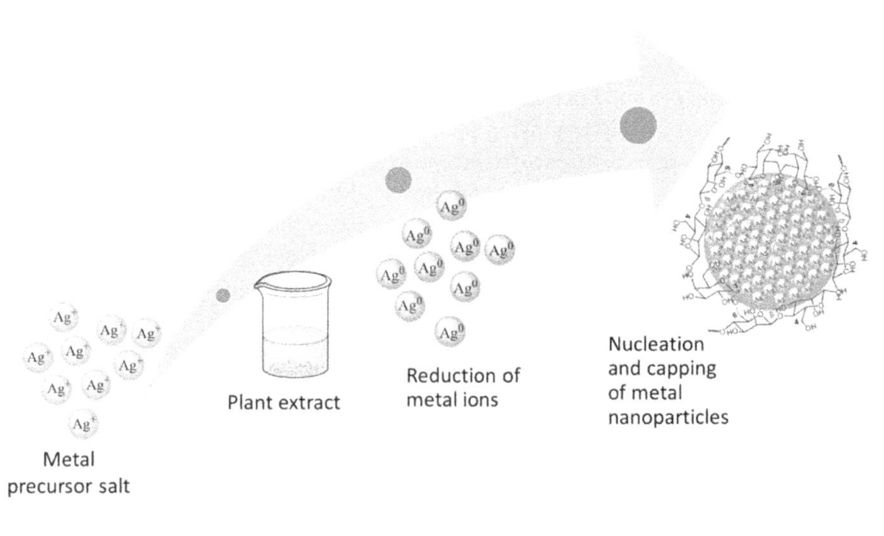

FIGURE 2.1 Biosynthesis of metal nanoparticles.

2.4 MECHANISM OF METAL NP SYNTHESIS

The issue of variable phytochemical composition plagues the synthesis of metallic NPs by the use of plant extracts in general. Therefore, it is extremely challenging to pinpoint the specific mechanism underlying the synthesis of metallic NPs using plant extracts. The role that polyphenols and flavonoids will play in the production and stabilization of metallic NPs, on the other hand, has only been hinted at in a small number of investigations. Determining the precise phytoconstituents responsible for the production of NPs in this situation is crucial. Therefore, in order to produce metallic NPs, researchers focused their efforts on isolating particular phytoconstituents that have antioxidant characteristics (Bao et al. 2021). In general, the formation of MNPs involves the reduction of MNPs to their atomic state, followed by nucleation and stabilization. In the first step of the process, metal ions are rapidly reduced, which results in the production of metal atoms. The second stage is initiated when these atoms begin to cluster together and form dimers and other extremely small groups. These clusters form in the third and final stage through the processes of aggregation and coalescence until they reach the desired particle size, at which time they are appropriately stabilized. This process continues until all of the clusters have been formed (Polte 2015).

2.5 FACTORS AFFECTING THE SYNTHESIS OF METAL NANOPARTICLES

The morphological and functional properties of the NPs are greatly influenced by various factors such as the choice of plant species, plant part, extraction method, the concentration of metal ions, and plant extract. Also, the reaction temperature, pH, and time are critical factors affecting the plant extract-mediated synthesis of MNPs. Careful optimization of these parameters is necessary to control the NP's properties and stability. The factors affecting NP synthesis are briefly discussed in the following sections.

2.5.1 Selecting the Right Plant Species for Nanoparticle Production

When it comes to producing MNPs using plants, selecting the right plant species is crucial. Not all plants are created equal when it comes to their ability to produce NPs. Some plants have higher levels of phytochemicals, which are natural compounds, and they may serve as reducing and stabilizing agents in the NP production process (Xia et al. 2013). Furthermore, the plant species chosen should also be easily available and cost-effective. Some popular plant species that have been used for NP production include green tea, neem, aloe vera, and garlic (Sun et al. 2014). It is important to note that the plant selection process should also take into account any potential environmental impact. Choosing invasive species or plants that are threatened or endangered could have negative consequences on the surrounding ecosystem.

Therefore, it is important to conduct thorough research and choose plant species that are sustainable and safe for the environment. Overall, selecting the right plant species for NP production is a critical step in the process and can ultimately impact the success and sustainability of the production.

2.5.2 Effect of pH

The pH of the solution medium plays a significant role in the plant-mediated biosynthesis of NPs. Numerous studies have indicated that the pH level of the solution directly affects the size, shape, and synthesis rate of the NPs. This phenomenon can be attributed to the formation of nucleation centers, which become more abundant as the pH increases. The presence of nucleation centers facilitates the reduction of metallic ions to MNPs, leading to an increase in their production rate. Simultaneously, the pH of the solution also influences the activity of functional groups present in the plant extract or biomass used for NP synthesis. Functional groups are specific chemical moieties that contribute to the reduction in metal salts. The activity of these functional groups can be modulated by variations in pH, thereby affecting the rate at which metal salts are reduced to form NPs (Iravani and Zolfaghari 2013; Nagar and Devra 2018).

The rate of AgNP synthesis has been found to increase with increased pH for various plant extracts and polysaccharides. With the increase in pH of the reaction mixture from acidic to more alkaline, the absorbance of the reaction mixture was found to increase with a concomitant increase in the color intensity of the colloidal solution. This could be because of more reduction of Ag ions to their atomic state and a subsequent increase in colloidal AgNPs (Iravani and Zolfaghari 2013). The synthesis of Cu NPs was observed to be significantly influenced by the pH of the reaction medium. Under acidic conditions, no Cu NPs were formed. However, as the pH increased, the formation of Cu NPs became apparent. The optimal pH for the synthesis was determined to be 6.6, as it resulted in the production of stable colloidal NPs. Furthermore, raising the pH beyond 6.6 further enhanced the rate of Cu NP formation. Nevertheless, this increase in pH also led to an increase in particle size, resulting in the aggregation of NPs (Nagar and Devra 2018).

Zan et al. synthesized AuNPs using *Cacumen Platycladi* leaf extract as a reducing and stabilizing agent and observed a significant impact of the pH on the synthesis of AuNPs (Zhan et al. 2011). The experimental findings revealed the presence of AuNPs exhibiting a variety of morphologies, including spheres, triangles, truncated triangles, and hexahedrons, at pH values of 4. At high pH levels, there was an accelerated reduction rate of Au^{3+} ions, facilitating homogeneous nucleation and suppressing anisotropic growth. Conversely, a slower reduction rate was observed under acidic conditions, leading to heterogeneous nucleation and the formation of small Au seed particles through secondary nucleation (Zhan et al. 2011). Moreover, at elevated pH levels, the AuNPs exhibited a higher negative charge due to the adsorption of hydroxide anions onto their surfaces. This resulted in electrostatic repulsion forces between the AuNPs, preventing their aggregation and stabilizing them.

2.5.3 Effect of Temperature and Pressure

The influence of elevated temperature on Ag, Au, and Cu nucleation was observed to significantly impact the reduction kinetics of Ag^{+}, Au^{3+}, and Cu^{2+} ions. This, in turn, affected the formation of NPs. Specifically, an increase in temperature expedited the transformation of Ag^{+}, Au^{3+}, and Cu^{2+} ions into Ag, Au, and Cu atoms by accelerating the rate of ion reduction. The heightened reduction rate facilitated the homogeneous nucleation of Ag/Au/Cu, where the atoms agglomerated to form stable nuclei, initiating the growth of NPs. The SPR peak intensity and the color of the colloidal NPs were found to increase with the increase in the temperature of the NP synthesis. Consequently, the utilization of higher temperatures, in combination with an efficient homogeneous nucleation process, resulted in the production of NPs characterized by a smaller average size and a narrower particle size distribution (Zhan et al. 2011).

2.5.4 Effect of Precursor Concentration

The concentration of precursor salts significantly influences the morphological characteristics of MNPs during synthesis. As the concentration of metal salts increases, more metal ions become available for reduction and subsequent nucleation and NP formation processes. This leads to an acceleration in the rate of NP formation. However, when the concentration surpasses a certain threshold, excessive nucleation occurs, forming larger NPs. Furthermore, NPs tend to aggregate at higher concentrations, causing their size destabilization and disrupting their colloidal behavior. Consequently, optimizing the precursor concentration is crucial to achieving the desired morphologies of the particles (Iravani and Zolfaghari 2013; Tagad et al. 2013a; Li et al. 2012).

2.6 METHOD OF METAL NP SYNTHESIS

The synthesis of MNPs using plant extract is a promising approach in the field of green nanotechnology. This method involves using plant materials, such as leaves, stems, or roots, as reducing and stabilizing agents for forming MNPs. Typically, the process begins by collecting plant samples and extracting their bioactive compounds. These extracts are then mixed with metal precursors and subjected to various synthesis conditions such as microwave irradiation, sonication, photoreduction, and autoclaving, initiating a reduction process mediated by the phytochemicals present in the plant extract. The reduction leads to the nucleation and growth of MNPs.

There are several techniques for synthesizing MNPs using plants, each with its own advantages and disadvantages. The choice of approach will rely on the particular requirements of the application. Each of these ways has benefits and drawbacks of its own. Regardless of the method used, synthesizing MNPs using plants is a promising approach that offers a more sustainable and eco-friendlier alternative to traditional methods. With further advancements in this field, it is expected that this approach will become even more widely adopted in NP synthesis for various applications (Sahu et al. 2016).

The commonly employed methods for synthesizing MNPs using plant extracts are summarized below.

2.6.1 Room Temperature Reduction

One of the most commonly used methods is the green synthesis method, which involves using plant extracts as reducing agents to facilitate the conversion from metal ions to NPs at room temperature. Green synthesis is a more eco-friendly approach using non-toxic and biodegradable chemicals, making it more sustainable (Shankar et al. 2004). Babak Sadeghi et al. synthesized AgNPs using *Ziziphora tenuior* (zt) extract as a reducing and stabilizing agent at room temperature (Sadeghi and Gholamhoseinpoor 2015). For the synthesis of AgNPs, equal volumes of 0.1 mM $AgNO_3$ and zt extract were mixed at room temperature, and the change in the color of the reaction mixture from colorless to yellow to brown was observed. The appearance of a characteristic yellow-brown color was attributed to the formation of AgNPs. The size of the AgNPs was found to be in the range of 8–40 nm as per the transmission electron microscopy (TEM) analysis (Sadeghi and Gholamhoseinpoor 2015). Several reports demonstrated the reduction of metal ions to their atomic state and the subsequent formation of NPs at room temperature. The ability to form NPs at room temperature depends on the extract type, the extract's phytoconstituents, etc. The rate of reaction at room temperature may be sluggish, resulting in the slow nucleation process and late stabilization of particles, leading to large aggregated particles. In order to form colloidal NPs with a narrow size distribution, the reactions can be performed at elevated temperatures.

2.6.2 Microwave Radiation

The microwave-assisted method involves using microwaves to accelerate the reduction process (Yin et al. 2004). Microwave irradiation represents a non-classical energy source that has demonstrated remarkable efficacy in facilitating uniform and rapid heating processes. This alternative energy delivery method has revolutionized the field of preparation and synthesis of NPs by significantly reducing overall reaction times (Yin et al. 2004).

This method is particularly useful for synthesizing NPs with a high degree of uniformity in size and shape. In this method, a reaction mixture containing precursor salts of the metal and the plant extract at optimal concentrations is exposed to microwave irradiation for a definite period. Microwave irradiation significantly enhances the formation of NPs with smaller sizes and a narrower particle distribution. Optimal concentrations of metal precursor salts were combined with the plant extract, and the resulting mixture was irradiated with microwaves. Microwave irradiation was carried out using a domestic microwave oven (Sharp R-219 T), operating at a power of 800 W and a frequency of 2,450 MHz (Vijayan et al. 2019). The appearance of a brownish color within 1 minute of microwave irradiation, with an SPR band at 441 nm, indicated the formation of AgNPs. By employing microwave irradiation, Tagad et al. could reduce the time required for AgNP synthesis from a few hours to a few seconds (Sharma et al. 2021).

2.6.3 Ultrasonication

Other methods of synthesizing MNPs using plants include the sonochemical method (Kumar et al. 2014). Because of its efficiency in generating uniformly dispersed and stable NPs, ultrasonication is one of the most common approaches used in NP synthesis (Sandhya et al. 2021; Alizadeh et al. 2019). Ultrasonication has been successfully employed for the synthesis of metal NPs using phytoconstituents from plant-based extracts. This method involves applying high-frequency sound waves to a liquid medium that already contains the precursor components. The ultrasonic waves cause sudden, significant pressure shifts within the liquid; this results in cavitation, the development, and subsequent bursting of minuscule gas bubbles (Sandhya et al. 2021). This cavitation process results in the nucleation and growth of NPs due to the generation of localized high temperatures and pressures. Ultrasonication generates significant shear pressures and turbulence, both of which help break down larger aggregates and encourage the production of smaller NPs with increased surface area and reactivity (Alizadeh et al. 2019; Sandhya et al. 2021). In addition, because both the strength and duration of ultrasonic treatment can be precisely controlled, ultrasonication enables rapid, effective control over the size and shape of a substance. Researchers can adjust the properties of NPs for a wide range of applications in disciplines such as catalysis, drug delivery, and energy storage using the ultrasonication method, providing a versatile, scalable approach to NP synthesis.

2.6.4 Hydrothermal Reduction

Hydrothermal synthesis is a method employed for the fabrication of NPs by subjecting an aqueous precursor solution containing metal ions or desired constituents to elevated temperatures and pressures. This process entails heating the precursor solution in an autoclave or an enclosed vessel to induce nucleation and the subsequent growth of NPs. The hydrothermal conditions create a supercritical environment that facilitates the formation of NPs with precise control over their dimensions, morphology, and composition. Manipulation of reaction parameters such as temperature, pressure, pH, and precursor concentration allows for tailored control over the size and shape of the resulting NPs. The hydrothermal technique enables the production of NPs exhibiting exceptional purity, uniform size distribution, and high crystallinity (Tagad et al. 2014; Tagad et al. 2013a,b). Kahandal et al. synthesized AgNPs using *Acacia nilotica* flower extract by autoclaving the reaction mixture at 120°C and 15 psi pressure. The appearance of a yellowish to brown color to the reaction mixture and the characteristic SPR peak at 405 nm indicated the formation of AgNPs (Kahandal et al. 2023). The functional groups of the biomolecules become more exposed and accessible to the metal ions for their reduction to their NP form (Tagad et al. 2013b).

2.6.5 Photoreduction

Another popular method of synthesizing MNPs using plants is the photochemical method, which involves the use of light to activate the reduction process. This technique harnesses the energy of light to drive the reduction of metal ions into their NP form. The photoreduction method has been employed previously to synthesize AuNPs by the living aquatic plant *Ulva armoricana* (Mukhoro et al. 2018). AuNPs were produced photochemically after transparent solutions containing *U. armoricana* extract were illuminated to sunlight. AuNPs were produced with a distinctive pink/purple hue, making it easy to see their quick growth. The ability of both live and aqueous extracts of *U. armoricana* to produce NPs was assessed (Mukhoro et al. 2018). In another report, Lopes et al. subjected the reaction mixture containing

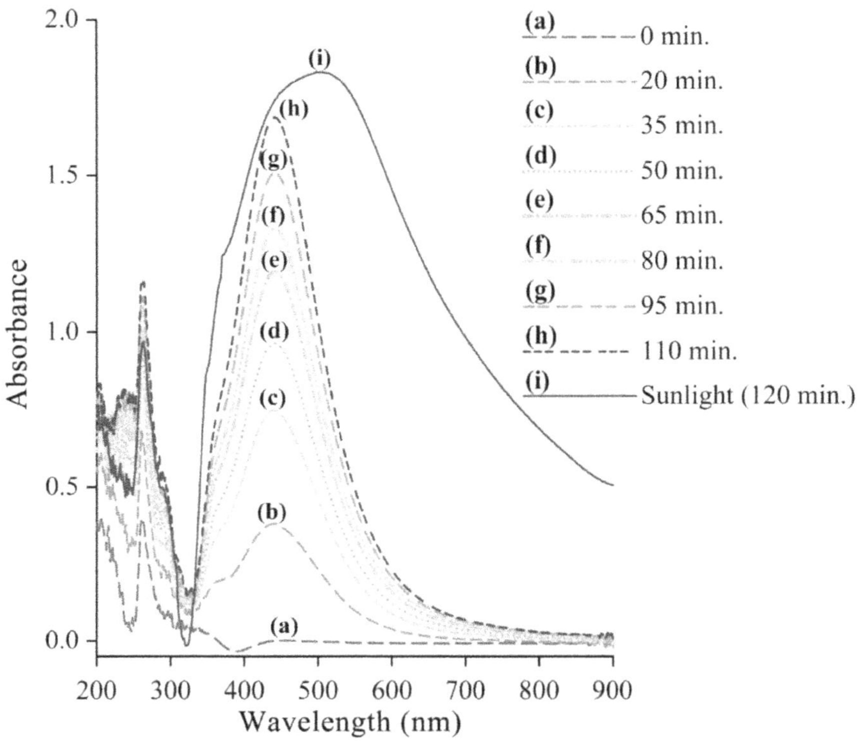

FIGURE 2.2 UV–vis spectra of plant extract–$AgNO_3$ mixtures exposed to varying reaction times. (*Source*: Ulug et al., 2015. Copyright @ 2015 Elsevier B.V.)

plant extract and the precursor salt to a 300-Watt Cermax xenon lamp for 2, 5, and 10 minutes for photoreduction of Ag^+ and subsequent formation of AgNPs (Lopes and Courrol 2018).

Bulent Ulug et al. studied the role of irradiation in the green synthesis of AgNPs mediated by *Ficus carica* leaf extract (Ulug et al. 2015). Figure 2.2 shows the UV–vis spectra of the AgNPs synthesized by irradiating the reaction mixture with 13.3 mW/cm^2 irradiance at 400 nm for different durations. The SPR peak intensity of the AgNPs in the reaction mixture increased remarkably with the photoirradiation time, indicating the consequence of photoirradiation to increase the rate of AgNP synthesis. The wavelength of light and the time or irradiation significantly affect the morphological properties of NPs (Ulug et al. 2015).

2.7 PHYTOCHEMICALS AS REDUCING AND CAPPING AGENTS FOR METAL NP SYNTHESIS

Phytochemicals, the naturally occurring bioactive compounds found in plants, have emerged as promising candidates for reducing and capping agents in the synthesis of NPs. These compounds possess inherent antioxidant and reducing properties, making them well-suited for mediating the reduction of metal ions into their NP form. Additionally, phytochemicals can serve as capping agents, providing a stabilizing layer around the synthesized NPs, preventing their agglomeration and maintaining their stability. Using phytochemicals as reducing and capping agents offers several advantages, including their abundance in nature, low toxicity, and eco-friendly nature compared to traditional chemical agents. Moreover, phytochemicals exhibit diverse structures and functional groups, allowing for a wide range of possibilities in tailoring the properties of the synthesized NPs.

Some of the most commonly used plant compounds for this purpose include polyphenols, flavonoids, terpenes, and alkaloids (Mashwani et al. 2016). These compounds are found in a wide range of plants, including grape seed extract, cinnamon, and turmeric (Ping et al. 2018). Polyphenols, for example, are known for their strong reducing properties, making them ideal for producing MNPs. On the other hand, flavonoids are excellent stabilizing agents, helping prevent NP aggregation and improve their overall stability (Sahu et al. 2016). By using these essential plant compounds in the NP production process, it is possible to create high-quality NPs that are both stable and biocompatible. This is particularly important for biomedical applications (Satyavani et al. 2011), where the safety and effectiveness of NPs are critical for their success. Overall, the use of plant compounds in the production of MNPs represents a promising new approach to NP synthesis that is both sustainable and environmentally friendly. As more research is conducted in this area, we can expect to see even more innovative uses for these plant-based NPs in a wide range of fields, from medicine to electronics and beyond (Kasana et al. 2017).

2.8 CHARACTERIZATION OF NANOPARTICLES

The successful utilization of NPs hinges on a thorough understanding of their structural, morphological, and chemical properties. Various analytical techniques play a crucial role in unraveling the intricacies of these nanomaterials. In this section, we delve into the characterization of NPs through X-ray diffraction (XRD), field emission scanning electron microscopy (FESEM), TEM, X-ray photoelectron spectroscopy (XPS), Fourier transform infrared spectroscopy (FTIR), and UV–vis spectroscopy.

XRD is a powerful tool employed for investigating the crystalline structure of NPs. When X-rays interact with the crystal lattice of a material, they undergo constructive interference, resulting in diffraction patterns. Analysis of these patterns allows for the determination of crystal phase, grain size, and lattice parameters. In the context of NPs, XRD aids in elucidating the size-induced effects on crystallographic properties. The broadening of peaks often observed in NP XRD patterns is attributed to the reduced size and increased surface energy.

FESEM and TEM are complementary techniques offering insights into the morphology and size distribution of NPs at different scales. FESEM provides high-resolution images of NP surfaces, enabling detailed examination of their external features. On the other hand, TEM provides cross-sectional views, offering a glimpse into the internal structure of NPs. These techniques are indispensable for assessing the size, shape, and distribution of NPs, crucial factors that influence their properties and applications.

XPS is a surface-sensitive technique employed to probe the elemental composition and chemical state of the outermost layers of nanomaterials. By analyzing the binding energies of electrons emitted upon X-ray irradiation, XPS provides information about the oxidation state and electronic environment of elements present in the NPs. In nanotechnology, XPS is particularly valuable for understanding the surface chemistry of NPs, helping researchers tailor their surfaces for specific functionalities.

FTIR is a spectroscopic technique utilized to identify the functional groups present in NPs. When infrared radiation interacts with a material, molecular vibrations occur, leading to characteristic absorption peaks. In the context of NPs, FTIR is instrumental in identifying the organic ligands or functional groups on their surfaces. This information is pivotal for understanding the chemical composition and potential interactions of NPs across various environments.

UV–vis spectroscopy is a versatile technique used to study the optical properties of NPs. The interaction of NPs with light in the UV–vis range results in electronic transitions, providing information about their band structure and SPR. SPR is a phenomenon arising from the collective oscillation of free electrons on the NP surface when exposed to light. Monitoring SPR via UV–vis spectroscopy enables researchers to tune the optical properties of NPs, a crucial aspect for applications such as sensing and imaging.

2.9 CHALLENGES AND FUTURE PROSPECTS

While the concept of producing MNPs using plants is certainly promising, it is important to understand the challenges and limitations of this approach. One of the main limitations is that not all plants can produce NPs. In fact, only a few species have been identified as having this ability, which can make it difficult to scale up production. Another challenge is the inconsistency of the process. The type of plant utilized, among other things, can have an impact on the NPs' size, shape, and purity, the concentration of metal ions, and the method of synthesis. This can make it difficult to produce NPs with the desired properties for specific applications. Furthermore, to fully comprehend the mechanisms, much research remains to be done on the green synthesis of NPs using plants. While this approach has shown promise as environmentally friendly and cost-effective, much remains unknown about how it works and how it can be optimized for commercial use. Despite these challenges and limitations, a promising field of research that has the potential to change manufacturing is the use of plants for the green production of MNPs the way we produce these materials. By continuing to explore this approach and addressing the challenges it entails, we may be able to develop innovative solutions for many applications.

2.10 CONCLUSION

The creation of NPs using green synthesis in the future has a bright future. Using plants to produce MNPs not only eliminates the need for toxic chemicals and reduces waste but also offers a sustainable option for meeting our growing demand for NPs. As more research is conducted in this field, we can expect to see advancements in the use of a broader range of plant species and the development of new techniques for producing NPs with specific properties. It is also important to note that while green synthesis offers many benefits, it is not a complete replacement for traditional methods. In certain cases, traditional methods may still be necessary to produce NPs with certain properties or in large quantities. Overall, the use of plants in NP synthesis is a promising development that offers a more sustainable and eco-friendly option for meeting our technological needs. As we continue to explore the possibilities of green synthesis, we can look forward to a future that is both innovative and environmentally responsible. This cutting-edge method of producing NPs is not only economical but also environmentally benign. With the help of plants, we can create high-quality NPs without the need for harmful chemicals or expensive equipment. As we become more conscious of our impact on the environment, it is exciting to see these types of sustainable solutions emerging.

REFERENCES

Ahmed, S., Ahmad, M., Swami, B.L., Ikram, S. 2016. A review on plants extract mediated synthesis of silver nanoparticles for antimicrobial applications: a green expertise. *Journal of Advanced Research* 7, 17–28. https://doi.org/10.1016/j.jare.2015.02.007

Alizadeh, S., Fallah, N., Nikazar, M. 2019. An ultrasonic method for the synthesis, control and optimization of CdS/TiO_2 core–shell nanocomposites. *RSC Advances* 9, 4314–4324. https://doi.org/10.1039/C8RA10155H

Anker, J.N., Hall, W.P., Lyandres, O., Shah, N.C., Zhao, J., Van Duyne, R.P. 2008. Biosensing with plasmonic nanosensors. *Nature Materials* 7, 442–453. https://doi.org/10.1038/nmat2162

Baig, N., Kammakakam, I., Falath, W. 2021. Nanomaterials: a review of synthesis methods, properties, recent progress, and challenges. *Materials Advances* 2, 1821–1831. https://doi.org/10.1039/D0MA00807A

Bao, Y., He, J., Song, K., Guo, J., Zhou, X., Liu, S. 2021. Plant-extract-mediated synthesis of metal nanoparticles. *Journal of Chemistry* 2021, 1–14. https://doi.org/10.1155/2021/6562687

Gajendiran, J., Rajendran, V. 2011. Synthesis of SnS_2 nanoparticles by a surfactant-mediated hydrothermal method and their characterization. *Advances in Natural Sciences: Nanoscience and Nanotechnology* 2, 015001. https://doi.org/10.1088/2043-6262/2/1/015001

Guzmán, M., Dille, J., Godet, S. 2012. Synthesis and antibacterial activity of silver nanoparticles against gram-positive and gram-negative bacteria. *Nanomedicine* 8, 37–45. https://doi.org/10.1016/j.nano.2011.05.007

Hussain, I., Singh, N.B., Singh, A., Singh, H., Singh, S.C. 2016. Green synthesis of nanoparticles and its potential application. *Biotechnology Letters* 38, 545–560. https://doi.org/10.1007/s10529-015-2026-7

Iravani, S., Zolfaghari, B. 2013. Green synthesis of silver nanoparticles using *Pinus eldarica* bark extract. *BioMed Research International* 2013, 1–5. https://doi.org/10.1155/2013/639725

Kahandal, A.W., Sharma, L., Sirdeshmukh, V., Kulkarni, A., Tagad, C.K. 2023. A sensitive image-based optical detection of heavy metal ions using green synthesized silver nanoparticles. *International Journal of Environmental Science and Technology* 20, 9077–9088. https://doi.org/10.1007/s13762-022-04539-4

Kasana, R.C., Panwar, N.R., Kaul, R.K., Kumar, P. 2017. Biosynthesis and effects of copper nanoparticles on plants. *Environmental Chemistry Letters* 15, 233–240. https://doi.org/10.1007/s10311-017-0615-5

Kumar, B., Smita, K., Cumbal, L., Debut, A. 2014. Green approach for fabrication and applications of zinc oxide nanoparticles. *Bioinorganic Chemistry and Applications* 2014, 1–7. https://doi.org/10.1155/2014/523869

Li, G., He, D., Qian, Y., Guan, B., Gao, S., Cui, Y., Yokoyama, K., Wang, L. 2012. Fungus-mediated green synthesis of silver nanoparticles using *Aspergillus terreus*. *International Journal of Molecular Sciences* 13, 466–476. https://doi.org/10.3390/ijms13010466

Lopes, C.R.B., Courrol, L.C. 2018. Green synthesis of silver nanoparticles with extract of *Mimusops coriacea* and light. *Journal of Luminescence* 199, 183–187. https://doi.org/10.1016/j.jlumin.2018.03.030

Mashwani, Z.-U.-R., Khan, M.A., Khan, T., Nadhman, A. 2016. Applications of plant terpenoids in the synthesis of colloidal silver nanoparticles. *Advances in Colloid and Interface Science* 234, 132–141. https://doi.org/10.1016/j.cis.2016.04.008

Mukhoro, O.C., Roos, W.D., Jaffer, M., Bolton, J.J., Stillman, M.J., Beukes, D.R., Antunes, E. 2018. Very green photosynthesis of gold nanoparticles by a living aquatic plant: photoreduction of AuIII by the seaweed *Ulva armoricana*. *Chemistry – A European Journal* 24, 1657–1666. https://doi.org/10.1002/chem.201704448

Nagar, N., Devra, V. 2018. Green synthesis and characterization of copper nanoparticles using *Azadirachta indica* leaves. *Materials Chemistry and Physics* 213, 44–51. https://doi.org/10.1016/j.matchemphys.2018.04.007

Ping, Y., Zhang, J., Xing, T., Chen, G., Tao, R., Choo, K.-H. 2018. Green synthesis of silver nanoparticles using grape seed extract and their application for reductive catalysis of direct orange 26. *Journal of Industrial and Engineering Chemistry* 58, 74–79. https://doi.org/10.1016/j.jiec.2017.09.009

Polte, J. 2015. Fundamental growth principles of colloidal metal nanoparticles – a new perspective. *CrystEngComm* 17, 6809–6820. https://doi.org/10.1039/C5CE01014D

Rai, M., Yadav, A., Gade, A. 2009. Silver nanoparticles as a new generation of antimicrobials. *Biotechnology Advances* 27, 76–83. https://doi.org/10.1016/j.biotechadv.2008.09.002

Sadeghi, B., Gholamhoseinpoor, F. 2015. Stability and green synthesis of silver nanoparticles using *Ziziphora tenuior* extract. *Spectrochimica Acta A* 134, 310–315. https://doi.org/10.1016/j.saa.2014.06.046

Sahu, N., Soni, D., Chandrashekhar, B., Satpute, D.B., Saravanadevi, S., Sarangi, B.K., Pandey, R.A. 2016. Synthesis of silver nanoparticles using flavonoids and their antibacterial effects and cytotoxicity. *International Nano Letters* 6, 173–181. https://doi.org/10.1007/s40089-016-0184-9

Sandhya, M., Ramasamy, D., Sudhakar, K., Kadirgama, K., Harun, W.S.W. 2021. Ultrasonication as an intensifying tool for preparation of stable nanofluids and influence on properties of graphene nanofluids – a systematic overview. *Ultrasonics Sonochemistry* 73, 105479. https://doi.org/10.1016/j.ultsonch.2021.105479

Satyavani, K., Gurudeeban, S., Ramanathan, T., Balasubramanian, T. 2011. Biomedical potential of silver nanoparticles synthesized from calli cells of *Citrullus colocynthis*. *Journal of Nanobiotechnology* 9, 43. https://doi.org/10.1186/1477-3155-9-43

Shankar, S.S., Ahmad, A., Sastry, M. 2003. Geranium leaf assisted biosynthesis of silver nanoparticles. *Biotechnology Progress* 19, 1627–1631. https://doi.org/10.1021/bp034070w

Shankar, S.S., Rai, A., Ahmad, A., Sastry, M. 2004. Rapid synthesis of Au, Ag and bimetallic Au–Ag nanoparticles using neem leaf broth. *Journal of Colloid and Interface Science* 275, 496–502. https://doi.org/10.1016/j.jcis.2004.03.003

Sharma, L., Gouraj, S., Raut, P., Tagad, C. 2021. Development of a surface-modified paper-based colorimetric sensor using AgNPs–alginate composite. *Environmental Technology* 42, 3441–3450. https://doi.org/10.1080/09593330.2020.1732471

Sun, Q., Cai, X., Li, J., Zheng, M., Chen, Z., Yu, C.-P. 2014. Green synthesis of silver nanoparticles using tea leaf extract and their antibacterial activity. *Colloids and Surfaces A* 444, 226–231. https://doi.org/10.1016/j.colsurfa.2013.12.065

Tagad, C.K., Dugasani, S.R., Aiyer, R., Park, S., Kulkarni, A., Sabharwal, S. 2013a. Green synthesis of silver nanoparticles for optical fiber based hydrogen peroxide sensor. *Sensors and Actuators B* 183, 144–149. https://doi.org/10.1016/j.snb.2013.03.106

Tagad, C.K., Kim, H.U., Aiyer, R.C., More, P., Kim, T., Moh, S.H., Kulkarni, A., Sabharwal, S.G. 2013b. A sensitive hydrogen peroxide optical sensor based on polysaccharide-stabilized silver nanoparticles. *RSC Advances* 3, 22940–22943. https://doi.org/10.1039/C3RA44547J

Tagad, C.K., Rajdeo, K.S., Kulkarni, A., More, P., Aiyer, R.C., Sabharwal, S. 2014. Green synthesis of polysaccharide-stabilized gold nanoparticles: catalytic and vapor sensing applications. *RSC Advances* 4, 24014–24019. https://doi.org/10.1039/C4RA02972K

Ulug, B., Turkdemir, M.H., Cicek, A., Mete, A. 2015. Role of irradiation in green synthesis of silver nanoparticles mediated by *Ficus carica* leaf extract. *Spectrochimica Acta A* 135, 153–161. https://doi.org/10.1016/j.saa.2014.06.142

Vijayan, R., Joseph, S., Mathew, B. 2019. *Costus speciosus* rhizome extract mediated synthesis of silver and gold nanoparticles and their biological and catalytic properties. *Inorganic and Nano-Metal Chemistry* 49, 249–259. https://doi.org/10.1080/24701556.2019.1661439

Xia, B., He, F., Li, L. 2013. Preparation of bimetallic nanoparticles via green synthesis and their application. *Langmuir* 29, 4901–4907. https://doi.org/10.1021/la400355u

Yaqoob, S.B., Adnan, R., Khan, R.M.R., Rashid, M. 2020. Gold, silver, and palladium nanoparticles: a chemical tool for biomedical applications. *Frontiers in Chemistry* 8, 1–16. https://doi.org/10.3389/fchem.2020.00376

Yin, H., Yamamoto, T., Wada, Y., Yanagida, S. 2004. Large-scale and size-controlled synthesis of silver nanoparticles under microwave irradiation. *Materials Chemistry and Physics* 83, 66–70. https://doi.org/10.1016/j.matchemphys.2003.09.006

Zhan, G., Huang, J., Lin, L., Lin, W., Emmanuel, K., Li, Q. 2011. Synthesis of gold nanoparticles by *Cacumen platycladi* leaf extract and insights into biosynthetic mechanism. *Journal of Nanoparticle Research* 13, 4957–4968. https://doi.org/10.1007/s11051-011-0476-y

Zhou, Y., Tang, R.-C. 2018. Eco-friendly fabrication of colored and bioactive silk materials using silver nanoparticles synthesized by flavonoids. *Polymers* 10, 404. https://doi.org/10.3390/polym10040404

3 Uptake, Translocation and Interaction of Nanoparticles in Plants

Sonika Sharma, Ashutosh Sharma, and Varsha Rattan

3.1 INTRODUCTION

Nanotechnology is a scientific discipline that refers to the study and application of materials whose particle size is reduced to the nanoscale (1–100 nm), exhibiting unique physicochemical behavior. At this extremely small scale, materials often demonstrate behavior and properties that differ markedly from their large bulk counterparts, which include enhanced strength, reactivity, and distinctive electrical and optical characteristics (Kumar and Singh 2025). Such nanoscale-specific traits have led to various transformative opportunities across multiple sectors and are reshaping modern technological development. Owing to its large scope, nanotechnology is inherently interdisciplinary, with applications extending into almost every domain of science and technology (Nasrollahzadeh et al. 2019; Sharma et al. 2025; Chauhan et al. 2023; Haleem et al. 2023). In medicine, nanoparticles (NPs) are engineered to achieve targeted and controlled drug delivery, particularly in the treatment of cancer, where they enhance the effectiveness of treatment, while reducing the side effects. In the energy sector, the performance of solar cells, fuel cells, and batteries is elevated by nanomaterials (NMs). Advances in electronics have made it possible to develop smaller, faster, and more energy-saving devices through nanoscale circuits and components. Additionally, nanobioremediation has emerged as a promising approach for environmental cleanup and pollution management (Bhardwaj et al. 2023).

Over the past few years, the rapid advancement of nanotechnology has demonstrated high potential in agriculture. Applications include NM-based fertilizers and pesticides for improved efficiency, nanosensors for real-time evaluation of soil and crop health parameters, and advanced nanocomposites for food packaging that extend their shelf life and preservation quality (Mishra et al. 2021; Islam et al. 2024). Beyond its small scale, nanotechnology has the ability to deal with various global challenges like sustainable agriculture, uncontaminated water, renewable energy, and advanced health services. Advances in imaging, fabrication, and analytical techniques have made it possible to do precise alteration of matter at the atomic state, which leads to the rapid expansion of nanoscience as a dynamic and interdisciplinary field. As a result, the expanding utilization of NMs in agricultural applications has highlighted the importance of understanding NP behavior in the environment and

 DOI: 10.1201/9781032659435-3

within plants. This chapter therefore explores the uptake, translocation, and interactions of NPs in plants, aiming to enhance crop productivity and stress resilience.

3.2 UPTAKE OF NANOMATERIAL IN PLANTS

NMs are readily taken up and translocated by plants. Several factors including particle size, exposure time, concentration, surface charge, and plant species influence this movement. Depending upon the route of exposure, NMs are introduced in plants either *via* roots from the soil or *via* leaves from foliar sprays. The route of NM exposure significantly impacts uptake and bioaccumulation of NMs by plants (Ma and Yan 2018). NM can readily infiltrate cell wall and plasma membrane of the root epidermal cells or pass into plant leaves through a polar pathway involving stomata, trichomes, hydathodes, and necrosis spots and a nonpolar path involving cuticle and the pores present in it (Ballikaya et al. 2023). Figure 3.1 illustrates "schematic representation of uptake of NMs in plants."

3.2.1 Nanomaterial Uptake by Roots

When present in soil, the NMs go through biogeotransformations, which eventually determine both bioavailability and toxicity of NMs. The first step in the

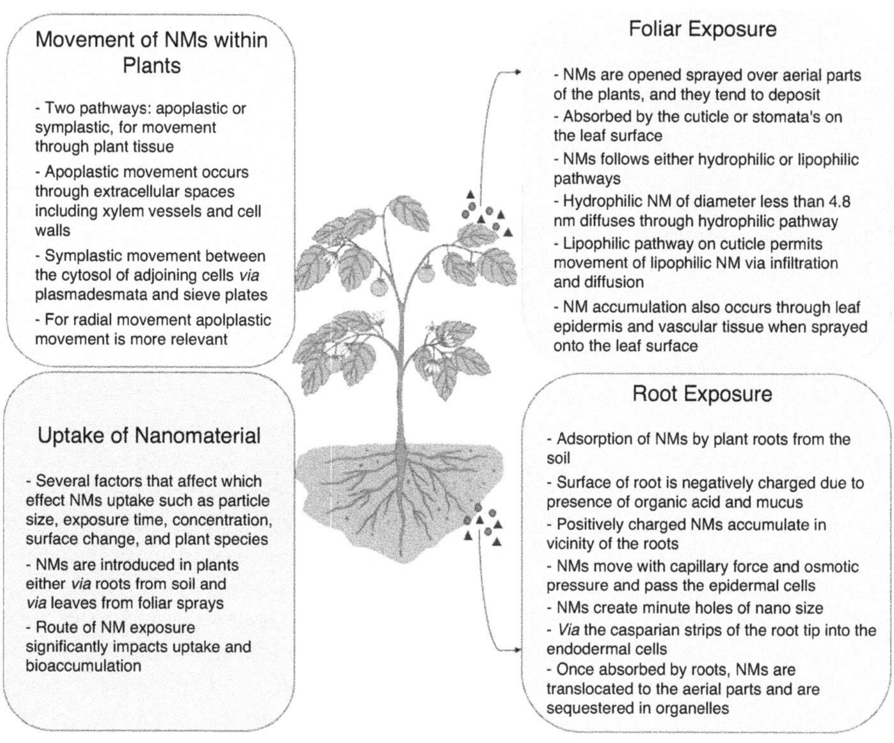

FIGURE 3.1 Schematic representation of uptake of nanomaterials in plants.

bioaccumulation of NMs is their adsorption by plant roots from soil (Wang et al. 2023). Since the root hairs release several chemicals including organic acids and mucus, the surface of the root is negatively charged, which makes NM possessing a positive charge to be readily accumulated in the vicinity of the roots and easily adsorbed onto the root surface (Lv et al. 2019). Lateral root formation further creates a new adsorption interface for NMs, thus facilitating the likelihood for NMs to gain entry into the root column (Peng et al. 2015). However, the root tip and root hair epidermis of both primary and secondary roots are not completely developed. When NM exposure occurs in these areas, it comes in contact with the semi-permeable cell wall of the epidermal cells and readily crosses it (Rajput et al. 2020).

The NMs with diameters ranging between 3 and 5 nm are observed to penetrate roots of the plants along with capillary forces and osmotic pressure, or they may pass right through the epidermal cells of the root. The cell wall of the root epidermis is semipermeable and contains minute pores that restrict the entry of large NMs (Ali et al. 2021). NM can also initiate the generation of membrane pores of nanosize in root cells that let them buildup in the plant roots. They can also move *via* the discontinuous areas in the Casparian strips of the root tip where the endodermal cells are immature by the crack-entry method and also at places where secondary root initiates. Even though there have been controversial statements on root uptake contingent with the type of NMs, time of exposure, and plant traits, it is commonly believed that NMs >100 nm are readily absorbed by plant roots regardless of size exclusion limits (<20 nm) (Ballikaya et al. 2023). Once absorbed by the plant roots, NMs translocate into aerial parts of the plant and are largely sequestered in cellular/subcellular organelles (Wang et al. 2023).

3.2.2 Nanomaterial Uptake by Leaves

During agricultural practices, NMs are often sprayed over aerial parts of the plant, where they tend to deposit and eventually are absorbed by the cuticle or the stomata on the leaf surface. The cuticle of the leaf epidermis is made up of cutin, pectin, and wax, which protects the plant from unnecessary water loss during its growth and eventually behaves as a natural barrier that precludes NMs from gaining entry into the leaves. However, to cross the barrier imposed by the cuticle, NM follows either the hydrophilic pathway or the lipophilic pathway. The channel diameters of hydrophilic and lipophilic pathways vary between 0.6 and −4.8 nm (Eichert et al. 2016). The hydrophilic pathways only allow the diffusion of hydrophilic NM of a diameter less than 4.8 nm (Banerjee et al. 2019), whereas the lipophilic pathway on cuticle surface permits the movement of only lipophilic NM through infiltration and diffusion (Wang et al. 2023). Hu et al. (2020) studied that carbon dots of size smaller than 2 nm enter the leaves of the cotton plant through the cuticle. Nonetheless, owing to the insufficient pore channel size of the cuticle, the NM absorption by plants *via* the epidermis is inadequate. Some researchers have indicated that NM accumulation occurs in leaf epidermis as well as the vascular tissue when sprayed onto the leaf surface.

3.2.3 Movement of NMs within the Plant

Once the NM enters the plant, it follows either of the two ways, apoplast or symplast, for moving through the plant tissues. Apoplastic transport occurs through extracellular spaces such as xylem vessels and cell walls of adjoining cells outside the plasma membrane through the extracellular spaces. On the other hand, symplastic transport encompasses water and NM movement between the cytosol of adjoining cells *via* specific structures known as plasmodesmata and sieve plates (Pérez-de-Luque 2017). Within the plant tissue, for radial movement, the apoplastic route is more relevant and permits NM to reach the central cylinder of the root and the vasculature for further upward movement into the aerial parts of the plant and this movement is largely dependent upon the surface charge and the particle size of the NM (Wang et al. 2023). Once the NM reaches the central cylinder, they move upward through the xylem, along with the transpirational stream. However, to reach xylem *via* roots requires a barrier crossover of the Casparian strips to the apoplast route, which must be done through the symplastic way *via* cells of the endodermis. In fact, certain NMs are stopped and cannot move further, leading to their accumulation in the Casparian strip (Lv et al. 2019). Furthermore, symplastic transport through the sieve tube elements of the phloem allows the distribution of NMs into the non-photosynthetic plant tissues and organs (Raliya et al. 2015). Several studies have revealed that NM with a size <50 nm is transported in plants through the symplast route, while most of the NMs with a size range between 50 and 200 nm are transported *via* the apoplast route (Raliya et al. 2015). It is of utmost importance to understand the way NM moves inside the plants, since it indicates in which part of the plant it will reach and accumulate. If NM transportation occurs through xylem and not through the phloem, they are more likely to travel in an upward direction from roots to shoot and leaves; thus, they are applied to the soil to get an even distribution within the plant. To the contrary, if NMs exhibit a better translocation via the phloem, they are applied in the form of foliar sprays (Pérez-de-Luque 2017).

3.2.4 Absorption of Nanomaterial by Plant Cells

Once NM gains entry into the plant tissue, they get absorbed in plant cells via different routes including endocytosis, ion pathway, binding with proteins of the cell membrane or physical damage (Lv et al. 2019). Invagination of the cell's plasma membrane leads to the formation of a vesicle that carries the NM to different cellular compartments and is referred to as endocytosis. Several studies have revealed that the protoplast of the plant cells can internalize materials of size <1 μm *via* endocytosis; therefore, the NMs absorbed through endocytosis exhibit no size specificity of the particle (Ma and Yan 2018). Carbon nanotubes were observed to be readily absorbed by *Catharanthus roseus* root cells by endocytosis (Liu et al. 2009). On the other hand, certain NMs can disorient the plasma membrane and induce the creation of pores in order to cross and gain entry into the cell and directly reach the cytoplasm without gaining entry into any cellular organelle (Wang et al. 2023). Certain NMs have the capacity to bind with cell membrane proteins, which act as carriers for the uptake and internalization of NMs into the cell. Another way by which NM gains entry into the plant cells is *via* plasmodesmata. For this, the NM must be present in the symplast, and once they cross the plasmodesmata, they are sequestered in the cytosol and eventually get transferred to the Casparian strips and endothelial layers in the plant. This mode of movement of NM is important for translocation *via* phloem (Miralles et al. 2012). Ion channels have also been suggested as a possible route for NM to gain entry into the plant cell. However, these channels have a size of about 1 nm making it difficult for the NM to effectively pass through them without significant alterations (Wang et al. 2023). How NM is internalized into the plant cells is of utmost importance since it determines the practicality of NM application. If NM is to be delivered into a certain plant cell organelle, then endocytosis seems to be the most relevant method. On the other hand, if the NM is to be delivered into the cytoplasm, pore formation in the plasma membrane is a more efficient and effective way for the same (Pérez-de-Luque 2017).

3.3 MECHANISM OF NANOPARTICLE UPTAKE AND TRANSLOCATION IN PLANTS

When NPs are applied to foliage, their entry and movement within plants depend on physicochemical interactions, leaf surface structure, and internal transport mechanisms. The movement of NPs is influenced by various factors such as droplet size, spray formulation, surfactants, humidity, wind, temperature, and plant physiological state (Banerjee et al. 2019). NPs first come in contact with the leaf cuticle, which is a hydrophobic, wax-rich barrier of cutin and waxes. At this stage, the adhesion and penetration of NP depend on charge, particle size, surface characteristics, and hydrophilic/hydrophobic nature. The smaller droplets of sprayed NP formulation having low surface tension spread more effectively over the entire leaf surface, while evaporation increases the concentration of NPs in the micro-cracks of the cuticle, around trichome bases, or near the stomatal opening. The cuticle comprises pathways mainly lipophilic and hydrophilic pathways, allowing the small hydrophobic or polar compounds to diffuse within wax layers or water-filled pores (Schreiber 2005). Certain engineered NPs like TiO_2 and ZnO are capable to locally modify the permeability of the cuticle, thereby facilitating their penetration (Navarro et al. 2008).

Stomata serve as a vital secondary route of entry. Under favorable environmental conditions, the stomata are usually open, which allows particles smaller than 500 nm to pass and enter via airflows or get deposited directly into the substomatal cavity (Eichert et al. 2008). Hydathodes, which are present at the leaf margins, may offer direct xylem entry during guttation (Li et al. 2018). Entry may additionally occur via lenticels, trichomes, or minor epidermal injuries.

Within leaf tissues, NPs move through the apoplast, cell walls, and intercellular spaces or the symplast, where they cross membranes via endocytosis, ion channels, aquaporins, and transport proteins. Apoplastic transport is often faster but may involve adsorption or dissolution influenced by wall chemistry, pH, and redox environment (Rai et al. 2018). Symplastic transfer through plasmodesmata, though size-restricted, bypasses some apoplastic limitations.

For systemic distribution, apoplast-moving NPs typically enter the xylem and move upward with transpiration, whereas ionic or soluble NPs may load into phloem

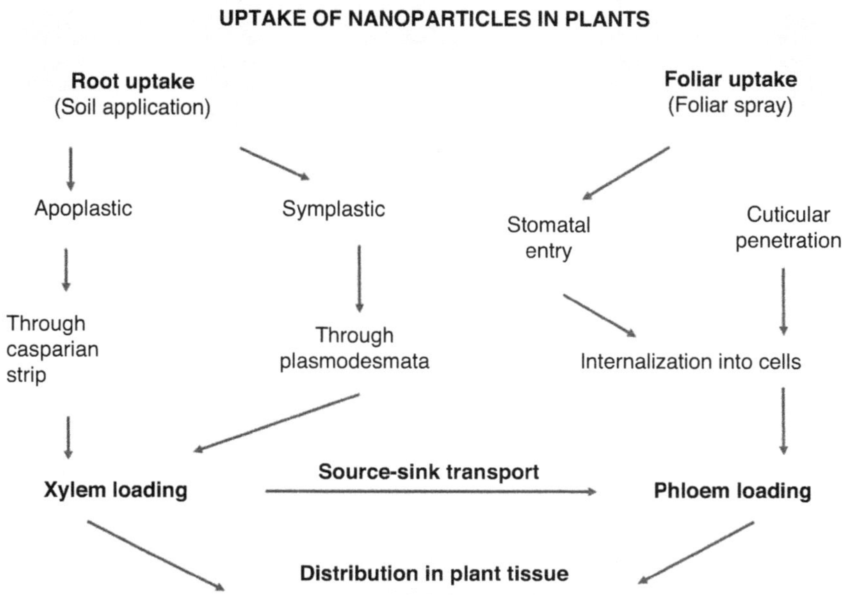

FIGURE 3.2 A flow chart showing the different modes of application, uptake, transport, accumulation, and distribution of nanoparticles in plants.

for bidirectional flow (Corredor et al. 2009). Inside tissues, NPs may dissolve, transform chemically, or become sequestered in cell walls or vacuoles. Overall uptake efficiency is determined by NP properties, species-specific leaf anatomy, environmental conditions, and application techniques (Figure 3.2).

3.4 TRANSLOCATION OF NANOPARTICLES IN PLANTS

After entering plant systems through roots or foliar surfaces, NPs are transported within tissues via vascular pathways (Figure 3.2) (Rodrigues et al. 2024). The xylem mainly facilitates upward movement from roots to aerial parts driven by transpiration pull, resulting in NP accumulation in stems, leaves, and other shoot structures (Wang et al. 2023). Conversely, the phloem allows bidirectional movement, particularly when NPs dissolve into ions or interact with organic molecules, enabling transport toward actively growing tissues, metabolic sinks, developing roots, and seeds (Wang et al. 2025). Movement through the apoplastic pathway (cell wall continuum) or symplastic pathway (through plasmodesmata and cytoplasm) depends on particle size, charge, and surface characteristics (Bhati et al. 2025). The NPs of smaller size usually less than 20 nm exhibit greater translocation potential and enhanced phloem loading efficiency (Corredor et al. 2009). Moreover, the transport efficiency is influenced by various factors like plant species, their tissue type, various developmental stages, and physicochemical properties of NPs (Khan et al. 2023). It is very important to understand these transport mechanisms for optimizing agricultural use of NPs and for evaluating food-chain safety risks associated with bioaccumulation.

The uptake of NPs by roots is influenced by various structural, biochemical, and physiological factors. Upon contact, the NPs first interact with the rhizoplane, where NPs are adsorbed through association with mucilage, organic acids, and sugars released by root exudates. Acidification of the surrounding rhizosphere by root hair further leads to an increase in the NP adhesion. The positively charged NPs exhibit stronger binding and uptake due to the negatively charged nature of the roots (Lev et al. 2019). Lateral root initiation zones also provide accessible entry sites (Peng et al. 2015). The easy penetration of NPs is facilitated by young root regions that lack well-developed suberin layers and Casparian strips compared to mature segments (Rajput et al. 2020). Cell wall pore size (3.5–20 nm) acts as a size-selective barrier controlling NP entry (Pérez-de-Luque 2017).

The NPs may travel through the apoplastic pathway toward the stele, in roots that lack fully formed barrier tissues (Péret et al. 2009). Their uptake into roots involves multiple pathways, including endocytosis (Ma and Yan 2018), ion channels, aquaporins, and metal transporters (Grillo et al. 2021). The NP penetration is enhanced through the damaged roots as compared to the young, actively growing roots, since the young roots show greater permeability (Péret et al. 2009). After reaching the stele, NPs are transported through the xylem and may later be redistributed into the phloem depending on the plant physiological state.

3.5 INTERACTION WITH SOIL

After entering the soil, NPs undergo various physical and chemical changes such as dissolution, aggregation, agglomeration, surface modification, growth, and aging, and all these changes determine their mobility, reactivity, persistence, and availability to organisms. The positive or negative interactions of NPs with soil microbes, plants, and minerals are affected by all these transformations. The soil complex matrix involving minerals, organic matter, water, and ions plays a significant role as it is responsible to either promote or slow down these processes, ultimately impacting nutrient cycling and contaminant behavior. Thus, the fate of NP in soil is regulated by parameters such as stability, aggregation/agglomeration, surface properties, dissolution, pH, point of zero charge (PZC), ionic strength, and zeta potential (Tarafdar and Adhikari 2015).

a. **Stability**: Various factors such as surface energy, size, shape, and solubility influence the stability of NPs. Materials with lower surface energy (e.g., anatase TiO_2) tend to be more stable, which determines whether NPs stay dispersed in the soil solution or undergo certain transformations in soil.
b. **Aggregation and Agglomeration**: The strong or loose aggregation of NPs is driven by the Brownian motion, electrostatic charge, and particle concentration. As the clusters increase in size, their sedimentation rate rises, which leads to a reduction in mobility of NPs in the soil matrix.
c. **pH**: The surface charge and zeta potential of NP are controlled by pH. Aggregation is more pronounced near PZC, where electrostatic repulsion is

minimal, while strong positive or negative surface charges promote dispersion and stability. Particles aggregate near PZC and remain stable at high positive/negative charges.

d. **Dissolution**: The smaller the NPs, the more is its reactivity and it usually dissolves faster, releasing ions (e.g., Ag^+, Zn^{2+}) depending on various factors like pH, redox conditions, and ligands, affecting toxicity and transformation.
e. **Surface Coating**: The surface charge and stability of NP depend on the surface coatings. The uncoated NPs are responsible for the easier aggregation near the isoelectric point per DLVO theory.
f. **Organic Matter**: The dispersion of NP is enhanced by soil organic matter coatings, improving dispersion by preventing aggregation. It also increases mobility and persistence within the soil matrix.
g. **Ionic Strength**: Elevated ionic strength compresses the electrical double layer around NP surfaces, reducing electrostatic repulsion and increasing NP aggregation, particularly for negatively charged particles.
h. **Zeta Potential**: NP with high absolute zeta potential values remains stable, whereas low zeta potential favors coagulation. The aggregation is typically greatest near the isoelectric pH (~6).

3.6 INTERACTION WITH RHIZOSPHERE MICROBES

Many microbes which live closely with plant roots, forming the rhizosphere microbiome, are often known as the plant's "second genome," which play an essential role in nutrient uptake, growth promotion, and stress tolerance. These microbes engage in various interactions ranging from mutualism to parasitism and operate within a root-influenced zone shaped by root exudates and the surrounding soil environment. When NPs enter this system, rhizosphere microorganisms exhibit distinct responses compared to free-living microbes because their behavior is influenced by plant-mediated processes and soil chemical characteristics. NPs can influence key microbial functions like nitrogen fixation, nutrient mineralization, and plant growth promotion; however, the magnitude and direction of these effects vary widely depending on both soil attributes and NP properties (Rajput et al. 2022).

Research indicates that the rhizosphere can adsorb or immobilize NPs, reducing their toxicity, though microbial sensitivity to these particles varies among species. For example, maize with arbuscular mycorrhizal fungi (AMF) tolerated 1 mg/kg Ag NPs, but AMFs were negatively affected (Wang et al. 2020). Reduced graphene oxide damages fungal hyphae; Ag and Ti NPs may stimulate fungal activity at low doses, whereas high ZnO NP levels suppress fungal colonization unless mitigated by $ZnSO_4$. CeO_2 NPs and MWCNTs lowered soybean root exudation, reducing rhizobial attraction and nodulation (Mortimer et al. 2020), and Ag, ZnO, and TiO_2 NPs reduced nodulation in other systems (Judy et al. 2015). Overall, low NP doses may be beneficial, but higher exposure typically harms rhizosphere microbes and plant–microbe interactions, stressing the need for cautious agricultural use.

3.7 USE OF NANOTECHNOLOGY TO CONTROL PLANT DISEASES AND INSECT PESTS

The overuse of chemical pesticides, fungicides, and herbicides in conventional agriculture has resulted in environmental contamination, pest resistance, and health hazards. Nanotechnology has become a viable substitute for crop protection in order to get around these restrictions. By destroying pathogen cell membranes, producing reactive oxygen species, or preventing insect feeding, NPs like silver, zinc oxide, titanium dioxide, silica, alumina, and chitosan-based biopolymers have strong antimicrobial, antifungal, and insecticidal qualities (Alimohamadian et al. 2022). The solubility, bioavailability, and gradual release of active ingredients are enhanced by nano-encapsulated fungicides and pesticides (Duhan and Gahlawat 2014; Kumar et al. 2025). Chitosan nanocarriers also activate plant defenses, strengthening systemic resistance (Elmer et al. 2018). While nanotechnology enhances efficacy and environmental safety, concerns about NP persistence, toxicity, and regulatory gaps remain, requiring careful evaluation for safe agricultural use (Ullah et al. 2020; Singh et al. 2021) (Table 3.1).

3.8 CHALLENGES

Nanotechnology has a huge potential in enhancing agricultural productivity, but its practical adoption is constrained by significant scientific and regulatory challenges. One of the major limitations is the problem in precise detection and determination of NPs in the plant tissues since their size is in nanoscale and resembles naturally occurring biogenic minerals. Conventional electron microscopy (TEM, SEM) provides qualitative insights, whereas advanced methods like synchrotron X-ray fluorescence and single-particle ICP-MS provide higher precision but are extremely expensive and technically demanding (Schwab et al. 2016).

The uptake and translocation of NPs differ widely among plant species and depend on the structure of the root/shoot, developmental stage, and mode of exposure (Ma et al. 2010). Various soil factors such as pH, organic matter, and microbial populations are also strongly determined by NP behavior, stability, and bioavailability (Wang et al. 2016), making universal application protocols difficult. Food safety concerns appear as NPs may gather in plant parts which are edible and potentially transfer into the food web (Rico et al. 2011). Long-term exposure to NPs can induce oxidative stress, DNA damage, metabolic disruption, and genotoxicity in plants (López-Moreno et al. 2010; Rizwan et al. 2019).

Another major bottleneck in nano-agriculture is the lack of internationally harmonized regulatory standards, which restricts farmer adoption and public acceptance (Kah et al. 2019). Future research directions emphasize biodegradable NMs such as chitosan, starch, and lignin, along with stimuli–responsive nanocarriers for safer agro-input delivery (Rastogi et al. 2017). The integration of advanced imaging and omics-based tools will further enable safe, efficient, and sustainable application of nanotechnology in agriculture (Kah et al. 2019).

TABLE 3.1
Use of Nanoparticles in the Management of Insect Pests

Nanoparticles/ Nanomaterials	Target Insect Pest	Effect/Mode of Action	Reference(s)
Silver nanoparticles (AgNPs)	*Tribolium castaneum* (red flour beetle) *Spodoptera litura*	Disrupts gut enzymes, induces oxidative stress, causes mortality	Bharani and Namasivayam (2017)
Zinc oxide nanoparticles (ZnO NPs)	*Callosobruchus chinensis* (pulse beetle) and *C. maculates*	Reduced fecundity, total development period, and complete mortality	Malaikozhundan et al. (2017); Mohammed and Aswd (2019)
Silica nanoparticles (SiO_2 NPs)	*Callosobruchus maculates* and *Sitophilus oryzae*	Cuticular abrasions, desiccation, and death	Ibrahim et al. (2024); Debnath et al. (2011)
Alumina nanoparticles (Al_2O_3 NPs)	*Sitophilus oryzae* (rice weevil)	Physical abrasion and disruption of the insect cuticle, leading to dehydration and mortality	Stadler et al. (2012)
Carbon nanotubes (CNTs)	*Plutella xylostella* (diamondback moth)	Efficacy of gene knockdown, leading to larval mortality	Ikawati et al. (2021)
Titanium dioxide nanoparticles (TiO_2 NPs)	*Musca domestica* (housefly)	Affected pupae weight, adult emergence, inhibiting adults, and sex ratio	Mohamadeen and Hassona (2024)
Iron oxide nanoparticles (FeNPs)	*S. granarius*	100% mortality	Sivapriya et al. (2018)
Silicon dioxide nanoparticles (SiO_2NPs)	*Cowpea weevil, C. maculatus*	100% mortality increased with time of exposure	Rouhani et al. (2013); Diagne et al. (2019)
Nanoemulsions	*T. castaneum*	Effective against both larvae and adults acutely toxic	Nenaah (2014); Giunti et al. (2019)

3.9 CONCLUSION AND FUTURE PERSPECTIVES

The uptake, movement, and interaction of NPs in crop plants define their potential for shaping the possible applications in sustainable agriculture practices. With their nanoscale properties, NPs play a significant role in enhancing nutrient delivery, strengthening plant defense mechanisms against pests and diseases, and reducing abiotic stress effects such as drought conditions, salinity, and heavy metal toxicity. Their ability to move through apoplastic and symplastic pathways and associate with cellular and biochemical processes provides them a distinctive advantage in influencing plant growth and physiology. Within the soil–plant–microbe continuum, NPs provide both opportunities and ecological concerns. They may enhance microbial activity, nutrient cycling, and soil health; however, inappropriate application can negatively impact the microbial balance, lead to bioaccumulation, and ultimately generate phytotoxicity. Agricultural adoption is progressing, majorly in India where nano-urea and nano-DAP have been commercialized by IFFCO and are being widely utilized. When used in combination with integrated nutrient management (INM), organic amendments, and beneficial microbes, NPs might lead to strengthening nutrient-use efficiency, productivity, and crop robustness.

For the application of nanotechnology to be safely incorporated in agriculture, it is very essential to be thorough with the toxicity evaluations, clearly defined exposure limits, and strong biosafety framework. Future progress will increasingly depend upon biodegradable, environmentally compatible NMs and collaborative, multidisciplinary research approaches to address challenges related to species-specific uptake behavior, detection difficulties, and uncertainties regarding long-term ecological impacts. Therefore, a responsible and balanced strategy that maximizes all benefits while minimizing potential hazards will enable nanotechnology to play a vital role in achieving global food security and sustainable agricultural development.

ACKNOWLEDGEMENT

The authors are thankful to the DAV University, Jalandhar, administration for providing all the infrastructural support required for the preparation of the manuscript.

REFERENCES

Ali, S., Mehmood, A., Khan, N. 2021. Uptake, translocation, and consequences of nanomaterials on plant growth and stress adaptation. *Journal of Nanomaterials* 2021, 1–17.

Alimohamadian, M., Aramideh, S., Mirfakhraie, S., Fourouzan, M. 2022. Silica nanoparticle: a potential new insecticide for beet armyworm, *Spodoptera exigua* control. *Agricultural Engineering International: CIGR Journal* 24(2), 1–9.

Ballikaya, P., Brunner, I., Cocozza, C., Grolimund, D., Kaegi, R., Murazzi, M.E., Schaub, M., Schönbeck, L.C., Sinnet, B., Cherubini, P. 2023. First evidence of nanoparticle uptake through leaves and roots in *Fagus sylvatica* and *Pinus sylvestris*. *Tree Physiology* 43, 262–276.

Banerjee, K., Pramanik, P., Maity, A., Joshi, D.C., Wani, S.H., Krishnan, P. 2019. Methods of using nanomaterials in plant systems: uptake, translocation, accumulation, biotransformation. *Advances in Phytonanotechnology* 2019, 123–152.

Bharani, R.A., Namasivayam, S.K.R. 2017. Biogenic silver nanoparticles mediated stress on developmental period and gut physiology of *Spodoptera litura*. *Journal of Environmental Chemical Engineering* 5, 453–467.

Bhardwaj, T., Khanna, K., Sharma, P., Bakshi, P., Devi, K., Madaan, I. et al. 2023. Nanobioremediation: a novel technology with phenomenal clean up potential for a sustainable environment. *Environmental Applications of Microbial Nanotechnology* 2023, 315–331.

Bhati, V.S., Singh, P.K., Parmar, Y.J., Rokins, P.G. 2025. Foliar uptake and translocation of nanoparticles in plants: a review. *Journal of Plant Nutrition* 49, 1–38.

Chauhan, A., Pallvi, Rattan, P., Ludarmani, Sharma, A. 2023. Effect of pre-soaking of potato tubers in nano-urea and nano-zinc on growth, quality and yield. *Journal of Pharmacy Innovation* 12(7), 980–995.

Corredor, E., Testillano, P.S., Coronado, M.J., González-Melendi, P., Fernández-Pacheco, R., Marquina, C., Ibarra, M.R., Rubiales, D., Pérez-de-Luque, A. 2009. Nanoparticle penetration and transport in living pumpkin plants: in situ subcellular identification. *BMC Plant Biology* 9, 45.

Debnath, N., Das, S., Seth, D., Chandra, R., Bhattacharya, S.C., Goswami, A. 2011. Entomotoxic effect of silica nanoparticles against *Sitophilus oryzae*. *Journal of Pest Science* 84, 99–105.

Diagne, A., Diop, B.N., Ndiaye, P.M., Andreazza, C., Sembene, M. 2019. Efficacy of silica nanoparticles on groundnut bruchid *Caryedon serratus*. *African Crop Science Journal* 27, 229–235.

Duhan, J.S., Gahlawat, S.K. 2014. Biogenesis of nanoparticles: a review. *African Journal of Biotechnology* 13(28), 1–12.

Eichert Faisal, M., Saquib, Q., Alatar, A.A., Al-Khedhairy, A.A., Ahmed, M., Ansari, S.M., Alwathnani, H.A., Dwivedi, S., Musarrat, J., Praveen, S. 2016. Cobalt oxide nanoparticle-induced DNA damage and cell death in eggplant. *Biological Research* 49, 1–13.

Eichert, T., Kurtz, A., Steiner, U., Goldbach, H.E. 2008. Size exclusion limits and lateral heterogeneity of the stomatal foliar uptake pathway. *Physiologia Plantarum* 134, 151–160.

Elmer, W., Ma, C., White, J. 2018. Nanoparticles for plant disease management. *Current Opinion in Environmental Science & Health* 6, 66–70.

Giunti, G., Palermo, D., Laudani, F., Algeri, G.M., Campolo, O., Palmeri, V. 2019. Repellence and acute toxicity of a nano-emulsion of sweet orange essential oil. *Industrial Crops and Products* 142, 111869.

Grillo, R., Mattos, B.D., Antunes, D.R., Forini, M.M.L., Monikh, F.A., Rojas, O.J. 2021. Foliage adhesion and interactions with particulate delivery systems. *Nano Today* 37, 101078.

Haleem, A., Javaid, M., Singh, R.P., Rab, S., Suman, R. 2023. Applications of nanotechnology in the medical field: a brief review. *Global Health Journal* 7, 70–77.

Hu, P., An, J., Faulkner, M.M., Wu, H., Li, Z., Tian, X., Giraldo, J.P. 2020. Nanoparticle charge and size control foliar delivery efficiency. *ACS Nano* 14, 7970–7986.

Ibrahim, S.S., Elbehery, H.H., Samy, A. 2024. Efficacy of green silica nanoparticles synthesized from rice straw. *Scientific Reports* 14, 8834.

Ikawati, S., Himawan, T., Abadi, A.L., Tarno, H. 2021. Toxicity of nanoinsecticide based on clove essential oil. *Journal of Pesticide Science* 46, 222–228.

Islam, F., Imran, A., Afzaal, M., Zahoor, T., Amjad, N., Wei, C.R. et al. 2024. Nanotechnology for improvement of food and agriculture. *Cogent Food & Agriculture* 10, 2398857.

Judy, J.D., McNear, D.H., Chen, C., Lewis, R.W., Tsyusko, O.V., Bertsch, P.M. et al. 2015. Nanomaterials in biosolids inhibit nodulation. *Environmental Science & Technology* 49, 8751–8758.

Kah, M., Tufenkji, N., White, J.C. 2019. Nano-enabled strategies to enhance crop nutrition. *Nature Nanotechnology* 14, 532–540.

Khan, Z., Thounaojam, T.C., Chowdhury, D., Upadhyaya, H. 2023. Role of selenium and nano-selenium in plants. *Plant Growth Regulation* 100, 409–433. https://doi.org/10.1007/s10725-023-00988-0

Kumar, R., Saharan, B.S., Rani, S., Duan, J.S. 2025. Nano-encapsulated pesticides for targeted pest control. In: Husen, A. (eds), *Handbook of Nanotechnology in Agriculture*, pp. 1–23. Springer Nature, Cham.

Kumar, S., Singh, N. 2025. Introduction to thematic issue on emerging trends in nanomaterials. *Indian Journal of Microbiology* 65, 1–2.

Lev, J., Christie, P., Zhang, S. 2019. Uptake, translocation, and transformation of metal nanoparticles in plants. *Environmental Science: Nano* 6, 41–59. https://doi.org/10.1039/C8EN00645H

Li, C., Wang, P., Lombi, E., Cheng, M., Tang, C., Howard, D.L., Menzies, N.W., Kopittke, P.M. 2018. Absorption of foliar-applied Zn fertilizers by trichomes. *Journal of Experimental Botany* 69, 2717–2729.

Liu, Q., Chen, B., Wang, Q., Shi, X., Xiao, Z., Lin, J., Fang, X. 2009. Carbon nanotubes as molecular transporters for walled plant cells. *Nano Letters* 9(3), 1007–1010.

López-Moreno, M.L., de la Rosa, G., Hernández-Viezcas, J.A., Castillo-Michel, H., Botez, C.E., Peralta-Videa, J.R., Gardea-Torresdey, J.L. 2010. Differential biotransformation of ZnO and CeO_2 nanoparticles. *Environmental Science & Technology* 44, 7315–7320.

Lv, J., Christie, P., Zhang, S. 2019. Uptake, translocation and transformation of metal nanoparticles in plants. *Environmental Science: Nano* 6, 41–59.

Ma, X., Geiser-Lee, J., Deng, Y., Kolmakov, A. 2010. Interactions between engineered nanoparticles and plants. *Science of the Total Environment* 408, 3053–3061.

Ma, X., Yan, J. 2018. Plant uptake and accumulation of engineered metallic nanoparticles. *Current Opinion in Environmental Science & Health* 6, 16–20.

Malaikozhundan, B., Vaseeharan, B., Vijayakumar, S., Thangaraj, M.P. 2017. *Bacillus thuringiensis* coated ZnO nanoparticles as biopesticides. *Journal of Photochemistry and Photobiology* B 174, 306–314.

Miralles, P., Church, T.L., Harris, A.T. 2012. Toxicity, uptake, and translocation of engineered nanomaterials. *Environmental Science & Technology* 46, 9224–9239. https://doi.org/10.1021/es202995d

Mishra, M., Pandey, A.K., Pandey, K., Dixit, S., Zohra, F., Seth, A., Singh, S. 2021. Nanotechnology in agriculture: current research and prospects. *Composite Materials* 7, 223–241.

Mohamadeen, F.T., Hassona, N.M. 2024. Activity of titanium dioxide nanoparticles on housefly stages. *BioNanoScience* 14, 3265–3273.

Mohammed, A.M., Aswd, S.A. 2019. Effect of nanoparticles on southern cowpea beetle. *Journal of Education and Science* 28, 188–199.

Mortimer, M., Li, D., Wang, Y., Holden, P.A. 2020. Physical properties of nanomaterials affect nodulation competitiveness. *Small* 16, e1906055.

Nasrollahzadeh, M., Sajadi, S.M., Sajjadi, M., Issaabadi, Z. 2019. Applications of nanotechnology in daily life. *Interface Science and Technology* 28, 113–143.

Navarro, E., Baun, A., Behra, R., Hartmann, N.B., Filser, J., Miao, A.J., Quigg, A., Santschi, P.H., Sigg, L. 2008. Environmental behavior and ecotoxicity of engineered nanoparticles. *Ecotoxicology* 17, 372–386.

Nenaah, G.E. 2014. Toxicity of essential oil nano-emulsions against *Tribolium castaneum*. *Industrial Crops and Products* 53, 252–260.

Peng, C., Duan, D., Xu, C., Chen, Y., Sun, L., Zhang, H., Yuan, X. 2015. CuO nanoparticle translocation in rice. *Environmental Pollution* 197, 99–107.

Péret, B., De Rybel, B., Casimiro, I., Benková, E., Swarup, R., Laplaze, L., Beeckman, T., Bennett, M.J. 2009. Arabidopsis lateral root development. *Trends in Plant Science* 14, 399–408. https://doi.org/10.1016/j.tplants.2009.05.002

Pérez-de-Luque, A. 2017. Interactions of nanomaterials with plants. *Frontiers in Environmental Science* 5, 12.

Rai, M., Ingle, A.P., Pandit, R., Paralikar, P., Shende, S., Gupta, I., Biswas, J.K., da Silva, S.S. 2018. Copper nanoparticles in pest and pathogen management. *Nanotechnology Reviews* 7, 303–315.

Rajput, V., Minkina, T., Mazarji, M., Shende, S., Sushkova, S., Mandzhieva, S., Burachevskaya, M., Chaplygin, V., Singh, A., Jatav, H. 2020. Accumulation of nanoparticles in soil–plant systems. *Annals of Agricultural Sciences* 65, 137–143.

Rajput, V.D., Faizan, M., Upadhyay, S.K., Kumari, A., Ranjan, A., Sushkova, S. et al. 2022. Influence of nanoparticles on rhizosphere microbiome. In: Rajput, V.D., Verma, K.K., Sharma, N., Minkina, T. (eds), *Nanoscience in Nutrient Use Efficiency*, pp. 83–102. Springer, Cham.

Raliya, R., Nair, R., Chavalmane, S., Wang, W.N., Biswas, P. 2015. Translocation and physiological impact of TiO_2 and ZnO nanoparticles. *Metallomics* 7, 1584–1594.

Rastogi, A., Zivcak, M., Sytar, O., Kalaji, H.M., He, X., Mbarki, S., Brestic, M. 2017. Impact of metal nanoparticles on plants. *Frontiers in Chemistry* 5, 78.

Rico, C.M., Majumdar, S., Duarte-Gardea, M., Peralta-Videa, J.R., Gardea-Torresdey, J.L. 2011. Interaction of nanoparticles with edible plants. *Journal of Agricultural and Food Chemistry* 59, 3485–3498.

Rizwan, M., Ali, S., Qayyum, M.F., Ok, Y.S., Adrees, M., Ibrahim, M., Al-Wabel, M.I. 2019. Effect of metal nanoparticles on food crops. *Journal of Hazardous Materials* 322, 2–16.

Rodrigues, S., Avellan, A., Bland, G.D., Miranda, M.C.R., Larue, C., Wagner, M. et al. 2024. Zinc phosphate shell effect on ZnO nanoparticle uptake. *Environmental Science & Technology* 58, 3213–3223. https://doi.org/10.1021/acs.est.3c08723

Rouhani, M., Samih, M.A., Kalantari, S. 2013. Insecticidal effect of silica and silver nanoparticles. *Journal of Entomological Research* 4, 297–305.

Schreiber, L. 2005. Polar diffusion paths across plant cuticles. *Annals of Botany* 95, 1069–1073.

Schwab, F., Zhai, G., Kern, M., Turner, A., Schnoor, J.L., Wiesner, M.R. 2016. Uptake and accumulation of nanomaterials in plants. *Nanotoxicology* 10, 257–278.

Sharma, D., Sharma, A., Gautam, H.R. 2025. Green synthesized nanoparticles for disease management. *Plant Nano Biology* 13, 100179.

Singh, H., Sharma, A., Bhardwaj, S.K., Arya, S.K., Bhardwaj, N., Khatri, M. 2021. Nano-agrochemicals for sustainable agriculture. *Environmental Science: Processes & Impacts* 23, 213–239.

Sivapriya, V., Azeez, A.N., Deepa, S.V. 2018. Phyto-synthesis of iron oxide nanoparticles for pesticidal activity. *Journal of Entomology and Zoology Studies* 6, 1050–1057.

Stadler, T., Buteler, M., Weaver, D.K., Sofie, S. 2012. Toxicity of nanostructured alumina to stored grain pests. *Journal of Stored Product Research* 48, 81–90.

Tarafdar, J.C., Adhikari, T.A. 2015. Nanotechnology in soil science. In: Rattan, R.K. (ed), *Soil Science: An Introduction*, pp. 775–807. Indian Society of Soil Science: New Delhi.

Ullah, R., Muhammad, Z., Inayat, N., Majeed, A. 2020. Nanoparticles in agriculture: challenges and opportunities. *New Frontiers in Stress Management* 721, 281–293.

Wang, F., Zhang, X., Zhang, S., Sun, Y. 2020. Interactions of microplastics and cadmium in soil–plant systems. *Chemosphere* 254, 126791.

Wang, P., Lombi, E., Zhao, F.J., Kopittke, P.M. 2016. Nanotechnology: a new opportunity in plant sciences. *Trends in Plant Science* 21, 699–712.

Wang, Q., Li, Q., Wang, L., Yang, L., Hu, Z., Wang, J., Yu, J., Hou, R. 2025. Optimizing mesoporous silica nano-delivery systems in tea plants. *Industrial Crops and Products* 227, 120789. https://doi.org/10.1016/j.indcrop.2025.120789

Wang, X.R., Xie, H., Wang, P., Yin, H. 2023. Nanoparticles in plants: uptake, transport and physiological activity. *Materials* 16, 3097.

4 Nanomaterials and Their Impact on Plant Growth and Development

Sukhmeen Kaur Kohli, Priyanka Sharma, and Harsimran Kaur

4.1 INTRODUCTION

Nanotechnology is serving as a high-tech innovative technology in this techno-science era and attracts innumerable disciplines including agriculture, energy, plant science, and nanomedicine and areas that are closely related to human welfare. As per the Food and Agriculture Organization latest predictions, the global population by year 2050 will increase to 9–10 billion, posing a need to increase food production by at least 25%–70% in comparison with its present condition. Currently, the agricultural sector is facing challenges such as production losses occurring from diseases, changing climate, and exhaustion of natural resources. Therefore, employing novel technologies in agriculture to maintain both productivity and sustainability to meet the demands of the growing population is pre-requisite. Thus, to feed the increasing population, new approaches are required to be deployed in the agricultural sector to ensure sustainability and productivity becomes a necessity (Aqeel et al. 2022). In agriculture, nanotechnological application includes the usage of nanopesticides, nanofertilizers, and nanoherbicides that augment plant nutrition while maintaining soil texture and protecting the plants from any microbial infections (Seleiman et al. 2023).

Nanomaterials (NMs) refer to organic, inorganic, or organo-metallic material that have physical, chemical, and/or electrical properties which alters as a function of shape and size of the material. Relative to the interior ones, these particles have high proportions of the surface atoms and owing to quantum confinement effects and NMs are subject to property disparity as a function of their size (Hochella et al. 2019). These nano-sized particle matter, present in earth, can readily move *via* different spheres of the environment (atmosphere, lithosphere, hydrosphere, and biosphere) within the biogeochemical cycle. They play an integral role in overall earth dynamics. Nanosize, high surface area-to-volume ratio, shape alterations, and ability to effectively deliver several chemicals into the cells are the main characteristics of NMs, making them an efficient candidate for delivering useful material to plants (Seleiman et al. 2023). However, NMs can negatively regulate the growth of plants and are known to induce phytotoxicity (Marslin et al. 2017). Thus, critical information about the biological, chemical, and environmental parameters that affect the fate of NM once applied to plants should be known.

 DOI: 10.1201/9781032659435-4

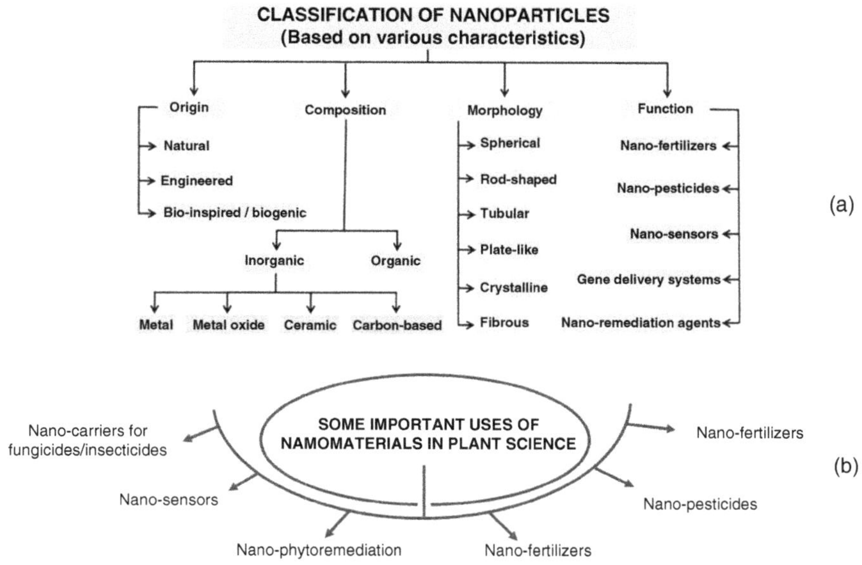

FIGURE 4.1 Classification of nanomaterials and key functions in plant science for increasing productivity, stress resilience, and environmental clean-up.

4.2 TYPES OF NANOMATERIALS

Nanoparticles (NPs) are materials characterized by at least one dimension ranging from 1 to 100 nm, exhibiting distinct biological and physicochemical properties that differ from their bulk counterparts. In plant science, NPs are increasingly explored for improving nutrient-use efficiency, crop protection, stress tolerance, gene delivery, and environmental monitoring. NMs can be classified based on their origin, composition, morphology, and function, which together determine their interaction with plant systems and their agricultural applicability (Figure 4.1a).

4.2.1 Based on Composition

Based on material makeup, NPs are broadly divided into organic, inorganic, and carbon-based NPs, each with distinct physicochemical and biological properties.

4.2.1.1 Organic Nanoparticles

In agriculture, organic NPs composed of lipids, polymers, proteins, or dendrimers are low-toxic and biodegradable carriers. Hydrophobic agrochemicals are protected, and controlled release is made possible by lipid NPs (Waghule et al. 2020). Polymeric NPs improve stability and solubility while lowering environmental loss (Pascoli et al. 2019). Micronutrients and highly biocompatible antimicrobials are encapsulated in protein-based NPs (Iravani and Shukla 2019). With high loading efficiency and minimal chemical consumption, dendrimers enable targeted delivery of genetic material and RNAi agents (Varatharaju et al. 2025).

4.2.1.2 Inorganic Nanoparticles

It consists of metals, metal oxides, or ceramics and is valued for its stability, tunable reactivity, and unique optical properties. Metal NPs such as Cu, Ag, and Au enhance plant growth, nutrient uptake, and antioxidant activity; AgNPs have potent antimicrobial effects, while CuNPs act as nanofertilizers that support photosynthesis and metabolism (Joshi et al. 2022). Metal oxide NPs, including Fe-, ZnO-, and TiO_2-based NPs, improve plant nutrition, chlorophyll synthesis, crop quality, enzyme activity, and tolerance to abiotic stress (Seleiman et al. 2023; Alabdallah et al. 2024). Ceramic NPs like alumina and nanosilica are thermally stable and chemically inert; nanosilica enhances water-use efficiency, strength of the cell wall, and resistance to drought and salinity, outperforming conventional silicon fertilizers under stress conditions (Surendar et al. 2024).

4.2.1.3 Carbon-Based Nanoparticles

These NPs, which have a large surface area, mechanical strength, and conductivity, are made of carbon arranged in tubular, spherical, or layered structures. Particle type, concentration, and species all affect how they interact with plants (Zhu et al. 2022).

Carbon nanotubes (CNTs), fullerenes, graphene oxide (GO), and carbon quantum dots (CQDs) are all extensively studied. While GO increases the effectiveness of nutrients and pesticides, CNTs facilitate gene delivery and improve seed germination. Because of their photoluminescent properties, CQDs are employed in plant bioimaging and stress sensing (Rao et al. 2021).

4.2.2 Based on Shape and Morphology

The morphology of NPs has a significant impact on biological activity and plant uptake. Spherical NPs, fibrous nanofibers, plate-like nanosheets, nanorods and nanotubes, and crystalline NPs are examples of common forms. While plate-like and fibrous forms facilitate controlled release and seed coating applications, high-aspect-ratio structures like CNTs improve penetration through plant cell walls (Singha Roy et al. 2022).

4.2.3 Based on Function

The functions of NPs in plant systems are highlighted by function-based classification (Figure 4.1b). Nanofertilizers improve uptake and lower losses by delivering nutrients in effective, slow-release forms (Punitha et al. 2023). According to Kah et al. (2018), nanopesticides reduce environmental contamination and application frequency by improving stability and targeted delivery of active ingredients. Real-time monitoring of nutrients, pathogens, and abiotic stress is made possible by nanosensors such as metal oxide NPs and quantum dots. Gene delivery, gene silencing, and genome editing are made possible without the use of transgenic vectors by NPs like CNTs, AuNPs, and polymeric carriers (Demirer et al. 2019). Furthermore, by enhancing heavy metal tolerance and contaminant degradation, nano-remediation agents like Fe_3O_4 and TiO_2 NPs improve phytoremediation.

4.3 IMPACT OF DIFFERENT NANOMATERIAL ON PLANT GROWTH AND DEVELOPMENT

The effect of NMs on the ecosystem functioning and mainly on the terrestrial plants has become an area of rigorous investigation. Plants exposed to various types of NMs exhibit responses at both molecular and physiological levels. The plant cellular interaction with NMs can either be positive or negative, depending upon the NM nature as well as the plant species. Concurrently, the size, chemical nature, concentration, and reactivity of NMs either inside or on the surface of the plant determine the effect of NM on plant systems (Ali et al. 2021). NM interacts with antioxidants and hormones, which further affect the patterns of plant growth that eventually lead to augment in growth and development of plants.

4.3.1 Impact of Gold Nanomaterial

Gold NMs (Au-NMs) owing to their specific properties, which influence the growth of plants, have gained significant attention. Au-NMs influence several plant processes including seed germination, antioxidant system, photosynthetic apparatus, and even gene expression. However, the growth-enhancing effects of these NMs vary from species to species. Besides the dosage of Au-NM, its surface charge, shape, size, and the capping agent influence their uptake, distribution, and their effects on plants (Bandi et al. 2023). Despite certain contradictory reports, Au-NM at concentrations above 100 mg/L poses detrimental effects on plants, and the toxicity increases as the particle size decreases from 5 nm in diameter. On the contrary, lower dosage of Au-NM can readily augment seed germination and content of chlorophyll and improve the productivity in different plants by enhancing the overall plant growth. It has been suggested that these NMs are initially oxidized to gold ions Au^{+}/Au^{3+} and then transported into the plant tissue, and consequently, they are reduced to Au-NMs (Siddiqi and Husen 2016). In recent years, numerous studies pertaining to plant growth and Au-NM exposure have been reported. Biosynthesized Au-NM positively improved seed germination rate and seedling growth in *Pennisetum glaucum* (Parveen et al. 2016). The foliar application of Au-NM on seedlings of *Brassica juncea* ameliorated the growth and seed yield besides augmenting the content of oil, total sugar, and reducing sugar in plants (Arora et al. 2012). An improvement in the redox status of the Au-NM-treated plants was also observed. Similar results were obtained by Abd El-Aziz et al. (2019) in lentil plants. However, they reported that only lower concentrations of Au-NM, i.e., 5 and 10 ppm, could successfully enhance physiological (seed germination, leaf number, plant height, and biomass) and biochemical (total soluble sugars, total chlorophyll, total soluble phenols, and protein content) attributes of the plant, but at higher concentration a reversal in the effect is observed. Nevertheless, producing Au-NM is economically not feasible since the cost of production is higher than the crop yield (Siddiqi and Husen 2016).

4.3.2 Impact of Silver Nanomaterial

Several studies reveal that plant response to silver (Ag)-based NM either augments or inhibits plant growth depending largely upon the dosage and shape of Ag-NM. In *Arabidopsis*, decahedral-shaped Ag-NMs of size 45 ± 5 nm encouraged a high degree of root growth promotion in comparison with triangular- and spherical-shaped Ag-NMs (Syu et al. 2014). Application of Ag-NMs is reported to stimulate root exudate production, thus facilitating plant–microbe interactions, which improve the overall plant growth (Wagi and Ahmed 2019). The Ag-NM impact is largely influenced by the species of the plant. In a study by Pallavi et al. (2016), the impact of Ag-NMs at the concentration of 50 and 75 ppm on growth parameters of wheat, cowpea, and *Brassica* sp. was studied, and their result revealed improved growth and root nodule formation in cowpea at 50 ppm along with improvement in shoot parameters of *Brassica* at 75 ppm concentration. However, in wheat, no such effect on shoot and root parameters was observed. Ag-NM-treated fenugreek seedlings exhibited enhancement in growth parameters including leaf number, shoot and root length, and wet weight besides showing remarked increase in levels of the major phytochemical diosgenin in comparison with the control (Jasim et al. 2017). The photostimulatory effect of Ag-NM synthesized by bacteria was observed in wheat as evident from improved shoot and root length, leaf number, and fresh weight (Wagi and Ahmed 2019). Several reports suggest Ag-NMs can upregulate the expression of genes implicated in ABA IAA biosynthesis, ABA signaling pathway, etc. Treatment of *Eruca sativa* with Ag-NM promotes root growth and induces *Jacalin-related lectin* (*JAL*) gene expression, which is responsible for the release of auxin synthesis precursor (Vannini et al. 2013).

4.3.3 Impact of Zinc Nanomaterial

Recently, zinc oxide (ZnO)-based NM has gained popularity and is widely applicable as a crop nutrient in areas deficient in zinc. They are considerably water soluble making their absorption and accumulation in plants easier in comparison with the conventional chemical zinc fertilizers (Sarkhosh et al. 2022). These ZnO-NMs not only encourage plant growth but also positively regulate plant physiological attributes (Yan et al. 2021). ZnO-NM are reported to be involved in the synthesis of chlorophyll, photosynthetic process, and starch formation, thereby elevating the levels of soluble carbohydrates (Wang et al. 2023). Application of ZnO-NM augments the chlorophyll content and antioxidant activity in cotton plant, escalates both the weight and number of cotton balls/plant, and improves the quality parameters of cotton fiber including fiber strength and uniformity (Vaghar et al. 2020). ZnO-NMs improve the yield of tomato plant by elevating nutrient absorption specifically of zinc and phosphorus (Faizan et al. 2021). Treating rice with ZnO-NMs not only improved the growth of rice plant, but it also reduced the accumulation of heavy metal arsenic in the plant. At the optimum dosage of 100 mg/L of ZnO-NM, the levels of arsenic in the root and shoot of rice seedlings reduced by 31.6% and

40.7%, respectively, in comparison with the control plant group (Yan et al. 2021). ZnO-NM not only leads to bioaugmentation of Zn in plants, but it also improves the overall nutritional value of plants. Foliar application of ZnO-NM to tomato plants led to a remarkable increase in the content of lycopene, a strong antioxidant, in comparison with the control plants (Raliya et al. 2015). ZnO-NM synthesized by *Aspergillus fumigatus* significantly improved plant biomass, shoot length, root length, root area, chlorophyll content, and total soluble leaf protein in clusterbean plants. Besides, an increase in the activity of several soil enzymes involved in the mobilization of phosphorous including phytase, alkaline phosphatase, and acid phosphatase (Raliya and Tarafdar 2013).

To the contrary, in most of the studies, high concentration of ZnO-NMs exhibits toxic effects. The negative impact of higher doses of ZnO-NMs on the growth of plants has been reported in several studies. ZnO-NM toxicity is a probable outcome of disturbed Zn homeostasis in plants, which further negatively impacts the homeostasis of other plants. High concentrations of ZnO-NMs adversely affect the growth of seedling in mung bean by eliciting oxidative stress. The levels of malondialdehyde, an indicator of lipid peroxidation and hydrogen peroxide in aerial plant parts, were considerably higher in comparison with the control plants (Sorahinobar et al. 2023). Thus, excessive accumulation of Zn upon application of Zn-based NM implicates oxidative stress in plants (Table 4.1).

4.3.4 Impact of Carbon Nanomaterial

Carbon-based NMs (CNMs) are a group of engineered NMs that includes fullerenes, nano-cones, nano-onions, nanohorns, CNTs, carbon dots, nano-beads, nano-diamonds, nanofibers, and grapheme (Mukherjee et al. 2016). CNMs are made of pure carbon and thus display high stability, low plant toxicity, and are environment-friendly (Zhao et al. 2021). Generally, CNMs are proven to be proficient at augmenting water uptake and water transport in plants, enhancing seed germination, and improving the photosynthetic apparatus and the antioxidant activities in plants besides lowering the levels of oxidative stress markers such as MDA and H_2O_2. Enhanced seed germination upon treatment with carbon nanohorns in different crop species including barley, rice, switchgrass, maize, wheat, tobacco, soybean, and tomato has been observed in different studies.

CNMs play a pivotal role in plummeting nutrient losses and enhancing fertilizer use proficiency. For instance, application of 200 mg/kg of CNM significantly improved corn growth as shown by enhanced plant height, nutrient uptake, nutrient-use efficacy, and biomass yield when applied to soil. The amended soils exhibited higher bioavailability of nitrogen and phosphorus in soil (Zhao et al. 2021). Similarly, CNM application on *Nicotiana tabacum* leads to augmented plant growth at different developmental stages in comparison with the use of conventional fertilizers and increases the levels of nitrogen and potassium in plant tissue. At maturity, the potassium levels in leaves amplified by at least 21%–37% (Liang et al. 2013).

TABLE 4.1
Dual Effect of Nanoparticles on Different Crop Plants

Nanoparticles	Positive Effect	Negative Effect	Crop(s)	References
ZnO nanoparticles	Increased chlorophyll content, enhanced growth, and grain yield	At high concentrations, they reduced root elongation and caused oxidative stress	Wheat (*Triticum aestivum* L.)	Rameshaiah et al. (2015)
TiO_2 nanoparticles	Improved germination, photosynthesis, and biomass	Excessive TiO_2 accumulation reduced chlorophyll content	Spinach (*Spinacia oleracea* L.)	Hong et al. (2005)
SiO_2 nanoparticles	Enhanced drought tolerance, improved water retention	Overuse caused silica deposition leading to reduced leaf flexibility	Rice (*Oryza sativa* L.)	Suriyaprabha et al. (2012)
Ag nanoparticles	Antimicrobial activity, reduced disease incidence	Toxic at higher doses, inhibited root growth and microbial diversity	Tomato (*Solanum lycoper sicum* L.)	Subhash (2017)
Chitosan nanoparticles	Controlled nutrient release, improved yield	Over-application caused nutrient imbalance and salt stress	Maize (*Zea mays* L.)	Abdel-Aziz et al. (2016)
CeO_2 nanoparticles	Increased tolerance to heat stress via antioxidant enzymes	Long-term exposure altered photosynthetic efficiency	Maize (*Zea mays* L.)	Rossi et al. (2017)
CuO nanoparticles	Strong antifungal activity, enhanced plant defense	Induced oxidative stress and reduced germination at high doses	Cotton (*Gossypium hirsutum* L.)	Singh et al (2021)
Carbon nanotubes (CNTs)	Faster germination, increased water uptake, and biomass	Caused mechanical damage to root cells at high levels	Tomato (*Solanum lycoper sicum* L.)	Lahiani et al. (2013)

4.3.5 Impact of Other Nanomaterials

Iron (Fe)-based NM can augment biomass and productivity in plants by upsurging the levels of protein, carbohydrates, and nutrients in plants (Wang et al. 2023). Fe_2O_3-NM application to peanut plants resulted in an increase in plant height, root length, and biomass. The Fe_2O_3-NMs promoted plant by modifying phytohormonal content and regulating activity of antioxidant enzymes (Rui et al. 2016). Foliar application of Fe_2O_3-NM and Fe_2O_3-NM considerably increased plant biomass, root development indices, and chlorophyll content in soybean plants. It also stimulated rhizobial nitrogen fixation in plants as supported by increased root nodule formation

(Yang et al. 2020). Similarly, Farahi et al. (2019) observed that TiO_2-NMs at different doses had stimulatory effects on shoot and root dry weight. The authors suggested that an increase in central cylinder and parenchyma density in roots upon exposure to TiO_2-NM was responsible for the higher dry weight of the roots. It is also believed that TiO_2-NM can encourage protein and chlorophyll formation by upregulating enzymes of nitrogen metabolism, thus enhancing plants' fresh and dry weight. Chen et al. (2018) suggested that an increase in the biomass of rice plants on the application of TiO_2-NM is related to higher uptake of minerals by treated plants. In a study by Mukherjee et al. (2023), CeO_2-NMs significantly reduced oxidative stress in *Vigna radiata* by inhibiting the generation of reactive oxygen species (ROS), hydrogen peroxide, and MDA and enhancing growth in plants grown under mercury stress (50 ppm). CeO_2-NMs with or without citric acid coating at the concentration of 500 mg/kg significantly increased stem length in *Solanum lycopersicum* seedlings by up to 13% and 9%, respectively (Barrios et al. 2016). Applying CeO_2-NM to soil promoted plant development in *Hordeum vulgare*, leading to 331% escalation in shoot biomass in comparison with the control. Besides, the NM treatment amended stress levels in plant leaves without causing any toxicity. Lower doses of the NM also promoted absorption and accumulation of mineral nutrients such as calcium, iron, and zinc. However, exposure of plants to the highest concentration of CeO_2-NM (500 mg/kg) inhibited grain formation (Rico et al. 2015).

4.4 EFFECTS OF NANOPARTICLES ON PLANTS

NPs, originally called "magic bullets," are now widely used in plant systems as nanofertilizers, pesticides, herbicides, and carriers for genes and growth regulators. Their nanoscale size enables precise delivery and improved plant performance. However, concerns about toxicity, bioaccumulation, and environmental persistence remain, highlighting the dual role of NPs as both beneficial tools and potential risks in agriculture.

4.4.1 Positive Effects of Nanoparticles

- **Enhanced Germination and Growth**: Metal oxide NPs (ZnO, TiO_2, Fe_3O_4) improve germination, root/shoot length, and seedling vigor by boosting water uptake and enzyme activity. TiO_2 accelerates photosynthesis and early growth (Singh et al. 2024).
- **Enhanced Nutrient Uptake**: Nanofertilizers (nano-Zn, nano-Fe, nano-Si) provide controlled nutrient release, biomass, enhancing root growth, and nutrient assimilation (Singh et al. 2024).
- **Stimulation of Photosynthesis**: TiO_2 and nano-CeO_2, which improve biomass and chloroplast stability by increasing chlorophyll, electron transport, and photosynthetic capacity (Rehman et al. 2025).
- **Stress Tolerance**: Nano-selenium, nanosilica, and nano-Zn/Cu enhance drought, salinity, and heavy metal tolerance via antioxidant enzymes and stress-responsive genes (Grodetskaya et al. 2022).
- **Hormone Regulation**: Ag NPs alter ABA under stress, controlling stomata; SiO_2 stimulates IAA synthesis, increasing root growth (Khalid et al. 2025).

4.4.2 Negative Effects of Nanoparticles on Plant Growth

- **Oxidative Stress and ROS Generation**: Excess metal NPs (Ag, CuO, ZnO) cause an excess of ROS, which can lead to lipid peroxidation, damage to proteins and DNA, disruption of signaling, and even cell death (Hatami and Ghorbanpour 2024).
- **Growth Inhibition**: By changing cell walls and vascular tissues, Ag NPs and Al_2O_3 NPs at toxic levels prevent cell division, limit root elongation, and prevent nutrient absorption (Kaveh et al. 2013). In *Arabidopsis* and *Spirodela polyrrhiza*, biomass decline and root/shoot abnormalities have been documented (Jiang et al. 2012).
- **Photosynthetic Damage**: High concentrations of nano-metal oxides damage chloroplast ultrastructure, reduce chlorophyll content, impair gas exchange, disrupt photosystem II (PSII) activity, and hinder carbon assimilation, ultimately limiting plant growth (Ghorbanpour et al. 2021).

4.5 EFFECT ON FUNCTIONING AND STRUCTURE OF PHOTOSYNTHETIC APPARATUS

Photosynthesis in plants is a fundamental vital process by which they convert the light energy from the sun into chemical energy. All forms of life depend either directly or indirectly on this process to obtain energy. Photosynthesis involves two sets of reaction procedures, light-dependent reactions and light-independent reactions, which eventually convert carbon dioxide (CO_2) into organic compounds, i.e., mainly sugars. During the light reaction, sunlight is converted into energy-trapping molecules (ATP and NADPH), whereas during the dark reaction, these byproducts of the light reaction form glucose by combining with CO_2 (Kataria et al. 2019). NM either augments the photosynthetic processes by enhancing the light-harvesting complex in the plants or obstructs the process by stalling the electron transport chain. Besides, they can also affect the rate of photosynthesis by modifying the expression of various genes and enzymes including phosphoenolpyruvate (PEP) carboxylase, ribulose-1,5-bisphosphate carboxylase/oxygenase (RUBISCO), and carbonic anhydrase (Poddar et al. 2020).

TiO_2 is a widely studied NM owing to its photocatalytic abilities and its potential to activate oxidation–reduction reactions that contributes for charge transfer between the NM and the light-harvesting complexes II. The effect of TiO_2-NM on spinach photosynthetic efficiency indicated that it can readily improve light absorption and accelerate transport and transformation of light energy. Furthermore, it prevents the aging of chloroplasts (Ali et al. 2021). Nanoanatase TiO_2 considerably enriched the oxygen-evolving and phosphorylating processes in plants along with the electron transport chain in spinach plants under ultraviolet and visible radiation. Raliya et al. (2015) in their study revealed an elevation in levels of chlorophyll in tomato plants upon aerosol foliar application of TiO_2-NM. Increase in the amount of TiO_2-NM during application leads to an increase in the absorption of photosynthetically active radiation (PAR), indicating NM can efficiently increase the photosynthetic rate in plants. In several studies in different plants, elevated levels of chlorophyll are commonly observed under the influence of NM treatment. Also, in ZnO-NM-treated mung bean plants, the content of soluble protein increased by 25% and chlorophyll content increased by 34.5% in comparison with the control plants (Raliya and Tarafdar 2013).

In *Arabidopsis thaliana*, TiO_2-NMs promoted light absorption capacity of the chloroplast, regulated the dispersal of light energy from PSI to PSII by augmenting LHCII, and accelerated the process of water photolysis, oxygen evolution, and transformation of light energy to electrical energy (Ze et al. 2011). Even under mild heat stress in tomato plants, TiO_2-NM increased the net photosynthetic rate, water conductance, and rate of transpiration. The NM application considerably reduced the relative electron transport and minimum chlorophyll fluorescence in tomato leaves. Besides, it enhanced regulated PSII energy dissipation and lowered the unregulated PS II energy dissipation, suggesting the promontory role of TiO_2-NM in photosynthesis (Qi et al. 2013). Carotenoid is a strong antioxidant that protects the chlorophyll molecules during stressful conditions from oxidative injury (Rasheed et al. 2022). Application of CuO-NM remarkably improved the content of carotenoid content in maize plants, which further prevented chlorophyll degradation during water stress and instead enhanced the concentration of chlorophyll, thereby improving the overall photosynthetic efficiency (Van Nguyen et al. 2022). Specific concentrations of ZnO-NM boosted the quantitative and qualitative characteristics of rice plants and resulted in higher PSII fluorescence kinetics and reduction in energy flux parameters (Singh et al. 2018). Additionally, CNTs uplifted the photosynthetic rate by at least three times by entering and amassing in the lipid envelope of plant chloroplast and elevating electron transport rate along with improved photoabsorption (Giraldo et al. 2014). An increment in the net photosynthetic rate is believed to be an outcome of an enhanced RUBISCO activity (Kataria et al. 2019). SiO_2-NM application results in replenishment in the process of photosynthesis by augmenting the activity of the enzyme carbonic anhydrase, which supplies CO_2 to RUBISCO, and the generation of photosynthetic pigments (Siddiqui and Al-Whaibi 2014). Similar results were obtained by Faizan et al. (2021) with ZnO-NM treatment in tomato plants. Escalation in plant growth and seed yield upon treatment with FeS-NM as an outcome of activation of genes of the RUBISCO small and large subunit was observed in *Brassica juncea* (Rawat et al. 2017). The authors suggested the growth-promoting role of FeS-NM is facilitated by activation of both nitrogen and carbon assimilatory pathways during particular growth stages.

On the other hand, certain metallic NMs can hinder photosynthesis by producing ROS or free radicals that can oxidize different components of the photosynthetic system (Poddar et al. 2020). TiO_2-NM downregulated the activity of both light-dependent and light-independent photosynthetic reactions and lowered the content of chlorophyll *a*, net photosynthetic rate, effective efficacy of PSII, transpiration rate, intercellular CO_2 concentration, starch content, and stomatal conductance in wheat plant, suggesting the toxic potential of the NM on sensitive biomarkers of photosynthesis (Dias et al. 2019). A study on *Arabidopsis thaliana* validated the harmful effects of ZnO-NM on gene expression of all the genes involved in chlorophyll synthesis including chlorophyll synthetase, chlorophyll A oxygenase, Mg-protoporphyrin IX methyltransferase, and Mg-chelatase subunit D. It was also revealed in the same study that genes related to PSI were also inhibited (Wang et al. 2016). A study in rice revealed that CuO-NM reduced thylakoid number per granum of the chloroplast, which remarkably reduced the photosynthetic capacity of the plant. Reduction in protein expression for PSI along with total demolition of PSII was reported with an increment in the dose of the NM (Da Costa and Sharma 2016).

4.6 NANOMATERIAL–PLANT INTERACTION PATHWAYS

NM can alter plant metabolic pathways by several ways such as by regulating genes, by regulating micronutrient supply or by interfering with several oxidative processes that may lead to oxidative burst. It is evident from several studies on NMs that when plants are exposed to an excess of NM, it leads to ROS formation, which further obstructs oxidative pathways. Once these NMs enter the plant cell, they interplay with the chloroplastic and mitochondrial electron transport chain, leading to oxidative burst (Jurkow et al. 2020). The concentration of ROS generated by NMs depends upon particle size, surface area, shape, and chemistry. It is well confirmed that NM of smaller size exhibits high-density surface imperfections and offers more free electrons and eventually creates holes that lead to ROS generation (Sorahinobar et al. 2023). The methods through which NMs encourage ROS generation and elicit cellular oxidative stress have been investigated in recent years. Ag-NMs activated ROS and Ca^{2+} signaling through oxidation of L-ascorbic acid in the apoplast and stimulation of the Ca^{2+}-permeable pores. *Arabidopsis thaliana* mutant *root hair defective 2* (*rhd2*) that lacks NADPH oxidase RBOHC exhibited significantly low levels of ROS formation in retortion to Ag-NMs in comparison with the wild-type *Arabidopsis* (Sosan et al. 2016), specifying that cellular ROS amassment in the apoplast is facilitated by NADPH oxidases (RBOH) enzymes bound to the plasma membrane. In another study, chloroplastic ROS formation in *S. polyrhiza* was observed as an outcome of inhibition of RUBISCO activity and PSII photo-protective ability (Jiang et al. 2017). Excessive ROS due to plant–NM interaction can damage different cellular components including proteins, lipids, and DNA that eventually causes cell death (Zia-ur-Rehman et al. 2022). In soybean, exposure to high concentrations of ZnO-NM (200 and 400 ppm) increased lipid peroxidation as evident from high levels of MDA in comparison with control plants (Hashemi et al. 2019). Atha et al. (2012) reported accumulation of mutagenic DNA lesions in radish, perennial ryegrass, and annual ryegrass on exposure to CuO-NM besides strong inhibition of plant growth.

The damaging or signaling nature of ROS largely depends upon an equilibrium between ROS generation and elimination. However, despite the destructive abilities, ROS also act as a signaling molecule role in varied cellular processes. Plant cells have established a robust antioxidant mechanism to control ROS levels. This antioxidant mechanism involves the formation of enzymatic and non-enzymatic antioxidants, the levels of which elevate during stress (Marslin et al. 2017). Exposure to low levels of NMs triggers the activation of ROS detoxification mechanisms by upregulating the signaling genes involved in the production of the antioxidant enzymes.

NMs can regulate several genes including *TAS14*, *DDF2*, *ZFHD*, *MAPK2*, and *P5CS* genes, which are intricated in the synthesis of non-enzymatic (ascorbic acid, jasmonic acid, proline) and enzymatic antioxidants (peroxidase, catalase, and superoxide dismutase). If the antioxidants fail to control ROS levels, oxidation of cellular molecules occurs, which eventually leads to cell death (Zia-ur-Rehman et al. 2022). Ag-NMs and Au-NMs in lavender plants augmented levels of ascorbate peroxidase and superoxide dismutase enzymes and increased the free radical scavenging capacity in the plant *in vitro* cultures. Metallic NMs, Ag-NM, and platinum-based NM (Pt-NM) increased levels of antioxidant glutathione in oakleaf lettuce seedlings (Jurkow et al. 2020).

The activities of glutathione S-transferase and guaiacol peroxidase increased in *Brassica napus* seedlings treated with Co_3O_4-NMs at the concentration of 250–1,000 mg/L. Another enzyme, catalase, that prevents cells from oxidative damage, significantly increased in wheat plant roots when treated with 500 mg/kg CuO-NMs. As disturbance in ROS homeostasis hinders growth and development in plants and maintenance of ROS concentrations within suitable levels encourages plant health, it is believed that stimulation of the NM-mediated antioxidant machinery promotes plant growth as long as the destructive ROS levels are not attained in plant cells. Once breached, it results in organelle dysfunction, membrane destruction, and finally phytotoxicity (Marslin et al. 2017). Even though NM-mediated plant responses depend upon factors like type of NM, exposure conditions, and plant species, several studies provide evidence regarding an underlying NM and phytohormone interactions. Enhanced contents of auxin, gibberellin, and abscisic acid under the influence of different NMs in several plants have been observed (Tripathi et al. 2022).

4.7 CONCLUSION AND FUTURE PROSPECTS

An evolutionary field of science, nanotechnology, has found new wide range of applications in the area of energy, medicine, electronics, life science, and agriculture. Even though several reports suggesting the negative toxicological consequences of NMs on plants are available, research aiming at understanding the positive outcomes of NMs on plant growth continues to grow. Studies on NMs have revealed significant scientific evidence of their capability of ameliorating the micronutrient content of the plants, as reflected by revamped growth parameters and improved physiological conditions including chlorophyll and carotenoid content, photosynthetic rate, transpiration rate, and metabolic pathways. This chapter thoroughly reviewed the uptake, transportation, and accumulation of the NMs and their positive as well as negative effects on plant growth and photosynthetic process on different plant species. However, there is a strong need to understand the biochemical, physiological, and molecular alterations implicated by NMs in plants. Besides, there is a need for additional studies to explore the mechanism of NMs action, their communication with plant biomolecules, and their influence on regulating gene expression within the plant cells.

REFERENCES

Abd El-Aziz, A.R.M., Al-Othman, M.R. 2019. Gold nanoparticles biosynthesis using *Zingiber officinale* and their impact on the growth and chemical composition of lentil (*Lens culinaris* medic). *Pakistan Journal of Botany* 51, 443–450.

Abdel-Aziz, H.M., Hasaneen, M.N., Omer, A.M. 2016. Nano chitosan-NPK fertilizer enhances the growth and productivity of wheat plants grown in sandy soil. *Spanish Journal of Agricultural Research* 14(1), e0902.

Alabdallah, N.M., Alluqmani, S.M., Almarri, H.M., Al-Zahrani, A.A. 2024. Physical, chemical, and biological routes of synthetic titanium dioxide nanoparticles and their crucial role in temperature stress tolerance in plants. *Heliyon*, 10(4), e26537.

Ali, S., Mehmood, A., Khan, N. 2021. Uptake, translocation, and consequences of nanomaterials on plant growth and stress adaptation. *Journal of Nanomaterials* 2021, 1–17.

Aqeel, U., Aftab, T., Khan, M.M.A., Naeem, M., Khan, M.N. 2022. Impacts of diverse nanoparticles on growth, development and physiological adjustments in plants. *Chemosphere* 291, 132672.

Arora, S., Sharma, P., Kumar, S., Nayan, R., Khanna, P.K., Zaidi, M.G.H. 2012. Gold-nanoparticle–induced enhancement in growth and seed yield of *Brassica juncea. Plant Growth Regulation* 66, 303–310.

Atha, D.H., Wang, H., Petersen, E.J., Cleveland, D., Holbrook, R.D., Jaruga, P. et al. 2012. Copper oxide nanoparticle–mediated DNA damage in terrestrial plant models. *Environmental Science & Technology* 46, 1819–1827.

Bandi, R., Dadigala, R., Alle, M. 2023. Emerging role of gold nanoparticles for healthier crop plant growth and enhanced yield. *Engineered Nanomaterials for Sustainable Agricultural Production* 2023, 125–143.

Barrios, A.C., Rico, C.M., Trujillo-Reyes, J., Medina-Velo, I.A., Peralta-Videa, J.R., Gardea-Torresdey, J.L. 2016. Effects of cerium oxide nanoparticles and cerium ions on tomato plants. *Science of the Total Environment* 563, 956–964.

Chen, R., Zhang, C., Zhao, Y., Huang, Y., Liu, Z. 2018. Foliar nano-silicon reduces cadmium in rice grains by inhibiting translocation. *Environmental Science and Pollution Research* 25, 2361–2368.

Da Costa, M.V.J., Sharma, P.K. 2016. Copper oxide nanoparticles affect growth and photosynthesis in *Oryza sativa. Photosynthetica* 54, 110–119.

Demirer, G.S., Zhang, H., Matos, J.L., Goh, N.S., Cunningham, F.J., Sung, Y. et al. 2019. High aspect ratio nanomaterials enable delivery of functional genetic material without DNA integration in mature plants. *Nature Nanotechnology* 14(5), 456–464.

Dias, M.C., Santos, C., Pinto, G., Silva, A.M.S., Silva, S. 2019. Titanium dioxide nanoparticles impair photochemical and non-photochemical phases of photosynthesis in wheat. *Protoplasma* 256, 69–78.

Faizan, M., Bhat, J.A., Chen, C., Alyemeni, M.N., Wijaya, L., Ahmad, P., Yu, F. 2021. Zinc oxide nanoparticles (ZnO-NPs) induce salt tolerance by improving the antioxidant system and photosynthetic machinery in tomato. *Plant Physiology and Biochemistry* 161, 122–130.

Farahi, S.M.M., Iranbakhsh, A., Mahmoodzadeh, H., Ebadi, M., Baharara, J. 2019. Effects of titanium dioxide nanoparticles (TiO_2) on germination and seedling growth of *Vitex agnus-castus. Journal of BioScience and Biotechnology* 8, 141–149.

Ghorbanpour, M., Movahedi, A., Hatami, M., Kariman, K., Bovand, F., Shahid, M.A. 2021. Insights into nanoparticle-induced changes in plant photosynthesis. *Photosynthetica* 59(4), 570–586.

Giraldo, J.P., Landry, M.P., Faltermeier, S.M., McNicholas, T.P., Iverson, N.M., Boghossian, A.A. et al. 2014. Plant nanobionics approach to augment photosynthesis and biochemical sensing. *Nature Materials* 13, 400–408.

Grodetskaya, T.A., Evlakov, P.M., Fedorova, O.A. 2022. Influence of copper oxide nanoparticles on gene expression of birch clones *in vitro* under stress caused by phytopathogens. *Nanomaterials* 12: 864.

Hashemi, S., Asrar, Z., Pourseyedi, S., Nadernejad, N. 2019. Investigation of ZnO nanoparticles on proline, anthocyanin, photosynthetic pigments, and lipid peroxidation in soybean. *IET Nanobiotechnology* 13, 66–70.

Hatami, M., Ghorbanpour, M. 2024. Metal and metal oxide nanoparticles-induced reactive oxygen species: phytotoxicity and detoxification mechanisms in plant cells. *Plant Physiology and Biochemistry* 213, 108847. https://doi.org/10.1016/j.plaphy.2024.108847

Hochella, M.F. Jr., Mogk, D.W., Ranville, J., Allen, I.C., Luther, G.W., Marr, L.C. et al. 2019. Natural, incidental, and engineered nanomaterials and their impacts on the earth system. *Science* 363, eaau8299.

Hong, R., Ren, Z., Ding, J., Li, H. 2005. Experimental investigation and particle dynamic simulation for synthesizing titania nanoparticles using diffusion flame. *Chemical Engineering Journal* 108(3), 203–212.

Iravani, S., Shukla, A.K. 2019. Plant protein-based nanoparticles and their biomedical applications. In: Husen, A., Iqbal, M. (eds), *Nanomaterials and Plant Potential*, pp. 177–191. Springer International Publishing, Cham.

Jasim, B., Thomas, R., Mathew, J., Radhakrishnan, E.K. 2017. Silver nanoparticles enhance growth and diosgenin in fenugreek. *Saudi Pharmaceutical Journal* 25, 443–447.

Jiang, H.S., Li, M., Chang, F.Y., Li, W., Yin, L.Y. 2012. Physiological analysis of silver nanoparticles and $AgNO_3$ toxicity to *Spirodela polyrhiza*. *Environmental Toxicology and Chemistry* 31(8), 1880–1886.

Jiang, H.S., Yin, L.Y., Ren, N.N., Zhao, S.T., Li, Z., Zhi, Y., Shao, H., Li, W., Gontero, B. 2017. Silver nanoparticles induce ROS via photosynthetic imbalance. *Nanotoxicology* 11, 157–167.

Joshi, S., Dar, A. I., Acharya, A., Joshi, R. 2022. Charged gold nanoparticles promote in vitro proliferation in *Nardostachys jatamansi* by differentially regulating chlorophyll content, hormone concentration, and antioxidant activity. *Antioxidants* 11(10), 1962.

Jurkow, R., Pokluda, R., Sękara, A., Kalisz, A. 2020. Metal nanoparticle foliar sprays alter antioxidant systems in lettuce. *BMC Plant Biology* 20, 1–12.

Kah, M., Kookana, R.S., Gogos, A., Bucheli, T.D. 2018. A critical evaluation of nanopesticides and nanofertilizers against their conventional analogues. *Nature Nanotechnology* 13(8), 677–684.

Kataria, S., Jain, M., Rastogi, A., Živčák, M., Brestic, M., Liu, S., Tripathi, D.K. 2019. Role of nanoparticles in photosynthesis. *Nanomaterials in Plants, Algae and Microorganisms* 2, 103–127.

Kaveh, R., Li, Y.S., Ranjbar, S., Tehrani, R., Brueck, C.L., Van Aken, B. 2013. Changes in *Arabidopsis thaliana* gene expression in response to silver nanoparticles and silver ions. *Environmental science and technology* 47(18), 10637–10644.

Khalid, F., Rasheed, Y., Ashraf, H., Asif, K. 2025. Nanoparticle-mediated phytohormone interplay: advancing plant resilience to abiotic stresses. *Journal of Crop Health* 77, 56. https://doi.org/10.1007/s10343-025-01121-z

Lahiani, M.H., Dervishi, E., Chen, J., Nima, Z., Gaume, A., Biris, A.S., Khodakovskaya, M.V. 2013. Impact of carbon nanotube exposure to seeds of valuable crops. *ACS Applied Materials and Interfaces* 5(16), 7965–7973.

Liang, T.B., Yin, Q.S., Zhang, Y.L., Wang, B.L., Guo, W.M., Wang, J.W. 2013. Carbon nanoparticles influence growth and nutrient accumulation in tobacco. *International Journal of Food, Agriculture and Environment* 11, 954–958.

Marslin, G., Sheeba, C.J., Franklin, G. 2017. Nanoparticles alter secondary metabolism via ROS. *Frontiers in Plant Science* 8, 257354.

Mukherjee, A., Majumdar, S., Servin, A.D., Pagano, L., Dhankher, O.P., White, J.C. 2016. Carbon nanomaterials in agriculture. *Frontiers in Plant Science* 7, 168246.

Mukherjee, S., Krishnamoorthy, S.B., Subrayan, R., Goswami, A., Mitra, S. 2023. Cerium oxide nanoparticles alleviate mercury stress in *Vigna radiata*. *Environmental Science and Pollution Research* 30, 73952–73963.

Pallavi, C.M., Mehta, R., Srivastava, R., Arora, S., Sharma, A.K. 2016. Silver nanoparticles impact plant growth and soil bacteria. *3 Biotech* 6, 1–10.

Parveen, A., Mazhari, B.B.Z., Rao, S. 2016. Impact of bio-nanogold on seed germination of *Pennisetum glaucum*. *Enzyme and Microbial Technology* 95, 107–111.

Pascoli, M., Lopes-Oliveira, P.J., Fraceto, L.F., Seabra, A.B., Oliveira, H.C. 2019. State of the art of polymeric nanoparticles as carrier systems with agricultural applications: a mini-review. *Energy, Ecology and Environment* 3(3), 137–148

Poddar, K., Sarkar, D., Sarkar, A. 2020. Nanoparticles and plant photosynthesis. In: Patra, J., Fraceto, L., Das, G., Campos, E. (eds), *Green Nanoparticles: Synthesis and Biomedical Applications*, pp. 273–287. Springer International Publishing, Cham.

Punitha, V.N., Vijayakumar, S., Alsalhi, M.S., Devanesan, S., Nilavukkarasi, M., Vidhya, E., Kumar, S., Pand Kim, W. 2023. Biofabricated ZnO nanoparticles as vital components for agriculture revolutionization–a green approach. *Applied Nanoscience* 13(9), 5821–5827.

Qi, M., Liu, Y., Li, T. 2013. Nano-TiO_2 improves photosynthesis of tomato leaves under mild heat stress. *Biological Trace Element Research* 156, 323–328.

Raliya, R., Nair, R., Chavalmane, S., Wang, W.N., Biswas, P. 2015. Mechanistic evaluation of translocation and physiological impact of TiO_2 and ZnO nanoparticles on tomato (*Solanum lycopersicum*). *Metallomics* 7, 1584–1594.

Raliya, R., Tarafdar, J.C. 2013. ZnO nanoparticle biosynthesis and its effect on phosphorous-mobilizing enzyme secretion in clusterbean (*Cyamopsis tetragonoloba*). *Agricultural Research* 2, 48–57.

Rameshaiah, G.N., Pallavi, J., Shabnam S. 2015. Nano fertilizers and nano sensors–an attempt for developing smart agriculture. *International Journal of Engineering Research and General Science* 3(1):314–320.

Rao, N., Singh, R., Bashambu, L. 2021. Carbon-based nanomaterials: synthesis and prospective applications. *Materials Today: Proceedings* 44, 608–614.

Rasheed, A., Li, H., Tahir, M.M., Mahmood, A., Nawaz, M., Shah, A.N. et al. 2022. Nanoparticles in plant biochemical, physiological, and molecular responses under drought stress: a review. *Frontiers in Plant Science* 13, 976179.

Rawat, M., Nayan, R., Negi, B., Zaidi, M.G.H., Arora, S. 2017. Physio-biochemical basis of iron-sulfide nanoparticle induced growth and yield enhancement in *Brassica juncea*. *Plant Physiology and Biochemistry* 118, 274–284.

Rehman, M., Salam, A., Ulhassan, Z., Ali, B., Haider, Z., Ahmad, I. et al. 2025. Titanium dioxide nanoparticles (TiO_2 NPs) in crop stress management: mechanisms, applications, and abiotic stress mitigation. *Plant Nano Biology* 14, 100207. https://doi.org/10.1016/j.plana.2025.100207

Rico, C.M., Barrios, A.C., Tan, W., Rubenecia, R., Lee, S.C., Varela-Ramirez, A., Peralta-Videa, J.R., Gardea-Torresdey, J.L. 2015. Physiological and biochemical response of soil-grown barley (*Hordeum vulgare*) to cerium oxide nanoparticles. *Environmental Science and Pollution Research* 22, 10551–10558.

Rossi, L., Sharifan, H., Zhang, W., Schwab, A. P., Ma, X. 2017. Mutual effects and in planta accumulation of co-existing cerium oxide nanoparticles and cadmium in hydroponically grown soybean (*Glycine max* (L.) Merr.). *Environmental Science: Nano* 5(1), 150–157.

Rui, M., Ma, C., Hao, Y., Guo, J., Rui, Y., Tang, X. et al. 2016. Iron oxide nanoparticles as a potential iron fertilizer for peanut (*Arachis hypogaea*). *Frontiers in Plant Science* 7, 815.

Sarkhosh, S., Kahrizi, D., Darvishi, E., Tourang, M., Haghighi-Mood, S., Vahedi, P., Ercisli, S. 2022. Effect of ZnO nanoparticles on seed germination characteristics in *Camelina sativa* and *Brassica napus*. *Journal of Nanomaterials* 2022, 1–15.

Seleiman, M.F., Ahmad, A., Alhammad, B.A., Tola, E. 2023. Exogenous ZnO nanoparticles improved antioxidant, photosynthetic and yield traits in salt-stressed maize. *Agronomy* 13, 2645.

Siddiqi, K.S., Husen, A. 2016. Engineered gold nanoparticles and plant adaptation potential. *Nanoscale Research Letters* 11, 1–10.

Siddiqui, M.H., Al-Whaibi, M.H. 2014. Role of nano-SiO_2 in germination of tomato (*Lycopersicum esculentum*). *Saudi Journal of Biological Sciences* 21, 13–17.

Singh Brar, R., Kumar, A., Kaur, S., Saha, S., Kumar, A., Kumar, S. 2021. Impact of metal oxide nanoparticles on cotton (*Gossypium hirsutum* L.): a physiological perspective. *Journal of Cotton Research* 4, 16.

Singh, A., Prasad, S.M., Singh, S. 2018. Impact of nano ZnO on metabolic attributes and fluorescence kinetics of rice seedlings. *Environmental Nanotechnology, Monitoring & Management* 9, 42–49.

Singh, A., Sharma, A., Singh, O., Rajput, V.D., Movsesyan, H., Minkina, T. et al. 2024. In-depth exploration of nanoparticles for enhanced nutrient use efficiency and abiotic stresses management: present insights and future horizons. *Plant Stress* 14, 100576. https://doi.org/10.1016/j.stress.2024.100576

Singha Roy, A., Kesavan Pillai, S., Ray, S.S. 2022. Layered double hydroxides for sustainable agriculture and environment: an overview. *ACS omega*, 7(24), 20428–20440.

Sorahinobar, M., Deldari, T., Nazem Bokaeei, Z., Mehdinia, A. 2023. Effect of zinc nanoparticles on growth and biofortification potential of mungbean (*Vigna radiata*) seedlings. *Biologia* 78, 951–960.

Sosan, A., Svistunenko, D., Straltsova, D., Tsiurkina, K., Smolich, I., Lawson, T. et al. 2016. Engineered silver nanoparticles modify physiology of *Arabidopsis thaliana. The Plant Journal* 85, 245–257.

Subhash, S.V. 2017. *Biosynthesis of silver nanoparticles (AgNPs) by using Trichoderma harzianum and its efficacy against soil borne plant pathogens of tomato* (Doctoral dissertation, MSc thesis).

Surendar, K.K., Raja, R.K., Sritharan, N., Ravichandran, V., Kannan, M. 2024. Impact of nano-silica on anatomical, morpho-physiological and yield characters of rice (*Oryza sativa* L.) for drought tolerance: a review. *International Journal of Botany and Horticulture Research* 2(2), 1–10.

Suriyaprabha, R., Karunakaran, G., Yuvakkumar, R., Rajendran, V., Kannan, N. 2012. Silica nanoparticles for increased silica availability in maize (*Zea mays*. L) seeds under hydroponic conditions. *Current Nanoscience* 8(6), 902–908.

Syu, Y.Y., Hung, J.H., Chen, J.C., Chuang, H.W. 2014. Impacts of size and shape of silver nanoparticles on *Arabidopsis* growth and gene expression. *Plant Physiology and Biochemistry* 83, 57–64.

Tripathi, D., Singh, M., Pandey-Rai, S. 2022. Crosstalk of nanoparticles and phytohormones regulates plant growth under abiotic and biotic stress. *Plant Stress* 6, 100107.

Vaghar, M.S., Sayfzadeh, S., Zakerin, H.R., Kobraee, S., Valadabadi, S.A. 2020. Foliar application of iron, zinc, and manganese nano-chelates improves soybean yield under water deficit stress. *Journal of Plant Nutrition* 43, 2740–2756.

Van Nguyen, D., Nguyen, H.M., Le, N.T., Nguyen, K.H., Nguyen, H.T., Le, H.M. et al. 2022. Copper nanoparticles enhance plant growth and grain yield under drought in maize. *Journal of Plant Growth Regulation* 41, 364–375.

Vannini, C., Domingo, G., Onelli, E., Prinsi, B., Marsoni, M., Espen, L. et al. 2013. Morphological and proteomic responses of *Eruca sativa* exposed to silver nanoparticles or silver nitrate. *PLoS One* 8, e68752.

Varatharaju, G., Arun Prasad, A.S., Narayanan, D.P., Sivaramakrishnan, R. 2025. Dendrimers in plant nanotechnology. In: Al-Khayri, J.M., Anju, T.R., Jain, S.M. (eds), *Plant Nanotechnology Fundamentals and Methodologies*, pp. 345–357. Springer Nature Switzerland, Cham.

Waghule, T., Rapalli, V.K., Gorantla, S., Saha, R.N., Dubey, S.K., Puri, A., Singhvi, G. 2020. Nanostructured lipid carriers as potential drug delivery systems for skin disorders. *Current Pharmaceutical Design* 26(36), 4569–4579.

Wagi, S., Ahmed, A. 2019. Green production of AgNPs and their phytostimulatory impact. *Green Processing and Synthesis* 8, 885–894.

Wang, X., Yang, X., Chen, S., Li, Q., Wang, W., Hou, C., Gao, X., Wang, L., Wang, S. 2016. ZnO nanoparticles affect biomass accumulation and photosynthesis in Arabidopsis. *Frontiers in Plant Science* 6, 1243.

Wang, X.R., Xie, H., Wang, P., Yin, H. 2023. Nanoparticles in plants: uptake, transport and physiological activity in leaf and root. *Materials* 16, 3097.

Yan, S., Wu, F., Zhou, S., Yang, J., Tang, X., Ye, W. 2021. ZnO nanoparticles alleviate arsenic toxicity and reduce arsenic accumulation in rice. *BMC Plant Biology* 21, 1–11.

Yang, X., Alidoust, D., Wang, C. 2020. Effects of iron oxide nanoparticles on mineral composition and growth of soybean. *Acta Physiologiae Plantarum* 42, 1–11.

Ze, Y., Liu, C., Wang, L., Hong, M., Hong, F. 2011. Regulation of TiO_2 nanoparticles on light-harvesting complex II and photosynthesis in *Arabidopsis thaliana*. *Biological Trace Element Research* 143, 1131–1141.

Zhao, F., Xin, X., Cao, Y., Su, D., Ji, P., Zhu, Z., He, Z. 2021. Carbon nanoparticles improve soil fertility and nutrient uptake in corn (*Zea mays*). *Nanomaterials* 11, 2717.

Zhu, L., Chen, L., Gu, J., Ma, H., Wu, H. 2022. Carbon-based nanomaterials for sustainable agriculture: their application as light converters, nanosensors, and delivery tools. *Plants* 11(4), 511.

Zia-ur-Rehman, M., Anayatullah, S., Irfan, E., Hussain, S.M., Rizwan, M., Sohail, M.I. et al. 2022. Nanoparticles-assisted regulation of oxidative stress and antioxidant enzyme system in plants under salt stress: a review. *Chemosphere* 314, 137649. https://doi.org/10.1016/j.chemosphere.2022.137649

5 Potential of Nanoparticles to Mitigate Abiotic Stress

*Indu Sharma, Sonika Sharma, Vikas Sharma, Ashish Sharma, Talwinder Kaur, and Ashutosh Sharma**

5.1 INTRODUCTION

The growth and yield of plants are directly dependent on the environmental conditions they are exposed to. During their life cycle, plants encounter various abiotic stresses like salinity, temperature extremes (both sides), drought or waterlogged conditions, pollutants including xenobiotics, heavy metals, pesticides, other agrochemicals, and nutritional deficiency. As a response to these abiotic stressors, plants produce reactive oxygen species (ROS) via ROS-generating systems like NADPH and glycolate oxidases (Kaur et al. 2014). Such an increased ROS accumulation induced by various abiotic stressors generally causes oxidative burst (Sharma et al. 2018, 2021, 2023), which is alleviated through the collaborative action of ROS-scavenging systems (nonenzymatic or enzymatic antioxidants) and a diverse range of plant growth regulators (Dhiman et al. 2023; Maurya et al. 2024; Sharma et al. 2010, 2011a, 2011b, 2014, 2023a, 2023b, 2024; Singh et al. 2023a, 2023b; Soni and Sharma 2023). Besides these, to remediate the harmful toxicants xenobiotics and agrochemicals, several other technologies such as phytoremediation and bioremediation are also being employed (Kour et al. 2020; Singh et al. 2024).

Moreover, new tools like nanotechnology are also used to alleviate the detrimental effects of abiotic stressors or the environmental contaminants (Bhardwaj et al. 2023). Nanotechnology employs the nano-size (10^{-9} of any value or unit) particles referred to as nanoparticles (NPs). The word 'Nanoparticle' originated from the Greek term 'Nanos' meaning 'dwarf', which is a natural or manmade substance with a dimension less than the size of 100nm from at least one side (Ioannou et al. 2020). Nanotechnology can also employ biological organisms like yeast, bacteria, fungi, and algae, to remove environmental contaminants and NPs produced by using such organisms are called biogenic NPs (Bhardwaj et al. 2023). Nano-remediation is a strategy combining bioremediation and nanotechnology to remediate contaminants, with applications in medicine, public health, textiles, sports, pharmaceuticals, cosmetics, food, industry, and technology (Bhardwaj et al. 2023; Kour et al. 2021a, 2021b). However, the use of metal-based NPs has raised some environmental and biosafety concerns. Khan et al. (2019) emphasized that those NPs, which involve toxic heavy metals like lead (Pb), mercury (Hg), and tin (Sn), can be harmful to the

DOI: 10.1201/9781032659435-5

environment, due to their exceptional physico-chemical characteristics (Kour et al. 2021a). The particle size of NP ranges from 1 to 100 nm (Khan et al. 2019). They are categorized into fullerenes, ceramic NPs, polymeric NPs, and metal NPs based on size, shape, and unique properties like toughness and reactivity potential and are typically synthesized using top-down or bottom-up techniques (Al-Khayri et al. 2023). The top-down approach produces NPs after the decomposition of larger molecules (Khan et al. 2019), whereas the bottom-up technique aggregates the simpler particles to form NPs through methods like spinning, sol–gel processes, condensation, and pyrolysis. The production of NPs from biomolecules or biological materials (such as microbes and plant extracts combined with metal salt solutions) referred to as green synthesis involves bottom-up techniques (Al-Khayri et al. 2023; Khan et al. 2019). The unique structural and optical characteristics of NPs make them a preferred candidate for several downstream uses including catalysis, imaging, medical research, environmental protection, and agriculture.

NPs can affect plant growth both ways (induce or reduce), subjected to their source, dose, application, and plant type (Kour et al. 2021a). Their implications in various industries are also subjected to their negative impact on living organisms. Further investigations focused on understanding the underlying methods of uptake, transport, and biotransformation of NPs would be more beneficial in determining the potential implications of NPs for plant stress management in plants (Singh et al. 2021). Moreover, the methodology, synthesis, and application of microbial NPs for bioremediation could be a better alternative to reduce the harmful effects of different abiotic stressors in plants. This chapter highlights the role of nanomaterials in agriculture, with special emphasis on abiotic stress management, diverse ways of treatments of nanomaterials, their types, and possible mechanisms of action for conferring tolerance against different abiotic stressors in plants.

5.2 NANOMATERIALS IN AGRICULTURE

The growing population and limited resources pose challenges to agriculture, particularly in developing nations (Ioannou et al. 2020). Among other environmental factors, climate change is an important aspect that causes remarkable loss in worldwide agricultural production, thereby necessitating the development of sustainable technologies. Nanotechnology offers valuable tools for improving growth, plant protection, and performance. In the agriculture sector (Figure 5.1), various nanomaterials are being used to improve crop production by using nanoscale products like nanofertilizers (e.g. nano-urea or nano-DAP), nanoherbicides, nano-pesticides, and nano-fungicides (Al-Khayri et al. 2023). These NPs are readily soluble and small in size. NPs also stimulate uptake in plants and improve both qualitative and quantitative characteristics of plants under adverse environmental conditions. Kumar et al. (2021) emphasized that transitioning sustainable practices using modern nanomaterials can be beneficial to maximize agricultural output from limited resources. Commonly used agrochemicals may enhance crop productivity, but at the same time, they also negatively impact the fertility of the soil. However, nanomaterial-based agrochemicals may positively regulate the crop physiology. Major applications of NPs in agriculture are discussed in Figure 5.1.

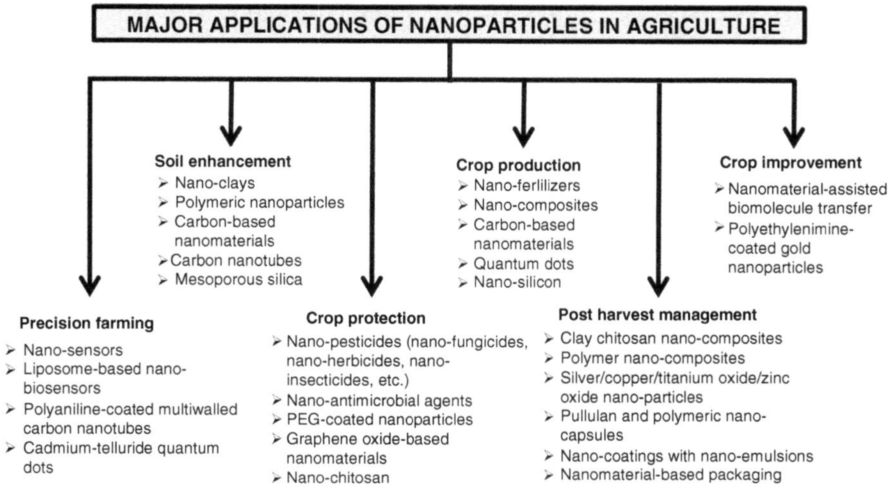

FIGURE 5.1 Major applications of nanoparticles in agriculture.

In addition to lowering the need for chemical fertilizers and pesticides, nanomaterials like nanocomposites are also helpful in energy storage, water purification, food processing, medicine delivery, illness detection, health monitoring, and pest control (Gupta 2018). Due to their special qualities, nanocomposites have several positive effects in plants (Kumar et al. 2021). Size exclusion limits, NP charge, and surface compositions all affect size-dependent translocation within different parts of plants. *In vitro* and *in vivo* studies have revealed that various NPs like nanopesticides, nanoherbicides, and nano-growth-promoting substances can be a better alternative to conventional agrochemicals (Chhipa 2017). Fertilizers containing nutrients (like phosphorus, potassium, nitrogen, iron, manganese, copper, zinc, and molybdenum) and carbon nanotube (C-NT) fertilizers are examples of nanofertilizers that exhibit improved release and targeted delivery efficiency. When compared to traditional pesticides, nanopesticides like copper (Cu), silver (Ag), silica (SiO_2), ZnO, and nanoformulations have superior broad-spectrum pest prevention efficacy.

Nanoformulations were created by coating urea with gelatine, calcium phosphate, gypsum powder, and calcium nitrate for sustainable and slow release of N (Khan et al. 2021). Gelatin-coated urea in sorghum improved nutrient loss, plant height, diameter, chlorophyll, dry matter yield, N absorption, and apparent N recovery compared to uncoated urea. Nanomaterials like metal-based NPs, C-NTs, liposomes, dendrimers, and nano-metal oxides are utilized in agricultural and food sectors, environmental management, plant disease management, packaging, taste development, pathogen detection, and various delivery methods (Kaphle et al. 2018). The treatment of uncoated C-NTs was found to enhance root length in onion and cucumber as compared to coated C-NTs (Mukherjee et al. 2016). C-NTs were found to be adsorbed onto the root surface, after that 'pierce' root epidermal cells, and then accumulate within the tissue upon exposure to wheat.

Recently, Pascoli et al. (2019) found that neem oil-based nanopesticides can be used in sustainable agriculture. Zein NPs coated with neem oil improved biopesticide properties. The neem oil-loaded zein NPs were safe for non-target organisms, demonstrating their eco-friendliness. C-NTs are being used in agriculture as nanomaterials for biosensors and pesticide systems (Chaud et al. 2021). They have positive impacts on plant biomass, plant height, and seed germination but may cause cell death and a decline in microbial communities in soil. Trypsin-modulating oostatic factor-conjugated hydrophilic polymers promote stability and limit metabolism, improving bioaccumulation and decreasing insect reproduction (Chaud et al. 2021; Parsons et al. 2018; Shen et al. 2009). Since they are inexpensive, simple to produce, easy to control size, biocompatible, nontoxic, and may be combined with a wide range of crosslinking agents, ionotropic reactions have drawn attention as pesticide nanocarriers. Chaudhry et al. (2018) suggested that nano-biosensors can detect and inhibit plant pathogens and monitor soil health.

5.3 ABIOTIC STRESS MANAGEMENT IN USING NANOMATERIALS

Nanotechnology has potential in managing abiotic stress, such as global warming, drought, floods, and extreme weather. Nanomaterials can improve crop growth and productivity under abiotic stresses by entering plant cells and affecting their life cycle activities. The interaction between plants and NPs can be beneficial or harmful, depending on their unique characteristics and species. Among various NPs, titanium dioxide (TiO_2)-NP, Zn-NP, ZnO-NP, cesium (Ce)-NP, cobalt (Co)-NP, Cu-NP, Mn-NP, Cu oxide (CuO)-NP, selenium (Se)-NP, Ag-NP, Si-NP, SiO_2-NP, iron oxide (FeO)-NP, $CaCO_3$-NP, magnesium (Mg)-NP, MgO-NP, Mo-NP, AlO_4-NPs, and C-NTs have been successfully employed to treat plants to overcome abiotic stresses (Khalid et al. 2022).

5.3.1 Different Modes of Nanomaterial Treatments

Advanced NPs and nano-polymers can be applied in plants through different modes (Figure 5.2), offering an alternative to conventional breeding and genetic modification (Ioannou et al. 2020). Two commonly employed methods, namely priming and foliar application of NPs, are discussed in the subsequent sub-sections.

5.3.1.1 Priming

Seed priming is done prior to sowing, which enhances seedling growth and capacity to withstand biotic and abiotic stressors (Zaytseva and Neumann 2016). It improves water use efficiency, nutrient uptake, germination rate, and shoot and root elongation. Seed priming regulates phytohormones, gene expression, metabolism, and expression of antioxidant enzymes, thereby ensuring cellular homeostasis and enhanced seed germination as well as mitigating environmental stressors. Seed nanopriming is a new technology using NPs to enhance germination, improve growth, and protect plants from various stresses (Acharya et al. 2019). It minimizes chemical input, increases plant uptake efficiency, and replaces high-dose herbicides and pesticides that cause soil pollution, thereby reducing the need for conventional materials

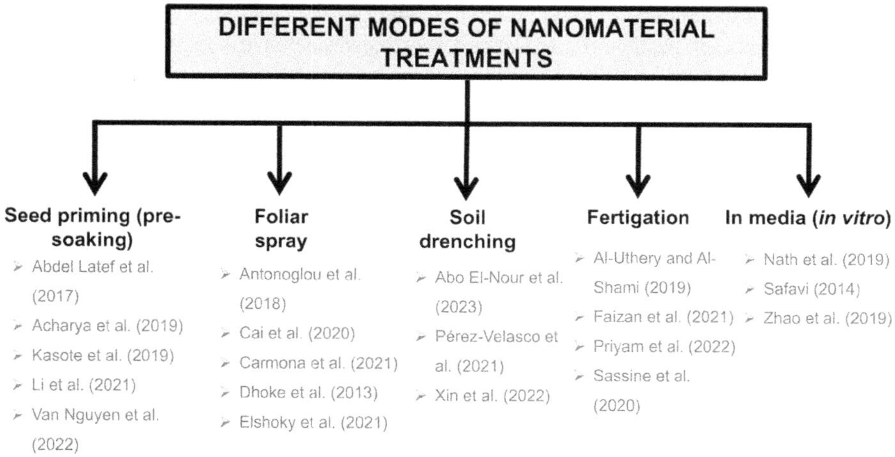

FIGURE 5.2 Different modes of nanomaterial treatments for plants.

(Acharya et al. 2019; Kasote et al. 2019). Two nanoemulsions made from turmeric and citrus oil were utilized for priming aged onion seeds (Acharya et al. 2019). Priming of seeds using Ag and Au NPs was observed to increase seed emergence and productivity of the onion. Seed priming with AuNPs significantly changed the emergence percentage (63.2%), and the average yield increased by 23.9%. Seed priming with these green-biosynthesized NPs was found to be safe and could ensure the sustainable usage of NPs in agriculture. Abdel Latef et al. (2017) found that priming of seeds with ZnO-NPs is an effective strategy for enhancing salt tolerance in lupine. Further, Kasote et al. (2019) reported that priming of watermelon seeds using Fe-NPs was effective in seedling health, antioxidant status, and the metabolic and hormonal profiles. Fe-NPs priming was found to be safe and had no adverse effect on germination, seedling development, or the biosynthesis of photosynthetic pigments. Further, Van Nguyen et al. (2022) also found that priming of maize seed using Cu-NPs was effective in conferring drought stress tolerance. Seed priming for 24 hours with ZnO-NPs and a plant growth regulator (24-epibrassinolide) could alleviate the adverse effects of salinity on germination and cellular homeostasis (Alhammad et al. 2023). Further, this study revealed that priming of seed using ZnO-NPs at a concentration of 100 mg/L was more effective than the 50 mg/L concentration. Another study by Li et al. (2021) emphasized that the priming of rice seeds with ZnO-NPs for 20 hours at 25°C in the dark was effective in alleviating Cd-induced phytotoxicity. Furthermore, a comprehensive investigation to develop a mathematical model for comparative analysis of abiotic stressors, NPs, and their optimal dose and time duration for seed incubation might pave the way for determining the potential efficacy of seed priming.

5.3.1.2 Foliar Spray

The NPs when applied through foliar spray were observed to significantly augment plant weights, activate antioxidants, and upregulate salicylic acid (SA) biosynthesis pathways, thereby inducing two important SA-responsive genes belonging to the category of pathogenesis-related (PR) genes, namely *PR1* and *PR2* (Cai et al. 2020). Foliar treatment of Fe_3O_4-NPs enhanced the resistance of tobacco plants through the stimulation of SA metabolism. The foliar spray with optimum concentrations of ZnO-NP, FeO-NP, and ZnCuFe-oxide NP in suspension form on hydroponically grown plants resulted in better growth (Dhoke et al. 2013). Further, it was established through inductively coupled plasma/atomic emission spectroscopy that NPs were absorbed by plant leaves. In tomato plants, the foliar spray of nanobrass CuZn–bimetallic-NPs was found to be an innovative and safe solution for controlling fungal infections and it showed enhanced antifungal activity against *Saccharomyces cerevisiae* (Antonoglou et al. 2018). Foliar treatment of ZnO and ZnO–Si-NPs at a concentration of 200–400 mg/L was found effective in pea plants under salt stress environment (Elshoky et al. 2021). Moreover, the foliar application of NPs has been suggested to be a better and faster mode of NP application as compared to fertilizers applied to the soil. Carmona et al. (2021) reported that roots uptake nutrients faster compared to leaves. Vine plants recognized and assimilated nutrients provided by calcium (ortho)phosphate (CaP)-NPs. Foliar and fertigation applications had similar effects compared to commercial fertilizers.

5.3.2 Different Nanomaterial Treatments Under Varied Abiotic Stress

Plants are exposed to various abiotic stress factors during development, including drought, salt, light, and low or high temperatures. The plant stress management efficiency of nanomaterials has been investigated by numerous researchers by exploring the effect of various treatments of NPs on plants subjected to abiotic stress(es). Nanomaterials, such as nanopesticides or nanofertilizers, have the capability to increase crop production and address some abiotic-related issues at the interface of soil and the plant. Ahmed et al. (2021) showed that engineered NPs like FeO-NPs and hydrogel-NPs can improve crop productivity, manage plant stress, and enhance water use efficiency in resource-deficient arable soils.

5.3.2.1 Salt Stress

Soil salinity, a global environmental factor, limits crop productivity due to osmotic imbalances, phytonutrition-related issues, and ionic toxicities (Elshoky et al. 2021). The effects of salt-induced toxicity in plants can be further enhanced by the oxidative damage due to increased ROS levels. It significantly influences photosynthetic efficiency due to structural changes in thylakoid membranes. High salinity stress inhibits the photosystem II (PSII) complex, causing a decrease in photochemical efficacy of PSII and the rate of electron transport. Treatment of salt stress–affected plants with NP was found to be effective in stress alleviation. ZnO-NP priming treatments significantly reduced germination time and daily germination speed, enhanced K^+ ion uptake, and increased the activity of α-amylase in maize seeds in saline environment (Alhammad et al. 2023). In *Mentha piperita* L., different concentrations (0, 10, 20, and 30 μm) of Fe_2O_3 NP under varying levels of salinity resulted in decreased accumulation of proline and ROS (Askary et al. 2017). The treatment of salt-stressed soya bean plants with CeO_2 (cerium oxide)-NPs was found effective in enhancing photosynthesis rates through an increase in the activity of the Rubisco (Cao et al. 2017).

At a 100 mg/kg dose of CeO_2-NPs, it improved growth and increased rate of photosynthesis; while it inhibited the activity of Rubisco and interfered with CO_2 diffusion routes, when applied at a dose of more than 500 mg/kg. The physiological effects of CeO_2-NPs were found to be dependent on both the concentrations and the surface properties of the NPs.

When *Capsicum annum* L. under salt stress was treated with different concentrations of Mn-NPs (0.1, 0.5, and 1 mg/L), it showed a redistribution of Mn, Na, K, and Ca contents in both shoot and root (Ye et al. 2020). Overall, Mn-NPs significantly enhanced the root growth in both the presence and absence of salt stress. Therefore, nanopriming has a massive potential to alter biomolecule composition and interactions, potentially aiding in managing plant salt stress and potentially promoting sustainable agriculture. An investigation by Elshoky et al. (2021) explored the use of ZnO-NPs in pea under a salt stress environment. It showed that the foliar application with ZnO–Si-NPs at a lower dose of 200 mg/L did not affect stomatal structure, photosystem functions, and membrane integrity; however, at a higher dose, i.e. 400 mg/L, it had a beneficial effects on PSII and also on overall photochemistry. Among the two NPs, ZnO–Si-NP application was found to have more salt stress protection potential in pea plants. The lupine seeds exposed to varying concentrations of ZnO-NPs showed reduced growth parameters and reduced photosynthetic pigments and catalase activity (Abdel Latef et al. 2017). However, salinity stress increased organic solutes, phenols, and enzyme activities in stressed plants. Seed priming using ZnO-NPs stimulated growth and enhanced salt resilience, particularly at a concentration of 60 mg/L ZnO-NPs.

5.3.2.2 Drought Stress

The treatment of Cu-NPs is shown to enhance growth and productivity in maize under drought conditions (Van Nguyen et al. 2022). Seed priming with Cu-NPs was observed to positively regulate the drought stress responses in maize. It enhanced drought tolerance by enhancing the contents of leaf water, anthocyanin, chlorophyll, carotenoid, and plant biomass in drought-stressed maize. Histochemical analyses showed decreased accumulation of ROS in Cu-NP-primed plants. Applying Cu-NPs increased seed yield under drought stress, suggesting these NPs as an important approach for crop production under drought conditions. Treatments of both silicate and Si-NPs to drought-stressed barley seedlings resulted in significant post-stress recovery performance (Ghorbanpour et al. 2020). Si-NP treatment led to the formation of Si-NP aggregates, large pores in roots, and rapid stomata closure in leaves. Shoot biomass increased significantly after recovery, especially in moderately stressed plants. The Si-NPs and silicate were associated with the changes in accumulation of osmolyte, changes in metabolism, cellular homeostasis, and antioxidant enzyme activities. Zhao et al. (2022) found that treating soya bean with a nanomaterial called graphene oxide (GO) improved drought resistance. GO-treated soya bean had higher water content, root parameters, and plant defence enzyme activities. Additionally, GO treatment decreased electrical conductivity, malondialdehyde content, and H_2O_2 accumulation. It also regulated drought-related hormones and gene expression in drought-affected soya bean. These findings suggest that GO treatment may offer new opportunities for developing effective soil water retention agents.

5.3.2.3 Heavy Metal Stress

Agricultural soil contains nanomaterials, which reduce heavy metal bioaccumulation in cucumber plants' roots, stems, and leaves, while also reducing macro- and micro-nutrient uptake (Song et al. 2019). MoS_2 nanomaterials showed promising potential in reducing heavy metal uptake but also had negative impacts in terms of altered nutrient uptake by the cucumber plants. Application of ZnO-NPs remarkably augmented the physiological attributes of rice, besides improving its seedling growth under Cd stress (Li et al. 2021). However, seed priming using ZnO-NPs could not be linked to any improvement in seed germination but promoted seedling weight, root–shoot ratio, and shoot weight under Cd toxicity. *Oryza sativa* exposed to ZnO-NPs at 5–25 mg/L concentration exhibited a reduced uptake of Cu and Pb (Akhtar et al. 2021). Application of both bacteria and ZnO-NPs at the lower application rates was effective in removing the heavy metals from polluted aquatic habitats. Seeds primed with bacteria in heavy metal–polluted water with ZnO-NP treatment reduced heavy metal uptake and reduced their bioaccumulation index, besides improving plant growth. Seed priming with ZnO-NPs decreased heavy metal toxicity, lowered bioaccumulation index, decreased metallothionine content, and improved plant growth in rice. Recently, Chen et al. (2022) studied the molecular mechanisms of multi-walled C-NTs in promoting the growth of *Solanum nigrum* under Cd and As stress conditions. It was found that 500 mg/kg multi-walled C-NT treatment promoted the growth of *S. nigrum* considerably, particularly in the root tissues. The multi-walled C-NTs also improved both P and Fe absorption in plant roots and the antioxidant enzyme activities. Transcriptomic analysis revealed that the application of multi-walled C-NTs led to the upregulation of the genes related to some beneficial biological processes, when the heavy metal(loid)s stress is prevalent and further emphasized the application of C-NTs in the reclamation and phyto-remediation of metal-contaminated areas.

5.3.2.4 Extreme Climatic Conditions

Climate change results in rapid fluctuations in the temperature that directly affect the crop productivity of staple crops (rice, wheat, and barley), particularly in developing countries (Zohra et al. 2021). To counteract these effects, Se, an essential trace element, can be applied. However, its higher concentrations are reported to be toxic, which results in overproduction of ROS, enhanced oxidative stress, disrupted cell structure, and reduced seed formation, but at lower levels, Se acts as an antioxidant agent, maintaining the structure and fluidity of the plastid membrane and chloroplast. Of late, implications of Se through nanotechnology as Se-NPs have been found to alleviate abiotic and biotic stresses in plants. In grain sorghum, at high temperatures (38°C/28°C), the foliar spray of Se-NPs was linked to alleviate high-temperature stress by enhancing antioxidant defence mechanisms, reducing oxidant levels, and improving the germination of pollens and increasing yield (Djanaguiraman et al. 2018). The study revealed that the Se-NPs could be translocated from plant roots to the shoots and 50% rate of cell mortality was recorded at 275 mg/L concentration of Se-NPs. The biogenic Ag-NPs were prepared using *Moringa oleifera* extract, and their effect was studied on wheat growth under heat (35°C–40°C) stress (Iqbal et al. 2019). The heat stress reduced biomass, height, leaf area, leaf number, and root number, but Ag-NP application improved these traits. In another experiment by Mohammadi et al. (2013), it was found that TiO_2-NPs may improve redox status and increase cold tolerance in chickpea, particularly sensitive genotypes. These findings could lead to further applications in agriculture.

5.3.2.5 Combination of Various Abiotic Stresses

Heavy metal (Cd) and drought are destructive abiotic stresses that negatively impact plant metabolism, photosynthesis, and nutrient use efficiency (Ahmed et al. 2021). In rice, treatment of FeO-NPs to mitigate both Cd and drought stresses resulted in increased plant biomass, photosynthesis performance, antioxidant status, and the ability for nutrient absorption, while decreasing ROS and Cd translocation. NP-treated rice plants also regulated the expression of genes involved in Cd transport. Therefore, the application of NP-based formulations might be an effective strategy for managing heavy metal pollution. Polymeric NPs have been utilized to lessen the toxicity of atrazine in the environment and to manage weeds (Chaud et al. 2021). These NPs have the potential to be used in atrazine-resistant crops since they have been modified with PLGA to boost bioavailability and decrease toxicity. Potential for atrazine-resistant crops, polymeric NPs made with PLGA have been utilized to suppress weeds and lessen environmental toxicity.

5.4 MECHANISM OF NANOMATERIAL-INDUCED ABIOTIC STRESS RESILIENCE IN PLANTS

The crop plants have evolved a wide array of molecular mechanisms that enable them to quickly detect environmental stressors and adjust their responses accordingly. When the plants and NPs interact, the interaction may lead to improvement in their performance by strengthening plant tolerance mechanisms against environmental stressors. The energy-dispersive spectroscopy and scanning electron microscopy (SEM)-based studies have suggested that Mn-NPs penetrate through the seed coat and then result in the formation of an NP–cornea complex (Ye et al. 2020). Under salt stress, the surface chemistries of Mn-NPs were observed to differ in water and salt conditions, affecting their affinity to proteins and altering N–H bonding. A hypothesis by Elshoky et al. (2021) emphasized that modifying the surface of some NPs like ZnO with Si may lead to an alteration of their charge, leading to increased stability and improved hydrophilicity compared to uncoated NPs. It might help in conferring salt stress alleviation in affected plants. Moreover, coating ZnO-NPs by a shell made up of Si would help in decreasing the metal-induced toxicity. Figure 5.3 represents various effects of NPs by which NPs act as abiotic stress alleviators in plants.

Plants synthesize and accumulate ROS more during abiotic stresses, which are managed through antioxidant defence systems. These enzymes mitigate the harmful effects of ROS. Ferromagnetic NPs can reduce superoxide radicals, H_2O_2 levels, and lipid peroxidation under stressed conditions. In plants, high concentration of heavy metal can affect plant metabolism (Zhou et al. 2020). Various strategies can help overcome stress including reducing soil heavy metal concentration, regulating heavy metal transport gene expression, enhancing antioxidant systems, and promoting

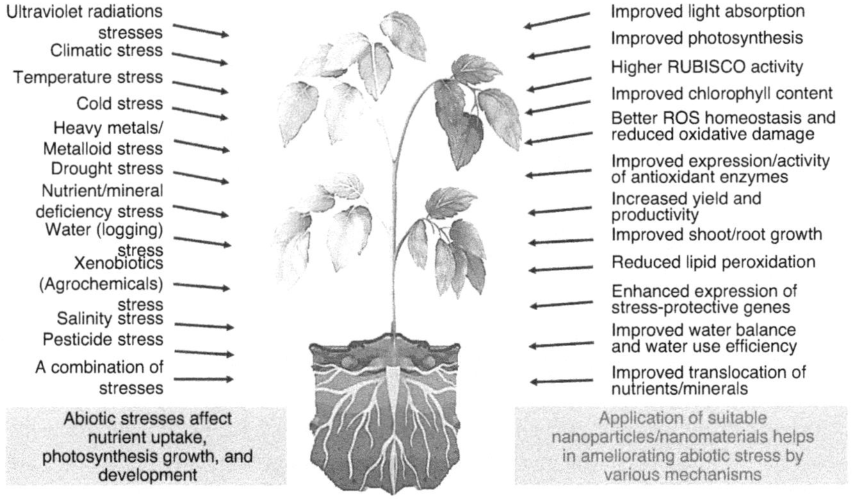

FIGURE 5.3 Mechanism of action of nanoparticles as abiotic stress mitigators in plants.

protective agent production. NPs alleviate heavy metal stress in soil and plant tissues by increasing soil pH, adsorbing and transforming heavy metal, regulating heavy metal transport genes, enhancing plant defence, promoting protective agents, and scavenging ROS. Apoplastic barriers present in roots can safeguard plants and regulate water, ions, and O_2 flow. NPs may contribute in the establishment of such barriers, reducing metal stress. However, these barriers alone may not be effective due to plant roots' ion and protein transporters and specific NPs regulating plant metal transport genes. NPs can also bind with heavy metals, making them unavailable and reducing biological activity. Additionally, NPs can enhance the production of structural protective agents.

5.5 CHALLENGES AND OPPORTUNITIES

NPs are being used in various industries and medical systems, but their uncontrollable use has raised concerns about health hazards due to their environmental discharge. Metallic NPs have been linked to pulmonary inflammation and immune system response, while tiny NPs can easily be ingested or inhaled by living organisms. The respiratory system is a unique target for NP toxicity. The knowledge of the harmful consequences of NP application in living organisms is limited, and the development of functionalized NPs is limited by data scarcity. A collaborative approach involving experts from various fields is needed to address these issues. Nanotechnology is being used to reduce the adverse effects of abiotic stressors in plants, but most investigations are limited to the laboratory stage. Further, more studies are required to understand the complex interactions of NPs with plants and soil, evaluate potential effects, and optimize green NP-based agricultural practices in a better way.

NOTE:

* Corresponding author.

REFERENCES

Abdel Latef, A.A.H., Abu Alhmad, M.F., Abdelfattah, K.E. 2017. Priming with ZnO nanoparticles mitigates salinity stress in lupine (Lupinus termis) plants. *Journal of Plant Growth Regulation* 36, 60–70.

Acharya, P., Jayaprakasha, G.K., Crosby, K.M., Jifon, J.L., Patil, B.S. 2019. Green-synthesized nanoparticles enhance seedling growth, yield, and quality of onion (*Allium cepa* L.). *ACS Sustainable Chemistry & Engineering* 7(17), 14580–14590.

Ahmed, T., Noman, M., Manzoor, N., Shahid, M., Abdullah, M., Ali, L. et al. 2021. Nanoparticle-based amelioration of drought stress and cadmium toxicity in rice via triggering stress-responsive genetic mechanisms and nutrient acquisition. *Ecotoxicology and Environmental Safety* 209, 111829.

Akhtar, N., Khan, S., Rehman, S.U., Rehman, Z.U., Khatoon, A., Rha, E.S., Jamil, M. 2021. Synergistic effects of zinc oxide nanoparticles and bacteria reduce heavy metal toxicity in rice (*Oryza sativa* L.). *Toxics* 9(5), 113.

Alhammad, B.A., Ahmad, A., Seleiman, M.F., Tola, E. 2023. Seed priming with nanoparticles and 24-epibrassinolide improves seed germination and enzymatic performance of *Zea mays* L. under salt stress. *Plants* 12(4), 690.

Al-Khayri, J.M., Rashmi, R., Ulhas, R.S., Sudheer, W.N., Banadka, A., Nagella, P., Aldaej, M.I., Rezk, A.A-S., Shehata, W.F., Almaghasla, M.I. 2023. Role of nanoparticles in plant responses to abiotic stress at physiological, biochemical, and molecular levels. *Plants* 12(2), 292.

Antonoglou, O., Moustaka, J., Adamakis, I-D.S., Sperdouli, I., Pantazaki, A.A., Moustakas, M., Dendrinou-Samara, C. 2018. Nanobrass CuZn nanoparticles as foliar spray nonphytotoxic fungicides. *ACS Applied Materials & Interfaces* 10(5), 4450–4461.

Askary, M., Talebi, S.M., Amini, F., Dousti Balout Bangan, A. 2017. Effects of iron nanoparticles on *Mentha piperita* L. under salinity stress. *Biologija* 63(1), 3476.

Bhardwaj, T., Khanna, K., Sharma, P., Bakshi, P., Devi, K., Madaan, I. et al. 2023. Nanobioremediation: a novel technology with exceptional cleanup potential for sustainable environments. In: Sharma, A., Kumar, V., Zheng, B. (eds), *Environmental Applications of Microbial Nanotechnology*, pp. 315–331. Elsevier, Amsterdam, Netherlands.

Cai, L., Cai, L., Jia, H., Liu, C., Wang, D., Sun, X. 2020. Foliar exposure to Fe_3O_4 nanoparticles in *Nicotiana benthamiana*: uptake, growth promotion, and elicitation of defense responses against plant virus. *Journal of Hazardous Materials* 393, 122415.

Cao, Z., Stowers, C., Rossi, L., Zhang, W., Lombardini, L., Ma, X. 2017. Physiological effects of cerium oxide nanoparticles on photosynthesis and water-use efficiency of soybean (Glycine max (L.) Merr.). *Environmental Science: Nano* 4(5), 1086–1094.

Carmona, F.J., Dal Sasso, G., Ramírez-Rodríguez, G.B., Pii, Y., Delgado-López, J.M., Guagliardi, A., Masciocchi, N. 2021. Urea-functionalized amorphous calcium phosphate nanofertilizers: optimizing synthetic strategies toward environmental sustainability and reduced manufacturing costs. *Scientific Reports* 11, 3419.

Chaud, M., Souto, E.B., Zielinska, A., Severino, P., Batain, F., Oliveira-Junior, J., Alves, T. 2021. Nanopesticides in agriculture: benefits, productivity enhancement, and toxicological risks to human health and the environment. *Toxics* 9(6), 131.

Chaudhry, N., Dwivedi, S., Chaudhry, V., Singh, A., Saquib, Q., Azam, A., Musarrat, J. 2018. Bio-inspired nanomaterials in agriculture and food: current status, emerging applications, and challenges. *Microbial Pathogenesis* 123, 196–200.

Chen, X., Wang, J., Wang, R., Zhang, D., Chu, S., Yang, X. et al. 2022. Growth-promoting effects of nanomaterials revealed through transcriptomics and metabolomics: molecular mechanisms of MWCNTs in enhancing hyperaccumulator plants under heavy metal(loid) stress. *Journal of Hazardous Materials* 439, 129640.

Chhipa, H. 2017. Nanofertilizers and nanopesticides for agriculture. *Environmental Chemistry Letters* 15, 15–22.

Dhiman, S., Ibrahim, M., Khanna, K., Bhardwaj, T., Devi, K., Sharma, I., Arora, U., Mir, B.A., Bhardwaj, R. 2023. Cross-talk between ROS, RNS, RCS, and RSS in plants under abiotic stresses. In: Singh, V.P., Singh, S., Tripathi, D.K., Prasad, S.M., Chauhan, D.K. (eds), *Nitric Oxide in Developing Plant Stress Resilience*, pp. 305–326. Academic Press, New York.

Dhoke, S.K., Mahajan, P., Kamble, R., Khanna, A. 2013. Effect of nanoparticle suspension on growth of mung (*Vigna radiata*) seedlings via foliar spray. *Nanotechnology Development* 3(1), e1–e1.

Djanaguiraman, M., Belliraj, N., Bossmann, S.H., Vara Prasad, P.V. 2018. Selenium nanoparticles alleviate high-temperature stress in grain sorghum. *ACS Omega* 3(3), 2479–2491.

Elshoky, H.A., Yotsova, E., Farghali, M.A., Farroh, K.Y., El-Sayed, Kh., Elzorkany, H.E. et al. 2021. Foliar application of zinc oxide nanoparticles enhances photosynthesis in *Pisum sativum* L. under salt stress. *Plant Physiology and Biochemistry* 167, 607–618.

Ghorbanpour, M., Mohammadi, H., Kariman, K. 2020. Nanosilicon-mediated recovery of barley (*Hordeum vulgare*) under drought stress. *Environmental Science: Nano* 7(2), 443–461.

Gupta, H. 2018. Role of nanocomposites in agriculture. *Nano Hybrids and Composites* 20, 81–89.

Ioannou, A., Gohari, G., Papaphilippou, P., Panahirad, S., Akbari, A., Dadpour, M.R., Krasia-Christoforou, T., Fotopoulos, V. 2020. Advanced nanomaterials in agriculture under a changing climate: future perspectives. *Environmental and Experimental Botany* 176, 104048.

Iqbal, M., Raja, N.I., Mashwani, Z.-U.-R., Hussain, M., Ejaz, M., Yasmeen, F. 2019. Silver nanoparticles enhance growth of wheat under heat stress. *Iranian Journal of Science and Technology, Transactions A: Science* 43, 387–395.

Kaphle, A., Navya, P.N., Umapathi, A., Daima, H.K. 2018. Nanomaterials for agriculture, food and environment: applications, toxicity and regulation. *Environmental Chemistry Letters* 16, 43–58.

Kasote, D.M., Lee, J.H.J., Jayaprakasha, G.K., Patil, B.S. 2019. Seed priming with iron oxide nanoparticles modulates antioxidant potential and defense-linked hormones in watermelon seedlings. *ACS Sustainable Chemistry & Engineering* 7(5), 5142–5151.

Kaur, G., Sharma, A., Guruprasad, K., Pati, P.K. 2014. Versatile roles of plant NADPH oxidases and emerging concepts. *Biotechnology Advances* 32(3), 551–563.

Khalid, M.F., Khan, R.I., Jawaid, M.Z., Shafqat, W., Hussain, S., Ahmed, T., Rizwan, M., Ercisli, S., Pop, O.L., Marc, R.A. 2022. Nanoparticles: the plant saviour under abiotic stresses. *Nanomaterials* 12(21), 3915.

Khan, I., Saeed, K., Khan, I. 2019. Nanoparticles: properties, applications and toxicities. *Arabian Journal of Chemistry* 12(7), 908–931.

Khan, O., Niazi, M.B.K., Shah, G.A., Hazafa, A., Jahan, Z., Sadiq, M., Sher, F. 2021. Green synthesis and evaluation of calcium-based nanocomposite fertilizers: a way forward to sustainable agriculture. *Journal of the Saudi Society of Agricultural Sciences* 20(8), 519–529.

Kour, J., Khanna, K., Bakshi, P., Sharma, I., Ohri, P., Mir, B.A., Kapoor, N., Bhardwaj, R. 2020. Role of beneficial microbes in the molecular phytotoxicity of heavy metals. In: Faisal, M., Saquib, Q., Alatar, A.A., Al-Khedhairy, A.A. (eds), *Cellular and Molecular Phytotoxicity of Heavy Metals*, pp. 227–262. Springer, Cham.

Kour, J., Khanna, K., Sharma, P., Arora, P., Dhiman, S., Kaur, R., Sharma, A., Ohri, P., Bhardwaj, R. 2021a. Variability, behaviour and impact of nanoparticles in the environment. In: Singh, V.P., Singh, S., Tripathi, D.K., Prasad, S.M., Chauhan, D.K. (eds), *Plant Responses to Nanomaterials: Recent Interventions, and Physiological and Biochemical Responses*, pp. 315–328. Springer, Cham.

Kour, J., Khanna, K., Sharma, P., Singh, A.D., Sharma, I., Arora, P. et al. 2021b. Hydrogen sulfide and phytohormones crosstalk in plant defense against abiotic stress. In: Singh, V.P., Singh, S., Tripathi, D.K., Prasad, S.M., Chauhan, D.K. (eds), *Hydrogen Sulfide in Plant Biology*, pp. 267–302. Academic Press, New York.

Kumar, A., Choudhary, A., Kaur, H., Mehta, S., Husen, A. 2021. Smart nanomaterials and nanocomposites with advanced agrochemical activities. *Nanoscale Research Letters* 16, 1–26.

Li, Y., Liang, L., Li, W., Ashraf, U., Ma, L., Tang, X., Pan, S., Tian, H., Mo, Z. 2021. ZnO nanoparticle-based seed priming modulates early growth and enhances physio-biochemical and metabolic profiles of fragrant rice against cadmium toxicity. *Journal of Nanobiotechnology* 19, 1–19.

Maurya, V., Singh, N., Sharma, I., Sharma, A., Kumar, R. 2024. Effect of melatonin in regulating salt stress responses in plants. In: Sharma, A., Ahammed, G.J. (eds), *Melatonin in Plants: Role in Plant Growth, Development, and Stress Response*, pp. 109–139. Springer Nature, Singapore.

Mohammadi, R., Maali-Amiri, R., Abbasi, A. 2013. Effect of TiO_2 nanoparticles on chickpea response to cold stress. *Biological Trace Element Research* 152, 403–410.

Mukherjee, A., Majumdar, S., Servin, A.D., Pagano, L., Dhankher, O.P., White, J.C. 2016. Carbon nanomaterials in agriculture: a critical review. *Frontiers in Plant Science* 7, 168246.

Parsons, K.H., Mondal, M.H., McCormick, C.L., Flynt, A.S. 2018. Guanidinium-functionalized interpolyelectrolyte complexes enabling RNAi in resistant insect pests. *Biomacromolecules* 19(4), 1111–1117.

Pascoli, M., Jacques, M.T., Agarrayua, D.A., Avila, D.S., Lima, R., Fraceto, L.F. 2019. Neem oil-based nanopesticide as an environmentally-friendly formulation for sustainable agriculture: an ecotoxicological perspective. *Science of the Total Environment* 677, 57–67.

Pérez-Velasco, E.A., Valdez-Aguilar, L.A., Betancourt-Galindo, R., Martínez-Juárez, J., Lozano-Morales, S.A., González-Fuentes, J.A. 2021. Gas exchange, fruit yield, quality, and nutrient status in tomato stimulated by ZnO nanoparticles with modified surface, morphology, and application form. *Journal of Soil Science and Plant Nutrition* 21(2), 991–1003.

Sharma, A., Pathania, A., Sharma, P., Bhardwaj, R., Sharma, I. 2023a. Role of glycine betaine in regulating physiological and molecular aspects of plants under abiotic stress. In: Sharma, A., Pandey, S., Bhardwaj, R. (eds), *The Role of Growth Regulators and Phytohormones in Overcoming Environmental Stress*, pp. 327–353. Academic Press, New York.

Sharma, A., Sharma, P., Kumar, R., Sharma, V., Bhardwaj, R., Sharma, I. 2021. Role of reactive oxygen species in the regulation of abiotic stress tolerance in legumes. In: Singh, V.P., Singh, S., Tripathi, D.K. (eds), *Abiotic Stress and Legumes*, pp. 217–243. Academic Press, New York.

Sharma, A., Sharma, S.K., Singh, N., Maurya, V., Kaur, S., Kumar, R., Sharma, I. 2024. Pesticides-mediated ROS generation in plants. In: Sharma, A., Kumar, V., Zheng, B. (eds), *Pesticides in a Changing Environment*, pp. 179–202. Elsevier, Amsterdam, Netherlands.

Sharma, I., Bhardwaj, R., Gautam, V., Kaur, R., Sharma, A. 2018. Brassinosteroids: occurrence, structure, and stress protective activities. *Frontiers in Natural Product Chemistry* 4, 204–239.

Sharma, I., Pati, P.K., Bhardwaj, R. 2010. Regulation of growth and antioxidant enzyme activities by 28-homobrassinolide in seedlings of *Raphanus sativus* L. under cadmium stress. *Indian Journal of Biochemistry & Biophysics* 47, 172–177.

Sharma, I., Pati, P.K., Bhardwaj, R. 2011a. Effect of 28-homobrassinolide on antioxidant defence system in *Raphanus sativus* L. under chromium toxicity. *Ecotoxicology* 20, 862–874.

Sharma, I., Pati, P.K., Bhardwaj, R. 2011b. Effect of 24-epibrassinolide on oxidative stress markers induced by nickel-ion in *Raphanus sativus* L. *Acta Physiologiae Plantarum* 33, 1723–1735.

Sharma, I., Sharma, A., Bhardwaj, R., Sirhindi, G. 2023b. *PGPR (Plant Growth Promoting Rhizobacteria) for Plant Stress Management*. Nova Science Publishers, Hauppauge, NY. https://doi.org/10.52305/MJDR1541

Sharma, N., Hundal, G.S., Sharma, I., Bhardwaj, R. 2014. 28-Homobrassinolide alters protein content and activities of glutathione-S-transferase and polyphenol oxidase in *Raphanus sativus* L. under heavy metal stress. *Toxicology International* 21(1), 44.

Sharma, S.K., Sharma, I., Bhushan, B. 2023. A review of melatonin metabolism with reference to plants. *International Journal of Advanced Research in Biological Sciences* 10(9), 1–10.

Shen, H., Brandt, A., Witting-Bissinger, B.E., Gunnoe, T.B., Roe, R.M. 2009. Novel insecticide polymer chemistry to reduce enzymatic digestion of a protein pesticide, trypsin modulating oostatic factor (TMOF). *Pesticide Biochemistry and Physiology* 93(3), 144–152.

Singh, A.D., Khanna, K., Kour, J., Dhiman, S., Ibrahim, M., Sharma, N., Sharma, I., Sharma, P., Mir, B.A., Bhardwaj, R. 2023a. Signaling pathways of gasotransmitters in heavy metal stress mitigation. In: Aftab, T., Corpas, F.J. (eds), *Gasotransmitters Signaling in Plants under Challenging Environment*, pp. 45–76. Springer, Cham.

Singh, A.D., Kour, J., Kumar, P., Sharma, N., Sharma, P., Madaan, I., Sharma, I., Kapoor, N., Singh, A.P., Bhardwaj, R. 2023b. Role of jasmonates in regulating physiological and molecular aspects of plants under abiotic stress. In: Sharma, A., Pandey, S., Bhardwaj, R. (eds), *The Role of Growth Regulators and Phytohormones in Overcoming Environmental Stress*, pp. 137–173. Academic Press, New York.

Singh, A.D., Sharma, I., Khanna, K., Handa, N., Kumar, P., Ibrahim, M., Ohri, P., Bhardwaj, R. 2021. Impact of cobalt oxide nanoparticles on morpho-physiological and biochemical responses in plants. In: Singh, V.P., Singh, S., Tripathi, D.K., Prasad, S.M., Chauhan, D.K. (eds), *Plant Responses to Nanomaterials: Recent Interventions, and Physiological and Biochemical Responses*, pp. 249–267. Springer, Cham.

Singh, N., Maurya, V., Sharma, I., Kumar, Y., Kumar, R., Sharma, A. 2024. Combinations of PGPR with other soil micro- and macro-organisms for improved stress tolerance in plants. In: Indu, S., Ashutosh, S., Renu, S. (eds), *PGPR (Plant Growth Promoting Rhizobacteria) for Plant Stress Management*, pp. 83–108. Nova Science Publishers, Hauppauge, NY.

Song, C., Ye, F., Zhang, H., Hong, J., Hua, C., Wang, B., Chen, Y., Ji, R., Zhao, L. 2019. Metal(loid) oxides and metal sulfides nanomaterials reduce heavy metals uptake in soil-cultivated cucumber plants. *Environmental Pollution* 255, 113354.

Soni, S., Sharma, I. 2023. Effect of indole-3-acetic acid and nano-urea on agronomic properties of radish crops. *International Journal of Plant & Soil Science* 35(20), 269–276.

Van Nguyen, D., Nguyen, H.M., Le, N.T., Nguyen, K.H., Nguyen, H.T., Le, H.M., Nguyen, A.T., Dinh, N.T.T., Hoang, S.A., Ha, C.V. 2022. Copper nanoparticle application enhances plant growth and grain yield in maize under drought stress conditions. *Journal of Plant Growth Regulation* 41(1), 364–375.

Ye, Y., Cota-Ruiz, K., Hernández-Viezcas, J.A., Valdes, C., Medina-Velo, I.A., Turley, R.S., Peralta-Videa, J.R., Gardea-Torresdey, J.L. 2020. Manganese nanoparticles control salinity-modulated molecular responses in *Capsicum annuum* L. through priming: a sustainable approach for agriculture. *ACS Sustainable Chemistry & Engineering* 8(3), 1427–1436.

Zaytseva, O., Neumann, G. 2016. Carbon nanomaterials: production, impact on plant development, agricultural and environmental applications. *Chemical and Biological Technologies in Agriculture* 3, 1–26.

Zhao, H., Liu, M., Chen, Y., Lu, J., Li, H., Qiao, S., Nechitaylo, G.S., Bogoslovskaya, O.A., Olkhovskaya, I.P., Glushchenko, N.N. 2019. Pepper plants response to metal nanoparticles and chitosan in nutrient media. *Australian Journal of Crop Science* 13(3), 433–443.

Zhao, L., Wang, W., Fu, X., Liu, A., Cao, J., Liu, J. 2022. Graphene oxide, a novel nanomaterial as soil water retention agent, dramatically enhances drought stress tolerance in soybean plants. *Frontiers in Plant Science* 13, 810905.

Zhou, P., Adeel, M., Shakoor, N., Guo, M., Hao, Y., Azeem, I., Li, M., Liu, M., Rui, Y. 2020. Application of nanoparticles alleviates heavy metals stress and promotes plant growth: an overview. *Nanomaterials* 11(1), 26.

Zohra, E., Ikram, M., Omar, A.A., Hussain, M., Satti, S.H., Raja, N.I., Mashwani, Z.-U.-R., Ehsan, M. 2021. Potential applications of biogenic selenium nanoparticles in alleviating biotic and abiotic stresses in plants: a comprehensive insight on mechanistic approach and future perspectives. *Green Processing and Synthesis* 10(1), 456–475.

6 Mechanism of Nanoparticles to Alleviate Biotic Stress

Mohd Ali, Deepak Kumar, Sandeep Kour, Parkirti, Roohi Sharma, Vikram, Tamanna Bhardwaj, Rajesh Kumari Manhas, Harish Changotra, Renu Bhardwaj, and Puja Ohri

6.1 INTRODUCTION

Under natural conditions, crops are exposed to a wide range of environmental stresses and these stresses may have an enormous effect on their productivity (Khan et al. 2023; Ali and Ohri 2023). It is believed that continuous reductions in agricultural land, water supply, global warming, and changing climatic conditions will cause a decline in crop production in the near future (Nguyen et al. 2023). Over the past few decades, population growth and increased number of environmental stressors have caused a noticeable decrease in agricultural productivity. Moreover, biotic stressors are significant barriers that negatively affect crop growth, development, and production. Biotic stressors consist of various disease-causing microorganisms like nematodes, fungi, bacteria, viruses, and herbivores (Hossain et al. 2023). To manage these pathogens, the use of new technologies and ongoing innovations is an essential advancement in the agricultural sector. While other technologies, such as genetic engineering, use of synthetic chemicals, and plant breeding, have increased agricultural output, they have dire consequences (Thakur et al. 2023). Hence, to meet the world's expanding food requirements without negatively affecting the economy, ecology, and human health, there is a compelling need for more nanotechnology-based discoveries (Ahmad et al. 2023). Nanotechnology is one of the most inventive and promising technologies available today and has been regarded as a "key enabling technology" by the European Commission (Valsami-Jones et al. 2022). Explorations in nanotechnology and its use in agronomy can be a vital alternative to meet the objective of sustainable agriculture in the face of climate change, environmental pollution, and world food shortages. Nanotechnology is based on the use of particles, having dimensions less than 100 nm, called as nanoparticles (NPs). They are bioactive substances that display hybrid quantum phenomena (Khan et al. 2022). Their physico-chemical characteristics like porosity, aggregating abilities, crystallinity, and catalytic activities allow them to pass through cellular barriers and influence biological processes, thus bestowing them with multiple applications (Kumar et al. 2023b).

DOI: 10.1201/9781032659435-6

In the last several years, nanotechnology has been significantly used in a number of fields including electronics, biotechnology, physics, chemistry, pharmacy, and life sciences. NPs as nanofertilizers greatly improve crop yield and soil enzymatic activity while preventing plant diseases by altering soil microbiota functions. Additionally, it has long been anticipated that NPs will be used actively in the fields of crop development and biotic stress management (Saeed et al. 2023).

6.2 SYNTHESIS OF NANOPARTICLE

NPs can be used to produce by physical, chemical, and biological methods. Physical processes use high-energy radiation, inert gas condensation, physical vapor deposition, laser pyrolysis, and melt mixing in order to produce NPs. Chemical method includes sol–gel method, microemulsion hydrothermal process, and polyol synthesis method (Kumari et al. 2023; Suroshe 2023). However, biological methods include NPs from plant extracts and microbes such as bacteria, viruses, and fungi. Such development of nanotechnology provides a promise to battle pathogenic organisms, and the search for a safe and alternative antimicrobial agent is essential in overcoming pathogen resistance (Noman et al. 2023).

6.3 MECHANISM OF NANOPARTICLES DURING BIOTIC STRESS

NPs appreciably enhance plant resistance toward different biotic stresses such as bacterial (*Pectobacterium carotovorum*, *Xanthomonas vesicatoria*, *Ralstonia solanacearum*), fungal (*Fusarium oxysporium*, *Magnaporthe grisea*, *Bipolaris sorokiniana*), and nematode (*M. incognita*) infection. Pathogens successfully invade the host plant and cause diseases. Application of NPs under such conditions can protect plants either by killing the causal organisms or by enhancing plant immunity. There are various mechanisms that NPs undergo while protecting plant from biotic stress. NPs may protect plant from invading pathogens by forming a layer on the plant surface. They may enhance phenolic content in leaves to give physical protection to plants or by elevating lignin content in plants, they decrease the risk of penetration of plant parasitic microbes. They also regulate the enzymatic activities of plant and reduce NO levels and reactive oxygen species (ROS) production (Cao and Wang 2022; Wang et al. 2022a). Application of bioengineered NPs enhanced plant immunity by reducing ROS with a 0.6–19.8-fold decrease in stress enzymes and autophagy inhibition. The proteomic analysis showed the overexpression in the production of abundant antimicrobial metabolite protein involved in early plant defense and protein biogenesis (Kumari et al. 2020). Thus, by enhancing the antioxidant activity of plants, NPs protect plants from oxidative stress by scavenging ROS. They also reduce lipid peroxidation in plants and induce plant growth and alter stress response by modulating their signaling pathways and hormonal levels. They also regulate plant gene expression. The molecular level, however, is still unclear, but NPs develop a stress-responsive gene cascade in plants. The application of NPs enhances cell division, cell growth, gene expression, and plant health (Wang et al. 2022a). The production of ROS, production of secondary metabolites, increased level of plant hormone, and rise in mitogen-activated protein kinase (MAPK)-dependent pathways are primary steps in the activation of the plant defense system. Calcium ions (Ca^{2+}) have been found to be a significant secondary messenger during stress conditions. Cytosolic Ca^{2+}activates Ca^{2+}-binding proteins (CaBPs), which eventually results in modification of gene expression and plant adaptation toward various stresses (Chandel et al. 2022). Complementary to this, the different mechanisms of NPs against pathogens involve ROS generation in pathogenic cells that cause disruption of enzymes, protein denaturation, DNA damage, electron transport chain interruption, and cell membrane damage (Tortella et al. 2023). NP-generated ROS in pathogens include hydroperoxyl, alkoxyl, carbon dioxide radicals, hydrogen peroxide, carbonates, super anions, and hydroxyl radical, along with nitric oxide, hypobromous acid, organic peroxides, and peroxide. NP accumulation in cells of pathogen causes injury of cell membrane, alteration in the permeability of cell membrane, cell damage, and lesions in pentose sugar and nitrogen base (Silva et al. 2022). Thus, NPs enter plants, kill pathogen, and modulate morphological, physiological, and biochemical attributes of plants to mitigate biotic stress (Figure 6.1). However, the significant response by NPs and the critical mechanism of stress signaling need to be developed with genomic and proteomic approaches.

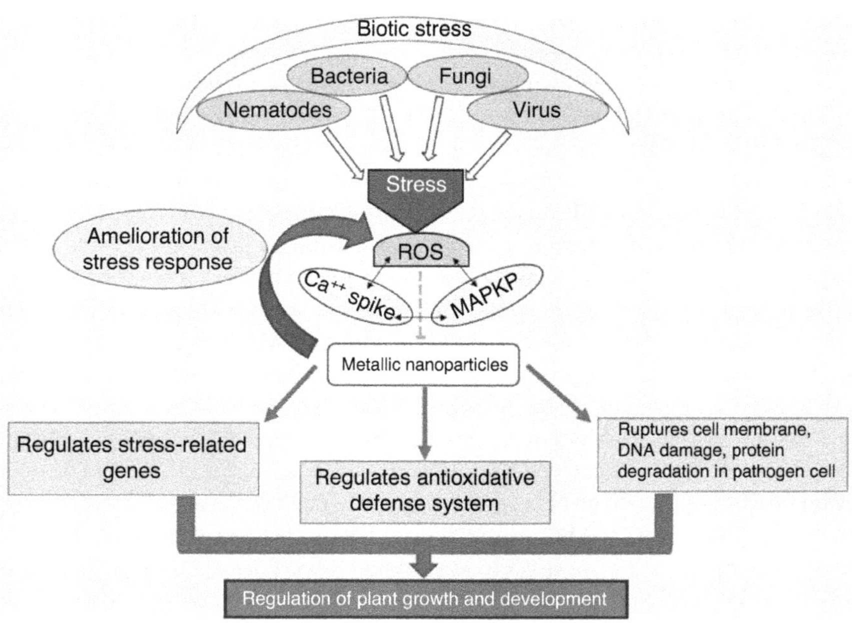

FIGURE 6.1 The mechanism of nanoparticles mitigating biotic stress in plants by regulating defense system of plants, by scavenging ROS, and by damaging cells of pathogens invaded in plants.

6.4 ROLE OF NANOPARTICLES DURING BIOTIC STRESS

Biotic stress in plants is caused by various pathogens such as viruses, bacteria, and fungi. NPs play a pivotal role in mitigating biotic stress in agriculture by bolstering plant resistance against pathogens. They act as delivery vehicles for bioactive compounds, induce systemic acquired resistance, and enhance plant defense mechanisms, offering a sustainable and targeted approach against diseases and pests and thereby improving crop yield and quality. Various studies prove that nanomaterials serve the purpose of mitigating biotic stress in crops. For example, copper oxide (CuO) NPs were synthesized from *Spirodela polyrhiza* and were tested for their nematicidal, antifungal, and antibacterial activities as well. The findings proved their potential to act as a safe alternative method to manage plant diseases caused by *Meloidogyne incognita*, *Phomopsis vexans*, *Pectobacterium carotovorum* (Khan et al. 2022). Similarly, zinc oxide NPs (ZnONPs) could generate reactive oxygen species (ROS) and showed their best antibacterial effect as ZnONPs that significantly reduced the population of bacteria, *Deinococcus radiodurans,* which was otherwise resistant toward radiation damage. ZnO enters into bacterial cell, and internally, it damages the cell membrane and DNA structure and downregulates various metabolic pathways (Singh et al. 2020). Titanium dioxide NPs (TiO_2-NPs) obtained from *Carum copticum* were predominant in inhibiting the growth of biofilm of both gram-positive and gram-negative bacteria, thus acting as antibiofilming agents (Altaf et al. 2021). Also, CuNPs synthesized from *Annona squamosa* seed extract were found effective against *Xanthomonas oryzae*, a pathogen of rice plant (Singh et al. 2021). Similarly, silicon NPs (SiNPs) biosynthesized from *Fusarium oxysporum* were effective against egg hatching and population growth of second-stage juveniles of *Meloidogyne incognita*, a disastrous pest of crop *Solanum melongena* (El-Ashry et al. 2022). SiNP application also decreased the threat of growth of branched broomrape, a root holoparasitic weed that drastically affects plant growth and yield, and improved the detoxification system of plant by decreasing ROS (Madany et al. 2020). The magnesium oxide NPs (MgO NPs) synthesized from strawberries were effective against two pathogens, i.e., *Ralstonia solanacearum*, a plant pathogenic bacterium, and *M. incognita*, a root-knot nematode (Khan et al. 2022). Thus, several NPs are involved in mitigating various kinds of biotic stress in plants.

6.4.1 Bacteria

Another major biotic constraint reducing crop production is plant pathogenic bacteria. It is well recognized that bacterial infections can trigger illness epidemics that result in significant economic losses in temperate, tropical, and subtropical locations across the globe. Bacteria are prokaryotic microorganisms with a simple structural form that exhibits diversity in their metabolic makeup (Sharma et al. 2023). The primary mechanism of bacterial infection is successful attachment or penetration by the use of growth regulators, toxins, extracellular enzymes, type secretion systems, and other chemicals that regulate plant defense, ultimately leading to successful colonization (Pfeilmeier et al. 2016). Because these bacterial diseases become resistant to medications, they are very challenging to control. As a result, efforts are undertaken to look for novel approaches to manage diseases (Pfeilmeier et al. 2016). Various studies have reported the antibacterial potential of NPs (Table 6.1). For instance, MgO, ZnO, and MnO_2NPs synthesized using rhizophytic bacteria *Paenibacillus polymyxa* were found to increase the rice biomass by reducing the bacterial late blight incidence. It was reported that these biogenic NPs had the potential to reduce the biofilm formation and inhibit the growth of *Xanthomonas oryzae* pv. *oryzae*, the causative agent of bacterial late blight in rice (Ogunyemi et al. 2020). In a study, it was reported that FeO NPs synthesized by using leaf extract of *Skimmia laureola* exhibit antibacterial properties against *Ralstonia solanacearum* causing bacterial wilt in tomato. During *in vitro* experiments, bacterial growth was significantly suppressed by the culture media containing 6 mg/mL of Fe_2O_3-NPs. Root zone treated with 6% w/v of Fe_2O_3-NPs substantially decreased the severity of the disease in in-planta conditions. Further, with Fe_2O_3-NP treatments, plant shoots, root length, and fresh biomass were also increased (Alam et al. 2019). In another study, chitosan NPs and ZnONPs were reported to exhibit antibacterial activity against *X. oryzae* pv. *oryzae*. Results revealed that both the NPs dramatically reduced the bacterial growth, which was attributed to the bacterial cell membrane damage, production of ROS, bacterial cell death, and reduced biofilm formation (Abdallah et al. 2020). Similar results were obtained by using AgNPs synthesized from sumac aqueous extract, where it was found that NPs exhibited inhibitory effects on *Pseudomonas syringae* pv. *syringae*, the causative agent of bacterial canker disease of stone fruit trees. Further, results from trials conducted in greenhouse demonstrated that varying concentrations of AgNPs lessened the disease severity of bacterial canker of stone fruits on two-year-old peach plants when compared to control plants (Shahryari et al. 2020).

6.4.2 Fungi

NPs also demonstrate efficacy in mitigating fungal stress by inhibiting fungal growth, disrupting cell membranes, and enhancing plant defense mechanisms. Serving as nano-fungicides, they offer a targeted and sustainable solution, contributing to improved crop health and resilience against fungal pathogens in agriculture. For example, some NPs were found to be effective against certain fungal pathogens like titanium dioxide NPs (TiO_2NPs), which were effective against *Puccinia striiformis,* a fungus that causes yellow rust or stripe in wheat crop. By eliciting biochemical modifications in plants, TiO_2NPs enhanced the immune system of plants against the disease (Satti et al. 2022). Similarly, magnetite NPs (Fe_3O_4NPs) obtained from spinach as a precursor were also effective against mycelial growth of *Fusarium oxysporum* f. sp. *lycopersici*, in tomato plants along with increasing enzymatic activities of plant (Ashraf et al. 2022). In another study on the watermelon plant having fusarium wilt disease, silver NPs (AgNPs) were effective in controlling mycelial growth of *F. oxysporium,* a fungus that causes fusarium wilt, and they also enhanced crop health (Chang et al. 2019). Copper oxide NPs (CuONPs) were effective in alleviating Fusarium infection in chickpea seedlings. These NPs also enhanced the root length, shoot length, protein, flavonoids, tannin, and phenolics in plants (Tiwari et al. 2024).

TABLE 6.1
Role of Nanoparticles During Biotic Stress

Nanoparticle (NP)	Nanoparticle Concentration	Biotic Stress	Size (nm)	Response	References
SiO_2 NPs	0.10 mg/mL	*Meloidogyne incognita*	5–15 nm	Reduced number of galls, reduced disease index, and increased growth of carrot plants	Ahamad et al. (2023)
Hydroxyapatite NPs	50, 150, and 200 ppm	*M. incognita*	15 nm	Increased juvenile mortality and inhibited egg hatching	Alamri et al. (2022)
FeNPs	75 mg/kg	*M. javanica*	–	Reduced density of females in soybean roots	Benedetti et al. (2023)
SeNPs	2, 6, and 100 ppm	*M. incognita*	50–500 nm	Reduced number of galls, egg masses, and improved growth and photosynthetic parameters of tomato plants	Daoush et al. (2023)
AgNPs	1, 2, and 3 µg/mL	*M. javanica*	15.71 nm	Reduced number of egg masses and infective juveniles and reproductive potential in concentration-dependent manner	Daramola et al. (2023)
AgNPs	250 µmol/L	*Pratylenchus brachyurus*	11.89 ± 9.28 nm	Reduced nematode population in soybean roots when applied twice	De Oliveira et al. (2023)
CuNPs	100, 125, 150, 175, 200, 225, and 250 ppm	*M. incognita*	23 nm	Increased number of dead infective juveniles and reduced egg hatching during *in vitro* tests	El-Ashry et al. (2024)
ZnO NPs	250, 500, 750, and 1,000 ppm	*M. incognita*	Diameter- 14–140 nm and length > 500 nm	Reduced infection in chickpea and improved plant growth parameters	Khan et al. (2023a)
CuO NPs	100 µg/mL	*M. incognita* and *P. carotovorum*	100 nm	Exhibited nematicidal and antibacterial properties	Khan et al. (2023b)

(Continued)

TABLE 6.1 (*Continued*)
Role of Nanoparticles During Biotic Stress

Nanoparticle (NP)	Nanoparticle Concentration	Biotic Stress	Size (nm)	Response	References
ZrO_2 NPs	0.10 g/L	*M. incognita* and *Fusarium oxysporum*	19.83 nm	Reduced disease indices and increased growth and photosynthetic pigments in bottle gourd	Siddiqui et al. (2023)
ZnO NPs	0.5 mg/mL	*M. incognita*	–	Increased juvenile mortality and reduced number of galls	El-Ansary et al. (2022)
SiNPs	50, 100, 150, 200, 250, 300, and 350 ppm	*M. incognita*	30–55 nm	Reduced juvenile emergence and increased mortality and improved growth of eggplant	El-Ashry et al. (2022)
MgO NPs	5, 50, and 100 µg/mL for nematicidal activity and 10, 20, 30, 40, and 50 µg/mL for antibacterial activity	*M. incognita* and *R. solanacearum*	100 nm	Inhibited growth of *R. solanacearum* and egg hatching of nematode	Khan et al. (2022)
MgO NPs	25, 50, and 100 ppm	*M. incognita*	–	Reduced number and size of galls and increased growth and chlorophyll content of cowpea plants	Tauseef et al. (2021)
AgNPs	20, 50, and 75 mL	*M. incognita*	2–20 nm	Increased growth of nematode-infested carrot plants	Fabiyi (2021)
AgNPs	5, 10, 15, 20, and 25 µg/mL	*Pectobacterium carotovorum*, *P. atrosepticum*, and *Ralstonia solanacearum*	33 ± 1 nm	Reduced bacterial growth	Elkobrosy et al. (2023)

(*Continued*)

TABLE 6.1 (*Continued*)
Role of Nanoparticles During Biotic Stress

Nanoparticle (NP)	Nanoparticle Concentration	Biotic Stress	Size (nm)	Response	References
ZnO NPs	0.5 mg/mL	*Pseudomonas syringae* pv.tabaci	18.6 nm	Disrupted biofilm formation and motility of bacteria and boosted tobacco plant growth and immunity	Fan et al. (2023)
AgNPs	50 µg/mL	*Dickeya dadantii*	48.5 nm	Reduced bacterial infection and improved growth characteristics of sweet potato plants	Hossain et al. (2023)
AgNPs	0.2%, 0.4%, and 0.6%	*Xanthomonas perforans*	–	Reduced incidence of bacterial spot disease in tomato	Mehmood et al. (2023)
Co_3O_4 NPs	200 µg/mL	*X. oryzae* pv. *oryzae*	34.9 nm	Exhibited antibacterial effect on bacterial biofilm formation and growth and enhanced growth and photosynthetic attributes of rice	Ogunyemi et al. (2023)
AgNPs	75 mg/L	Huanglongbing disease	74 nm	Restored health of Huanglongbing-diseased "Kinnow" mandarin plants	Umair Raza et al. (2023)
AgNPs	–	*Acidovorax avenae* subsp. *avenae*	35 nm	Reduced disease severity and controlled red stripe disease in sugarcane	Guerrero et al. (2022)
CuO, FeO, and ZnO NPs	500 mg/kg	*R. solanacearum*	20–39 nm, 19–27 nm, and 16–31 nm, respectively	Improved growth and physiological characteristic of tomato wilt-infected tomato plants	Jiang et al. (2022)
Chitosan-derived NPs	0, 10, 50, 100, 150, 200, 250, and 500 mg/kg	*R. solanacearum*	98.6 nm	Protected tomato plants against pathogen-induced wilt disease	Narasimhamurthy et al. (2022)
Thymol-loaded chitosan NPs	100–600 µg/mL	*X. campestris pv. campestris*	54–250 nm	Inhibited bacterial growth and biofilm formation and damaged bacterial cell membrane	Sreelatha et al. (2022)

(*Continued*)

TABLE 6.1 (*Continued*)
Role of Nanoparticles During Biotic Stress

Nanoparticle (NP)	Nanoparticle Concentration	Biotic Stress	Size (nm)	Response	References
Silica NPs	–	*R. solanacearum*	18.99 ± 0.68 nm	Decreased oxidative stress and increased activity of antioxidative enzymes in infected plants	Wang et al. (2022)
ZnO NPs	18 µg/mL	*R. solanacearum*	8.9–32.6 nm	Inhibited bacterial growth and improved tomato growth attributed by decreasing disease severity and bacterial population in soil	Khan et al. (2021)
AgNPs	25, 50, 75, and 100 mg/L	*X. oryzae*	18–39 nm	Reduced bacterial leaf blight diseases and increased biomass and activity of antioxidants in rice plants	Ahmed et al. (2020)
Hybrid Cu/Zn NPs	100, 200, 500, and 1,000 µg/mL	*Xanthomonas* spp.	–	Reduced bacterial population and disease severity of bacterial spot in tomato plants	Carvalho et al. (2019)
AgNPs	64 and 128 µmol/L	*R. solanacearum*	˂ 100 nm	Decreased bacterial growth and reduced wilt incidence in tomato	Santiago et al. (2019)
AuNPs	0.034 mg/100 mL	Barley yellow dwarf virus-PAV (BYDV-PAV)	30 nm	Damaged and disappeared the virus	Alkubaisi and Aref (2017)
ZnONPs	100 µg/mL	Tobacco mosaic virus (TMVs)	20 nm	Elevation in plant growth parameters	Cai et al. (2019)
NiO	150 µg/L	Cucumber mosaic virus (CMV)	20 nm	Lowered the degree of illness and the concentration of CMV	Derbalah and Elsharkawy (2019)

(*Continued*)

TABLE 6.1 *(Continued)*
Role of Nanoparticles During Biotic Stress

Nanoparticle (NP)	Nanoparticle Concentration	Biotic Stress	Size (nm)	Response	References
AgNPs	0.0338 g/100 mL	Bean yellow mosaic virus (BYMV)	13.7–40 nm	Showed higher expression of genes for enhanced growth parameters	Abdelkhalek et al. (2022)
TiO_2NPs	20, 40, 60, and 80 mg/L	*Puccinia striiformis*	100 nm	Enhanced the immune system of the whole plant	Satti et al. (2022)
Fe_3O_4NPs	15 µg/mL	*Fusarium oxysporum*	20 nm	Enhanced the enzymatic activity of tomato plant	Ashraf et al. (2022)
AgNPs	150 and 200 ppm	*F. oxysporium*	–	Enhanced crop health	Chang et al. (2019)
CuONPs	10, 25, and 50 ppm	*F. oxysporium*	85 nm	Enhanced root length, shoot length, protein, flavonoids, tannin, and phenolics in chickpea	Tiwari et al. (2024)

6.4.3 Virus

NPs prove instrumental in alleviating virus stress by inhibiting viral replication, enhancing antiviral plant defenses, and reducing disease severity. Engineered NPs, acting as antiviral agents, provide a targeted and sustainable approach, contributing to enhanced crop protection and productivity in the face of viral infections, thus offering potential solutions for sustainable agriculture. Research on using NPs to treat plant illnesses has been conducted; however, studies on using them to treat viral infections are still in their early phases. *In vitro* studies using ZnONPs and SiO_2NPs have been shown to have antiviral action against tobacco mosaic virus (TMV) *in vitro* (Cai et al. 2019). The findings of their study indicated that metallic NPs directly inactivate TMV through their interaction with envelope glycoproteins, resulting in direct damage to TMV shell proteins, TMV aggregation, and even breakage. Similarly, AgNPs were also found effective against bean yellow mosaic virus (BYMV) that causes disease in the faba bean plant. Even at low concentration, the NPs were effective in controlling disease by inducing defense-related enzymes in plants (El Gamal et al. 2022). Using TEM, it was demonstrated that AuNPs harm the virus-like particles (VLPs) of the barley yellow dwarf virus (PAV), seeing both damaged and disappearing particles and inflated and degraded VLPs coated with AuNPs (Alkubaisi and Aref 2017). Additionally, in vivo investigations have been carried out to determine how well MeNPs work against viral infections in the plants belonging to families Solanaceae, Asteraceae, Poaceae, Fabaceae, and Cucurbitaceae. In a study, the impact of AgNPs from *Bacillus pumilus, Bacillus persicus*, and *Bacillus licheniformis* on yellow bean mosaic virus was examined. Post-infection treatment after 24 hours halted the virus effects, but concurrent infection and AgNP treatment showed mild symptoms. Pre-infection therapy 72 hours before inoculation had no effect on virus presence or disease severity (Elbeshehy et al. 2015).

6.4.4 Nematodes

Plant–parasitic nematodes (PPNs) form the major biotic factor that reduces crop productivity. In addition to directly lowering agricultural productivity, PPNs have a major negative influence on disease complexes involving other harmful species. Sometimes crop plants show signs of nematode damage, but diagnosis becomes difficult due to the lack of markers (Kumar et al. 2023a). To feed the world's expanding population, the most efficient method to increase agricultural output is to employ sustainable methods of management against these phytopathogens. Applying the chemical nematicides that are currently being used in the market is one of the most successful management techniques (Grabau et al. 2020). However, because requirements for food safety and environmental protection have increased, several extremely poisonous nematicides are no longer appropriate for intensive cultivation (Kumar and Ohri 2023). So, the use of nanotechnology such as use of NPs is a safer and more ecologically friendly alternative method to manage PPNs in light of the major health and environmental risks associated with pesticide use (Tortella et al. 2023). Various studies have reported the nematicidal potential of different NPs (Table 6.1). For instance, AgNPs synthesized from the latex extract of *Euphorbia tirucalli* exhibited nematicidal potential against *M. incognita*. *In vitro* experiments revealed that AgNPs induced mortality of infective juveniles and reduced egg hatching. Further, during *in-planta* experiments, *M. incognita*–infested tomato plants exhibited better growth and reduced number of gall and egg masses after treatment with AgNPs (Kalaiselvi et al. 2019). In another study, the effectiveness of different preparations of silver-botanical NPs, such as *Lantana camara*, cloves, and *Conyza dioscoridis*, as nematicide alternatives against the root-knot nematode (*M. incognita*) was assessed in both laboratory and greenhouse settings. Results revealed that NP formulations exhibited better results compared to the normal extract in terms of reduced egg hatching and increased mortality. Similarly, the infective stage of *M. incognita* showed considerable mortality after being exposed to 0.0005 µg of AgNPs for 1 minute, and within 24 and 48 hours, almost all the nematodes were found inactive, demonstrating NPs as a possible green substitute for control.

6.5 CONCLUSION

In order to accomplish sustainable agriculture, the scientific community needs to find suitable environment-friendly strategies to deal with agricultural yield loss caused by biotic stress. Improving agricultural productivity and quality, developing the farming industry, and controlling the world's food demand may all be accomplished with the help of nanotechnology. A lot of research is being done on the possible roles that various NPs may play in reducing damage caused by biotic stress, enhancing crop output, and promoting plant growth. NPs that activate plant defense mechanisms through the stimulation of ROS production and phytotoxicity might lessen the adverse effects of biotic stress. Because of their small size, NPs may also easily enter the plant tissue and thus can have a favorable impact on the morphological, physiological, and biochemical processes of the plant (Figure 6.1). Additionally, the vast surface area of NPs enhances the delivery and absorption of certain targeted nutrients. However, the development of NP and their application for crop enhancement and sustainable agriculture is still in its infancy and demands further exploration to completely endorse the role of NPs as a sustainable alternative.

REFERENCES

Abdallah, Y., Liu, M., Ogunyemi, S.O., Ahmed, T., Fouad, H., Abdelazez, A., Yan, C., Yang, Y., Chen, J., Li, B. 2020. Bioinspired green synthesis of chitosan and zinc oxide nanoparticles with strong antibacterial activity against rice pathogen *Xanthomonas oryzae* pv. *oryzae*. *Molecules* 25, 4795. https://doi.org/10.3390/molecules25204795

Abdelkhalek, A., Yassin, Y., Abdel-Megeed, A., Abd-Elsalam, K.A., Moawad, H., Behiry, S.I. 2022. *Rhizobium leguminosarum* bv. *viciae*-mediated silver nanoparticles for controlling Bean Yellow Mosaic Virus (BYMV) infection in faba bean plants. *Plants* 12(1), 45. https://doi.org/10.3390/plants12010045

Ahamad, L., Siddiqui, Z.A., Hashem, A., Abd_Allah, E.F. 2023. Use of AM fungus *Rhizophagus irregularis* and silicon dioxide nanoparticles for the management of *Meloidogyne incognita*, Alternaria dauci and Rhizoctonia solani and growth of carrot. *Archives of Phytopathology and Plant Protection* 56(6), 466–488. https://doi.org/10.1080/03235408.2023.2207958

Ahmad, N., Youjin, L., Žiković, S., Belyaeva, Z. 2023. The effects of technological innovation on sustainable development and environmental degradation: evidence from China. *Technology in Society* 72, 102184. https://doi.org/10.1016/j.techsoc.2022.102184

Ahmed, T., Shahid, M., Noman, M., Niazi, M.B.K., Mahmood, F., Manzoor, I. et al. 2020. Silver nanoparticles synthesized by using *Bacillus cereus* SZT1 ameliorated the damage of bacterial leaf blight pathogen in rice. *Pathogens* 9(3), 160. https://doi.org/10.3390/pathogens9030160

Alam, T., Khan, R.A.A., Ali, A., Sher, H., Ullah, Z., Ali, M. 2019. Biogenic synthesis of iron oxide nanoparticles via *Skimmia laureola* and their antibacterial efficacy against bacterial wilt pathogen *Ralstonia solanacearum*. *Materials Science and Engineering: C* 98, 101–108. https://doi.org/10.1016/j.msec.2018.12.067

Alamri, S., Nafady, N.A., El-Sagheer, A.M., El-Aal, M.A., Mostafa, Y.S., Hashem, M., Hassan, E.A. 2022. Current utility of arbuscular mycorrhizal fungi and hydroxyapatite nanoparticles in suppression of tomato root-knot nematode. *Agronomy* 12(3), 671. https://doi.org/10.3390/agronomy12030671

Ali, M., Ohri, P. 2023. Deciphering the synergistic effect of jasmonic acid and spermine in mitigating root-knot nematode stress in tomato plants through enhancing growth and activity of antioxidant enzymes. *South African Journal of Botany* 161, 21–35. https://doi.org/10.1016/j.sajb.2023.01.010

Alkubaisi, N.A., Aref, N.M. 2017. Dispersed gold nanoparticles potentially ruin gold barley yellow dwarf virus and eliminate virus infectivity hazards. *Applied Nanoscience* 7, 31–40. https://doi.org/10.1007/s13204-016-0540-0

Altaf, M., Zeyad, M.T., Hashmi, M.A., Manoharadas, S., Hussain, S.A., Abuhasil, M.S.A., Almuzaini, M.A.M. 2021. Effective inhibition and eradication of pathogenic biofilms by titanium dioxide nanoparticles synthesized using *Carum copticum* extract. *RSC Advances* 11(31), 19248–19257. https://doi.org/10.1039/D1RA02876F

Ashraf, H., Batool, T., Anjum, T., Illyas, A., Li, G., Naseem, S., Riaz, S. 2022. Antifungal potential of green synthesized magnetite nanoparticles (black coffee–magnetite nanoparticles) against wilt infection by ameliorating enzymatic activity and gene expression in *Solanum lycopersicum* L. *Frontiers in Microbiology* 13, 754292.

Benedetti, T., Sordi, E., Antoniolli, Z.I., Santos, J.R.P., Bortoluzzi, E.C. 2023. Iron nanoparticle and biological nematicide agent for the management of *Meloidogyne javanica* in soybean. *Nematropica* 53, 1–10.

Cai, L., Liu, C., Fan, G., Liu, C., Sun, X. 2019. Preventing viral disease by ZnO nanoparticles through directly deactivating TMV and activating plant immunity in *Nicotiana benthamiana*. *Environmental Science: Nano* 6(12), 3653–3669.

Cao, X., Wang, Z. 2022. Application of nano-agricultural technology for biotic stress management: mechanisms, optimization, and future perspectives. *Environmental Science: Nano* 9(12), 4336–4353.

Carvalho, R., Duman, K., Jones, J.B., Paret, M.L. 2019. Bactericidal activity of copper–zinc hybrid nanoparticles on copper-tolerant *Xanthomonas perforans*. *Scientific Reports* 9(1), 20124.

Chandel, S.S., Shree, B., Mondal, K., Mali, S. 2022. Mechanism of nanoparticles-mediated alleviating biotic and abiotic stresses in agricultural crops: recent advances and future perspectives. In: Rajput, V.D., Verma, K.K., Sharma, N., Minkina, T. (eds), *The Role of Nanoparticles in Plant Nutrition under Soil Pollution: Nanoscience in Nutrient Use Efficiency*, pp. 117–139. Springer, Cham.

Chang, T.H., Liu, Y.W., Lin, Y.H., Lin, J.J., Huang, J.W., Hussain, A., Chang, P.F.L. 2019. Silver nanoparticles on nanoscale silica platelets (AgNP/NSP) inhibit the development of *Fusarium oxysporum* f. *sp. niveum*. *ACS Applied Bio Materials* 2(11), 4978–4985.

Daoush, A.S.M., Hendawey, M.H., Yaseen, R., El-Nuby, A.S.M., Bedair, T.M., Alwutayd, K.M., Al-Hoshani, N., Shaaban, A., Bashir, A., Li, L. 2023. Effect of biosynthesized nanoselenium on controlling tomato root-knot nematode *Meloidogyne incognita. Agronomy* 13(7), 1668.

Daramola, F., Lewu, N., Nkiko, J., Lewu, F. 2023. Nematicidal effects of silver nanoparticles on the root-knot nematode associated with Swiss chard. *Helminthologia* 60(2), 189–195.

De Oliveira, L.S., Furtado, L.L., Diniz, F.A.S., Mendes, B.L., Araújo, T.R., Silva, L.P., Santiago, T.R. 2023. Eco-friendly silver nanoparticles synthesized from a soybean by-product with nematicidal efficacy against Pratylenchus brachyurus. *Nanomaterials* 14(1), 101.

Derbalah, A.S.H., Elsharkawy, M.M. 2019. A new strategy to control Cucumber mosaic virus using fabricated NiO nanostructures. *Journal of Biotechnology* 306, 134–141.

El Gamal, A.Y., Tohamy, M.R., Abou-Zaid, M.I., Atia, M.M., El Sayed, T., Farroh, K.Y. 2022. Silver nanoparticles as a viricidal agent to inhibit plant-infecting viruses and disrupt their acquisition and transmission by their aphid vector. *Archives of Virology* 167, 1–13.

El-Ansary, M.S.M., Hamouda, R.A., Elshamy, M.M. 2022. Using biosynthesized zinc oxide nanoparticles as a pesticide to alleviate toxicity in banana infested with parasitic nematodes. *Waste and Biomass Valorization* 13, 1–11.

El-Ashry, R.M., El-Saadony, M.T., El-Sobki, A.E.A., El-Tahan, A.M., Al-Otaibi, S., El-Shehawi, A.M., Saad, A.M., Elshaer, N. 2022. Biological silicon nanoparticles maximize the efficiency of nematicides against biotic stress induced by *Meloidogyne incognita* in eggplant. *Saudi Journal of Biological Sciences* 29(2), 920–932.

El-Ashry, R.M., Nader, M.M., Shami, A., Alduwish, M.A., Ahmed, A.E., Alamoud, S.A. et al. 2024. Ecofriendly synthesis and nematicidal application of copper nanoparticles fabricated from *Bacillus subtilis* AM18 against root-knot nematode of cucumber. *European Journal of Plant Pathology* 168(1), 53–81.

Elbeshehy, E.K., Elazzazy, A.M., Aggelis, G. 2015. Silver nanoparticles synthesis mediated by new isolates of *Bacillus* spp., nanoparticle characterization and their activity against Bean Yellow Mosaic Virus and human pathogens. *Frontiers in Microbiology* 13(6), 453, doi: 10.3389/fmicb.2015.00453

Elkobrosy, D., Al-Askar, A.A., El-Gendi, H., Su, Y., Nabil, R., Abdelkhalek, A., Behiry, S. 2023. Nematicidal and bactericidal activities of green synthesized silver nanoparticles mediated by *Ficus sycomorus* leaf extract. *Life* 13(5), 1083.

Fabiyi, O.A. 2021. Sustainable management of *Meloidogyne incognita* infecting carrot (*Daucus carota*): green synthesis of silver nanoparticles using *Cnidoscolus aconitifolius. Vegetos* 34(2), 277–285.

Fan, G., Xiao, Q., Li, Q., Xia, Y., Feng, H., Ma, X., Cai, L., Sun, X. 2023. Antimicrobial mechanisms of ZnO nanoparticles against phytopathogen *Pseudomonas syringae*: damage of cell envelope, suppression of metabolism, biofilm and motility, and stimulation of stomatal immunity in host plant. *Pesticide Biochemistry and Physiology* 194, 105455.

Grabau, Z.J., Mauldin, M.D., Habteweld, A., Carter, E.T. 2020. Nematicide efficacy at managing and non-target effects on free-living nematodes in peanut production. *Journal of Nematology* 52(1), 1–10.

Guerrero, D.S., Bertani, R.P., Ledesma, A., de los Angeles Frías, M., Romero, C.M., Dávila Costa, J.S. 2022. Silver nanoparticles synthesized by the heavy metal resistant strain *Amycolatopsis tucumanensis* and their application in controlling red strip disease in sugarcane. *Heliyon* 8(5), e09463.

Hossain, A., Luo, J., Ali, M.A., Chai, R., Shahid, M., Ahmed, T. et al. 2023. Synergistic action of biosynthesized silver nanoparticles and culture supernatant of *Bacillus amyloliquefaciens* against the soft rot pathogen *Dickeya dadantii. Plants* 12(9), 1817.

Jiang, H., Lv, L., Ahmed, T., Jin, S., Shahid, M., Noman, M., Osman, H.E.H., Wang, Y., Sun, G., Li, X. 2022. Nanoparticles-mediated amelioration of tomato bacterial wilt disease by modulating the rhizosphere bacterial community. *International Journal of Molecular Sciences* 23, 414.

Kalaiselvi, D., Mohankumar, A., Shanmugam, G., Nivitha, S., Sundararaj, P. 2019. Green synthesis of silver nanoparticles using latex extract of *Euphorbia tirucalli*: a novel approach for the management of root-knot nematode, *Meloidogyne incognita*. *Crop Protection* 117, 108–114.

Khan, A., Khan, A.A., Jameel, M., Khan, M.F., Khan, M., Khan, A., Ahmad, F., Alam, M. 2023. Grass-shaped zinc oxide nanoparticles synthesized by the sol–gel process and their antagonistic properties towards the biotrophic parasite *Meloidogyne incognita*. *Bioinorganic Chemistry and Applications* 2023, 1–12.

Khan, A.U., Khan, M., Khan, A.A., Parveen, A., Ansari, S., Alam, M. 2022. Effect of phyto-assisted synthesis of magnesium oxide nanoparticles (MgO-NPs) on bacteria and the root-knot nematode. *Bioinorganic Chemistry and Applications* 2022, 1–11.

Khan, M., Khan, A.A., Parveen, A., Min, K., Yadav, V.K., Khan, A.U., Alam, M. 2023b. Mitigating the growth of plant pathogenic bacteria, fungi and nematodes using plant-mediated synthesis of copper oxide nanoparticles (CuO NPs). *Green Chemistry Letters and Reviews* 16(1), 2177520.

Khan, M.S.S., Ahmed, S., Ikram, A.U., Hannan, F., Yasin, M.U., Wang, J., Zhao, B., Islam, F., Chen, J. 2023a. Phytomelatonin: a key regulator of redox and phytohormone signaling against biotic and abiotic stresses. *Redox Biology* 60, 102805.

Khan, R.A.A., Tang, Y., Naz, I., Alam, S.S., Wang, W., Ahmad, M. et al. 2021. Management of *Ralstonia solanacearum* in tomato using ZnO nanoparticles synthesized through *Matricaria chamomilla*. *Plant Disease* 105(10), 3224–3230.

Kumar, D., Manhas, R.K., Ohri, P. 2023a. Deciphering the growth-promoting and stress-curtailing role of nitric oxide in *Meloidogyne incognita* infested *Solanum lycopersicum* seedlings. *Scientia Horticulturae* 319, 112147.

Kumar, D., Ohri, P. 2023. Say "NO" to plant stresses: unraveling the role of nitric oxide under abiotic and biotic stress. *Nitric Oxide* 130, 36–57.

Kumar, V., Kaushik, N.K., Tiwari, S.K., Singh, D., Singh, B. 2023b. Green synthesis of iron nanoparticles: sources and multifarious biotechnological applications. *International Journal of Biological Macromolecules* 236, 127017.

Kumari, M., Pandey, S., Mishra, S.K., Giri, V.P., Agarwal, L., Dwivedi, S., Pandey, A.K., Nautiyal, C.S., Mishra, A. 2020. Omics-based mechanistic insight into the role of bioengineered nanoparticles for biotic stress amelioration by modulating plant metabolic pathways. *Frontiers in Bioengineering and Biotechnology* 8, 242.

Kumari, S., Raturi, S., Kulshrestha, S., Chauhan, K., Dhingra, S., Kovács, A., Thu, K., Khargotra, R., Singh, T. 2023. A comprehensive review on various techniques used for synthesizing nanoparticles. *Journal of Materials Research and Technology* 25, 5930–5957.

Madany, M.M.Y., Saleh, A.M., Habeeb, T.H., Hozzein, W.N., AbdElgawad, H. 2020. Silicon dioxide nanoparticles alleviate the threats of broomrape infection in tomato by inducing cell wall fortification and modulating ROS homeostasis. *Environmental Science: Nano* 7(5), 1415–1430.

Mehmood, B., Abbas, M., Hussain, S., Rahim, N., Shafique, S., Khan, R.T. et al. 2023. Exploring the potential of green silver nanoparticles from *Berberis vulgaris* against bacterial spot of tomato and its surveillance in Poonch District. *International Journal of Phytopathology* 12(1), 49–61.

Narasimhamurthy, K., Udayashankar, A.C., De Britto, S., Lavanya, S.N., Abdelrahman, M., Soumya, K., Shetty, H.S., Srinivas, C., Jogaiah, S. 2022. Chitosan and chitosan-derived nanoparticles modulate enhanced immune response in tomato against bacterial wilt disease. *International Journal of Biological Macromolecules* 220, 223–237.

Nguyen, T.T., Grote, U., Neubacher, F., Do, M.H., Paudel, G.P. 2023. Security risks from climate change and environmental degradation: implications for sustainable land use transformation in the Global South. *Current Opinion in Environmental Sustainability* 63, 101322.

Noman, M., Ahmed, T., Ijaz, U., Hameed, A., Shahid, M., Li, A.D., Song, F. 2023. Microbe-oriented nanoparticles as phytomedicines for plant health management: an emerging paradigm to achieve global food security. *Critical Reviews in Food Science and Nutrition* 63(25), 7489–7509.

Ogunyemi, S.O., Xu, X., Xu, L., Abdallah, Y., Rizwan, M., Lv, L. et al. 2023. Cobalt oxide nanoparticles as an effective growth promoter of Arabidopsis and nano-pesticide against bacterial leaf blight pathogen in rice. *Ecotoxicology and Environmental Safety* 257, 114935.

Ogunyemi, S.O., Zhang, M., Abdallah, Y., Ahmed, T., Qiu, W., Ali, M.A., Yan, C., Yang, Y., Chen, J., Li, B. 2020. Biosynthesis of three metal oxide nanoparticles (ZnO, MnO_2 and MgO) and their antibacterial activity against the bacterial leaf blight pathogen. *Frontiers in Microbiology* 11, 588326.

Pfeilmeier, S., Caly, D.L., Malone, J.G. 2016. Bacterial pathogenesis of plants: future challenges from a microbial perspective. *Molecular Plant Pathology* 17(8), 1298–1313.

Raza, M.U., Abasi, F., Shahbaz, M., Ehsan, M., Seerat, W., Akram, A., Raja, N.I., Mashwani, Z.U.-R., Hassan, H.U., Proćków, J. 2023. Phytomediated silver nanoparticles embellish the antioxidant defense system, ameliorating HLB-diseased 'Kinnow' mandarin plants. *Molecules* 28(5), 2044.

Saeed, A., Javed, H., Alserae, H., Nawaz, R., Farooqi, Z.U.R., Riaz, S., Nawaz, H. 2023. Nanotechnology in management of environmental contaminants. In: Ansari, A., Gill, S., Gill, R., Lanza, R.G., Newman, L. (eds), *Phytoremediation: Management of Environmental Contaminants*, Vol. 7, pp. 383–401. Springer, Cham.

Santiago, T.R., Bonatto, C.C., Rossato, M., Lopes, C.A.P., Lopes, C.A., Mizubuti, E.S.G., Silva, L.P. 2019. Green synthesis of silver nanoparticles using tomato leaf extract and their entrapment in chitosan nanoparticles to control bacterial wilt. *Journal of the Science of Food and Agriculture* 99(9), 4248–4259.

Satti, S.H., Raja, N.I., Ikram, M., Oraby, H.F., Mashwani, Z.-U.-R., Mohamed, A.H., Singh, A., Omar, A.A. 2022. Plant-based titanium dioxide nanoparticles trigger biochemical and proteome modifications in *Triticum aestivum* L. under biotic stress of *Puccinia striiformis*. *Molecules* 27(13), 4274.

Shahryari, F., Rabiei, Z., Sadighian, S. 2020. Antibacterial activity of synthesized silver nanoparticles by sumac aqueous extract and silver–chitosan nanocomposite against *Pseudomonas syringae* pv. *syringae*. *Journal of Plant Pathology* 102, 469–475.

Sharma, N., Khanna, K., Jasrotia, S., Kumar, D., Bhardwaj, R., Ohri, P. 2023. Metabolites and chemical agents in plant roots: an overview of their use in plant-parasitic nematode management. *Nematology* 25(3), 243–257.

Siddiqui, Z.A., Khan, M.R., Aqib. 2023. Comparative efficacy of zirconium dioxide nanoparticles and AM fungus against wilt disease complex of bottle gourd and upregulation of biochemical and physiological processes. *International Journal of Vegetable Science* 29(3), 243–260.

Silva, S., Dias, M.C., Silva, A.M.S. 2022. Titanium- and zinc-based nanomaterials in agriculture: a promising approach to deal with biotic stresses. *Toxics* 10(4), 172.

Singh, P., Singh, K.R.B., Singh, J., Das, S.N., Singh, R.P. 2021. Tunable electrochemistry and efficient antibacterial activity of plant-mediated copper oxide nanoparticles synthesized using *Annona squamosa* seed extract for agricultural utility. *RSC Advances* 11(29), 18050–18060.

Singh, R., Cheng, S., Singh, S. 2020. Oxidative stress-mediated genotoxic effect of zinc oxide nanoparticles on *Deinococcus radiodurans*. *3 Biotech* 10(2), 66.

Sreelatha, S., Kumar, N., Yin, T.S., Rajani, S. 2022. Evaluating the antibacterial activity and mode of action of thymol-loaded chitosan nanoparticles against the plant bacterial pathogen *Xanthomonas campestris* pv. *campestris*. *Frontiers in Microbiology* 12, 792737.

Suroshe, J.S. 2023. Chemical methods for the synthesis of nanomaterials. In: Sharma, A., Oza, G. (eds), *Nanochemistry*, pp. 77–87. CRC Press, Boca Raton, FL.

Tauseef, A., Hisamuddin, Khalilullah, A., Uddin, I. 2021. Role of MgO nanoparticles in the suppression of *Meloidogyne incognita*, infecting cowpea and improvement in plant growth and physiology. *Experimental Parasitology* 220, 108045.

Thakur, N., Nigam, M., Mann, N.A., Gupta, S., Hussain, C.M., Shukla, S.K., Shah, A.A., Casini, R., Elansary, H.O., Khan, S.A. 2023. Host-mediated gene engineering and microbiome-based technology optimization for sustainable agriculture and environment. *Functional & Integrative Genomics* 23(1), 57.

Tiwari, V., Bambharoliya, K.S., Bhatt, M.D., Nath, M., Arora, S., Dobriyal, A.K., Bhatt, D. 2024. Application of green synthesized copper oxide nanoparticles for effective mitigation of *Fusarium* wilt disease in roots of *Cicer arietinum*. *Physiological and Molecular Plant Pathology* 131, 102244.

Tortella, G., Rubilar, O., Pieretti, J.C., Fincheira, P., Santana, B.M., Fernández-Baldo, M.A., Benavides-Mendoza, A., Seabra, A.B. 2023. Nanoparticles as a promising strategy to mitigate biotic stress in agriculture. *Antibiotics* 12(2), 338.

Valsami-Jones, E., Cassee, F.R., Falk, A. 2022. From small to clever: what does the future hold for the safety and sustainability of advanced materials? *Nano Today* 42, 101364.

Wang, L., Ning, C., Pan, T., Cai, K. 2022a. Role of silica nanoparticles in abiotic and biotic stress tolerance in plants: a review. *International Journal of Molecular Sciences* 23(4), 1947.

Wang, L., Pan, T., Gao, X., An, J., Ning, C., Li, S., Cai, K. 2022b. Silica nanoparticles activate defense responses by reducing reactive oxygen species under *Ralstonia solanacearum* infection in tomato plants. *NanoImpact* 28, 100418.

7 Recent Advances in Nanofertilizers and Nanopesticides

Savita Bhardwaj, Dhriti Sharma, Usha Thakur, and Dhriti Kapoor

7.1 INTRODUCTION

World populace with being over 7.7 billion now has been estimated to reach the mark of 9.8 billion by 2050 (Kah et al. 2019). To satisfy the hunger requirements of this ever-soaring number, sustainable maintenance of crop productivity becomes the dire need of the hour (Tang et al. 2019). However, agricultural business is overworked on account of multiple ecological and biological complexities (Yin et al. 2018). Therefore, the precise management of finite resources and sustainable enhancement of crop yields make all the more sense (Giraldo et al. 2019). The famous green revolution, which was incepted in the sixties realized its objective of increment in crop productivity even with limited land use, but with the passage of time, it ended on account of an enhancement in agricultural inefficiency and unsustainability, which, in turn, paved the way for the second green revolution. This new transformative phase is centered around the usage of novel strategies and technologies to achieve the goal of fulfilling future food requirements in a sustainable manner (Lowry et al. 2019). Among them, nanotechnology has succeeded in standing out as a rapidly growing arena dealing with nanoscale processing of materials (Wang et al. 2016; Yin et al. 2018; Moulick et al. 2020). This outreach of nanotechnology in the sphere of plants could possibly give rise to a novel area of research in the form of plant nano-science.

In order to realize the aim of betterment in agricultural production, heed must be paid toward furnishing the essential nutrients along with managing the pests, which otherwise hamper plant growth. However, the popularly applied fertilizers with instant solubility affect the environment adversely and also possess low usage efficacy (<30%) (Song et al. 2019). Additionally, a large chunk of these fertilizers is lost in the process of application, resulting in defiling of the ecosystem (Kumar et al. 2019). All this makes it more pertinent to shift the focus toward developing new propositions to deal with the abovementioned issues. Fertilizers enabled with nanotechnology, i.e., nanofertilizers, and pesticides, i.e., nanopesticides, have the requisite attributes of being highly biocompatible, durable, and efficient when applied to the crop field (Deepa et al. 2015; Adhikari et al. 2016; Sassine et al. 2020). They are reported to be an excellent replacement for traditionally used agrochemicals, which can efficiently deal with nutritional inadequacies and pests (Zhao et al. 2020).

DOI: 10.1201/9781032659435-7

Improvement in both the crop yield and quality of produce has been achieved by the usage of nanofertilizers and nanopesticides, but at the same time a fundamental solution of these problems is needed in the form of preparing plants with innate high yield potential with pest resistance. Lately, genetic manipulation of plants whereby the gene of interest is incorporated or existing genetic material is regulated has been largely employed to make them more productive and stress resistant (Li et al. 2012; Jat et al. 2020). Application of nanomaterials as a delivery vehicle of desired genes in plants has seen some breakthroughs on account of their more surface area for chemical alterations and protection against cellular metabolism; however, plants show some recalcitrance by their cell walls to the internalization of these nanocarriers (Wang et al. 2019b; Demirer et al. 2020; Zhang et al. 2020). Nevertheless, safety concerns must also be taken into consideration for long-term and large-scale agricultural applications of nanofertlizers and nanopesticides for which comprehensive investigative works are still required.

This book chapter summarizes the applicability of nanotechnology in the form of nanofertilizers and nanopesticides along with their recent advancements as a breakthrough for realizing novel agri-tech solutions so that more systematic, safe, and sustainable practices could be adopted to cope with future food and energy security crises.

7.2 ROLE OF NANOFERTILIZERS AND NANOPESTICIDES IN AGRICULTURE

7.2.1 Nanofertilizers

A key constituent of agricultural production, fertilizers are still essential on larger scales to enhance the crop production and meet global food demand. Normally, uptake of required nutrients by the plants is inefficient, which proves to be cost-ineffective to the farmers. Emerging field of nanotechnology has led to various innovative usages in the arena of plant nutrition to upgrade the process of plant growth, and nanofertilizers are one of them. These nanotechnology-enabled fertilizers have been proven to promote nutrient delivery and thereby cut the farmer's costs by minimizing nutrient losses to the environment (Liu and Lal 2015). Such fertilizers can be incorporated via applying them directly onto the roots or through foliar path and are more target-oriented for enhancing the plants' efficacy of nutrient utilization (Adisa et al. 2019). Nanofertlizers incredibly reduce the over-usage of ordinary composts, and their steady release offers a chance to the plants for utilizing the nutrients as per the natural requirements. This makes the process of plant development more profound and enables the plants to combat the stresses of both biotic and abiotic nature. Beneficial role of nanofertilizers and nanopesticides is depicted in Figure 7.1

Generally, the nature of nanomaterials along with the nutrients in use is employed as a basis of categorizing the nanofertilizers into four classes of (i) macronutrient (mostly N, P, K) nanofertilizers, (ii) micronutrient nanofertilizers, (iii) nanomaterial-loaded nanofertilizers, and (iv) other miscellaneous nanofertilizers. Out of these, the first two are nutritive in nature and make the nutrient absorption of plants more efficient, whereas the third one although lacking any kind of essential nutrients

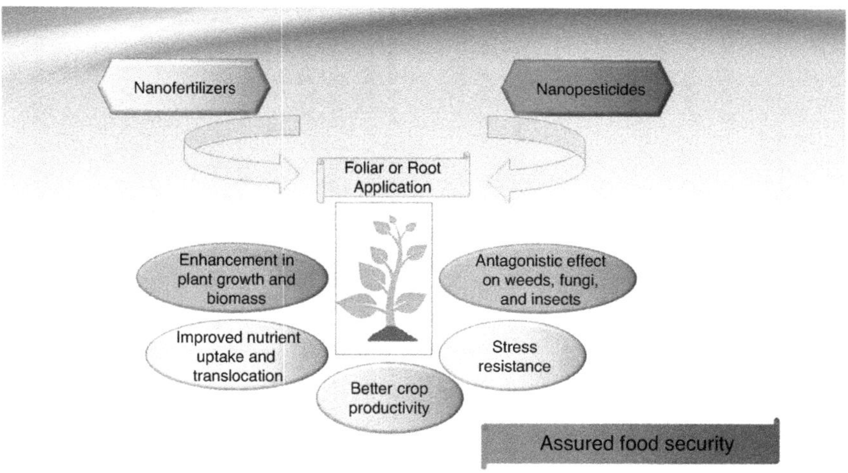

FIGURE 7.1 Positive impact of nanofertilizers and nanopesticides in agriculture at a glance.

acts as a nanocarrier for releasing the nutrients slowly. The last class of nanofertilizers does not fall into the category of nutritionally required biomolecules but still casts beneficial impacts on the growth and productivity of plants (Adisa et al. 2019). These nanofertilizers possess a very small diameter with high solubility and upscale growth-related attributes and are also reported to confer stress resistance. Some currently employed nanofertilizers in agriculture and their impact are briefly summarized in Table 7.1.

7.2.2 Nanopesticides

Agricultural procedures have a high requirement for managing the problem of pests effectively. However, in accordance with an estimate, volatilization, photolysis, or degradation occurring either after or during the process of application results in a loss of applied chemical pesticides by 90%. This could negatively impact the well-being of plants and humans alike via gaining an entry into the food chain (Liang et al. 2017). Additionally, long-term and widespread applications of pesticides have led to an enhancement in pesticide resistance in unwanted plants and insects plus pathogens. Therefore, the need arises for resorting to more ecologically safe and target-specific pest control practices (Irfan et al. 2018). Since numerous experimental works have substantiated the role of nanotechnology in improving plant growth and nutrient utilization, its attribute of dealing with pests and pathogens has also been gaining interest among the members of the scientific community and agri-business (Kookana et al. 2014). Nanopesticides are the pesticidal formulations enriched with nanomaterials of biocidal nature, which can either be utilized directly as pest killers or indirectly, whereby they protect other pesticides and upscale their release at the desired site of action. So, two major classes of nanopesticides can be ascribed as (i) nanomaterials with direct usage as nanopesticides and (ii) the ones acting as nanocarriers for other pesticides.

TABLE 7.1
Some Nanofertilizers of Current Usage in the Agricultural Field

Nanomaterial	Type	Plant	Concentration	Fertilization Technique	Impact on Growth	Reference/s
Nitrogen-doped carbon dots (N-CDs)	Macronutrient	*Vigna radiata* (mung bean)	0.2 mg/L	Nutrient solution	A 200% improvement in growth rates (root + shoot length) was noticed	Wang et al. (2019a)
Apatite nanoparticles (NPs)	Macronutrient	*Glycine max* (soybean)	21.8 mg/L	Soil	A respective enhancement of 20.4% and 32.6% in seed yield and growth rates was achieved. This increment was 18.2% and 41.2% in biomass production for above- and below-ground parts, respectively	Liu and Lal (2014)
Monopotassium-phosphate (MKP)	Macronutrient	*Lycopersicum esculentum* (tomato)	3 g/L	Foliar and soil	Different growth parameters were recorded to improve even under salt-induced stress	Sassine et al. (2020)
Calcium oxide nanoparticles (CaO NPs)	Macronutrient	*Arachis hypogea* (groundnut)	500 mg/L	Foliar	Enhancement in both seed germination efficiency and growth rates was recorded	Deepa et al. (2015)
Sulfur NPs	Macronutrient	a. *Cucurbita pepo* (pumpkin)	100–400 mg/L	Soil	The number of leaves and branches, height per plant, stem diameter, and healthy plant increased	Salem et al. (2015)
		b. *Lycopersicum esculentum* (tomato)	300 mg/L	Soil	A concentration-dependent positive impact was noted whereby root and shoot growth both were enhanced	Salem et al. (2016)
Zinc oxide (ZnO)NPs	Micronutrient	a. *Oryza sativa* (rice) var. PR-121	5 g/L	Foliar	Different growth and yield depicting parameters along with microbial count plus enzymatic action of dehydrogenase recorded an enhancement	Bala et al. (2019)
		b. *Coffea arabica* (coffee)	10 mg/L	Foliar	37% (roots) and 95% (leaves) enhancement in fresh weight together with 28% (roots) and 85% (leaves) in dry weight was documented, while net photosynthetic rate was improved by 55%	Rossi et al. (2019)

(Continued)

TABLE 7.1 (*Continued*)
Some Nanofertilizers of Current Usage in the Agricultural Field

Nanomaterial	Type	Plant	Concentration	Fertilization Technique	Impact on Growth	Reference/s
Gamma-ferric oxide (γ-Fe_2O_3) NPs	Micronutrient	*Brassica napus* (rapeseed)	2 mg/mL	Soil	The content of oxidative stress markers like MDA and H_2O_2 was lowered down to 26 mm/g and 83 µM/g, respectively; foliar growth rate and photosynthetic efficiency got enhanced by 50% and 52%, respectively	Palmqvist et al. (2017)
Nanopriming with zero valent iron (nZVI)	Micronutrient	*Oryza sativa* (Rice)	100 mg/kg	Soil	Increment of 47.1%–55% in the grain yield was recorded along with a decrease of 83.6%–86.2% in grain PCP content, while removal of PCP from soil became rapid after adding these nanofertilizers	Liu et al. (2021)
Copper oxide (CuO) NPs	Micronutrient	*Zea mays* (maize)	8 mg/L	Foliar	Improvement of about 51% was registered in plant growth	Adhikari et al. (2016)
Manganese (Mn) NPs	Micronutrient	*Vigna radiata* (mung bean)	0.05 mg/L	Foliar	An increment of length (root + shoot) and biomass (fresh + dry) was documented	Pradhan et al. (2013)
Molybdenum (Mo) NPs	Micronutrient	*Cicer arietinum* (chickpea)	8 mg/L	Seed	The microbial action and growth parameters of seed were recorded to be improved	Taran et al. (2014)
Chitosan solution-methylacrylic acid (CS-MMA) NPs	Nanomaterial-loaded	a. *Triticum aestivum* (wheat)		Foliar	Notable enhancement in the content of total sugars, phosphorus, and potassium inside wheat grains was observed in contrast to commonly used fertilizers	Abdel-Aziz et al. (2018)
		b. *Pisum sativum* (Pea) var. Master B		Foliar	Upregulation of mitotic cell divisions along with the expression levels of important proteins was induced by these nanofertilizers	Khalifa and Hasaneen (2018)

(*Continued*)

TABLE 7.1 (*Continued*)
Some Nanofertilizers of Current Usage in the Agricultural Field

Nanomaterial	Type	Plant	Concentration	Fertilization Technique	Impact on Growth	Reference/s
Titanium dioxide (TiO_2) NPs	Other	*Vigna radiata* (mung bean)		Foliar	Foliar treatment brought about significant promotion of root + shoot length, chlorophyll measures, and total soluble protein in leaves	Raliya et al. (2015)
Multi-walled carbon nanotube (MWCNT)	Other	*Avena sativa* (oat)		Seed priming	Elevation in growth rates of xylar cells with that of chlorophyll measures and photosynthetic potential	Joshi et al. (2018)

7.2.2.1 Nanomaterials with Direct Application as Nanopesticides

The broad range of antifungal or antibacterial attributes of nanomaterials enables them to be applied directly to different plant parts (leaves, roots, or seeds) in order to protect them from pests and pathogens. Additionally, they also mitigate the problem of the development of microbial resistance as observed after long-term usage of traditional pesticides. These nanopesticides extensively use metal-based nanoparticles (NPs) like Cu, Zn, Ti, Mg, and Ag, while Si and S are also reported to be of antimicrobial nature and much safer for both flora and fauna.

Out of these, silver (Ag) NPs belong to the category of broad-spectrum germicidal agent, which can harm bacterial and fungal species attacking plants. The most calamitous pathogen of rice crop, i.e., *Xanthomonas oryzae* pv. oryzae, along with the attack of nematodes has been reported to be effectively dealt with by the application of Ag NPs (Cromwell et al. 2014; Ahmed et al. 2020). Additionally, Cu ions or CuO-based nanopesticides can also act against pathogenic bacteria and fungi as reported in Bt cotton plants where the crop resistance was enhanced further by elevating the expression levels of Bt toxin proteins inside both the foliar and rhizal regions (Sidhu et al. 2017). To overcome the problem of agglomeration, the pure forms of both the Ag and Cu-based nanopesticides are now being converted into nanocomposites by combining them with reduced graphene oxide (rGO), which makes them more stable plus soluble.

Among other metal-based nanopesticides, usage of Mg or MgO-based NPs is advantageous due to their non-toxic nature, easy preparation, low cost, and great efficacy against bacteria. Treatment of tomato roots with suspension of MgO NPs can strongly inhibit the occurrence of bacterial wilt (Imada et al. 2016). Akin to this, magnesium hydroxide ($Mg(OH)_2$) NPs are reported to have equivalent bactericidal properties and less phytotoxic impact on both the leaves and seeds of tomato plants (Huang et al. 2018). Moreover, Zn and Ti-backed NPs can also play a role in the effective management of pests as observed in the case of *Xanthomonas citri* ssp. citri (citrus canker of orange) and *X. perforans*, respectively (Paret et al. 2013). Further, the disease severity of bacterial spot in tomatoes under greenhouse conditions can be notably reduced if Zn NPs doped with TiO_2 are supplemented.

Besides being utilized as nanopesticides, Cu, Zn, and Mg are nutritive in nature and their application can also upgrade the process of growth by making the enzymatic action and physiological processes of vital importance better. Sulfur-based nanomaterials (SNPs) also being a macronutrient effectively inhibit the growth of different microbes as observed in the case of foliar supplementation in tomato plants, which curbs the spread of *Fusarium solani* and *Venturia inaequalis* and rhizome treatment of ginger plants, which combats *Fusarium oxysporum* causing soft rot in them (Rao and Paria 2013; Athawale et al. 2018).

Despite the low phytotoxicity, the long-term efficacy of these nanopesticides still requires to be thoroughly investigated as the impact of their accumulation in the food chain and, in turn, on the human health need generation of more relevant information.

7.2.2.2 Nanomaterials Utilized as Nanocarriers for Pesticides

It is quite pertinent to regulate the release of pesticides with more efficiency in low concentrations in the field of pest management (Khandelwal et al. 2016). Nanocarriers, being constituted either of a membranal structure or a shell so that a pesticidal chemical could be encapsulated inside them, have proven remarkably effective in protecting the pesticides from adverse environmental conditions. Nanocarrier makes the pesticides more chemically stable, dispersible, and wettable, which reduces the risk of their runoff (Kumar et al. 2019). These substances formulate a smart delivery system for pesticides, which is more target-specific with high penetrability. Some major examples include metal organic frameworks (MOFs), solid lipid NPs, polymeric nanomaterials, and Si NPs. Out of these, polymeric nanomaterials are widely used as nanocarriers on account of their enhanced biocompatibility and less side effects on non-targeted sites (Kumar et al. 2017), whereas solid lipid NPs are emulsion like in nature and are a promising option for a varied array of lipophilic substances and also prevent the pesticides from chemical degradation as in the case of deltamethrin, which is otherwise prone to photodegeneration (Borel and Sabliov 2014; Wei et al. 2017). Of late, MOFs, a type of hybrid between organic and inorganic materials, have secured a significant position as nanocarriers due to their regular porosity, extensive surface area, and more stability (Wang et al. 2020). Moreover, they are biodegradable in nature and so pose no threat to the well-being of ecological components as nanocarriers of pesticides. Further, Si NPs are easy to synthesize and are available in a variety of shapes and sizes, which enables them to be utilized as nanocarriers for pesticides sensitive to UV light degradation (Mody et al. 2014). The inner region of Si NPs loads pesticides, while its outer shell prevents the entry of UV radiations, which, in turn, regulate the release of pesticides to the target-specific regions.

In comparison with traditional systems, nanocarriers possess peculiar benefits of being stable, biocompatible, target-specific, and environment-friendly. However, for realizing the goal of sustainable agricultural production, the mechanism of uptake and translocation of these nanocarrier-encapsulated pesticides still needs a strong and comprehensive backing of long-term field trials.

7.3 MITIGATION OF ABIOTIC STRESSES BY THE USE OF NANOFERTILIZERS

Nano-enabled approaches offer a promising tool in agriculture to control nutrient deficiency, increase crop yields, and manage plant stresses from the environment (Dimkpa et al. 2017; White and Gardea-Torresdey 2018). Their unique properties allow for precise and targeted delivery of nutrients, improving plant growth and yields. Their use is a promising development that can address challenges in the industry such as food security and sustainability (Irshad et al. 2021; Shah et al. 2021).

7.3.1 Drought Stress

Zinc NPs (ZnNPs) improved plant height, biomass, root/turnip length, photosynthetic pigments, i.e., chl and carotenoid content, activities of antioxidative enzymes (SOD (superoxide dismutase), POD (peroxidase), and CAT (catalase)), and content of secondary metabolites (total phenolics, flavonoids, and ascorbic acid), in the leaves of *Brassica rapa* L. under drought stress (Li et al. 2023). Calcium NPs (Ca-NPs) improved the plant biomass, gas exchange attributes, chlorophyll content, PSI and PSII efficiency, level of secondary metabolites, functioning of antioxidative enzymes such as SOD, POD, CAT, and APX, and content of non-enzymatic components in *Brassica napus* under drought stress (Ayyaz et al. 2022). Yttrium-doped-stabilized γ-Fe_2O_3 NPs under drought stress increased the leaf length and iron uptake and reduced the lipid peroxidation level by reducing the MDA level in *Brassica napus* (Palmqvist et al. 2017). Foliar spraying with Zn nanochelate improved the leaf area index, total dry matter, crop growth rate, relative growth rate, yield attributes, oleic acid, and linoleic acid in canola plants under drought stress (Saffari et al. 2023). A nanopesticide was prepared by loading salicylic acid (SA)-functionalized mesoporous silica NPs with emamectin benzoate (EB), which enhanced the salt stress resistance of plants by increasing the SOD and POD activities by reducing the MDA content (Zhong et al. 2023). Increase in chlorophyll content, leaf area index, number of capsules, seed weight, oil percentage, and grain yield was observed in potassium nanofertilizers applied to sesame plants under drought stress (Mizani et al. 2022). Drought-exposed basil plants showed an increase in height, dry leaf weight and stem, yield, and absorption and supply of essential elements for plant growth by the application of nanofertilizer (Sohrabi et al. 2023) (Table 7.2).

TABLE 7.2
Role of Nanofertilizers in Various Plant Species Under Drought Stress

Nanofertilizer	Plant Species	Results	References
Nano-chelated nitrogen fertilizer	Wheat (*Triticum aestivum* L.)	Increase in RWC, ion leakage, protein content, phosphorus, potassium, and photosynthesis rate was observed	Astaneh et al. (2021)
Nanochelated silicon, boron, and zinc fertilizers	Wheat	Increase in RWC, cell membrane stability index, protein, WUE, yield components, and activities of antioxidative enzymes, i.e., SOD and CAT	Ahmadian et al. (2021)
Nano-micro-fertilizer and nano-macro-fertilizer	Sorghum (*Sorghum bicolor* L. Moench)	Protein and tannin content and yield were noticed to be improved under drought stress conditions	Kazemi et al. (2022)
Iron nanofertilizer	Dill (*Anethum graveolens* L.)	Morphological attributes, seed yield, percentage, and yield of essential oil were increased under drought stress conditions	Gholinezhad (2017)

(*Continued*)

TABLE 7.2 (*Continued*)
Role of Nanofertilizers in Various Plant Species Under Drought Stress

Nanofertilizer	Plant Species	Results	References
Potassium nanochelate	Sesame (*Sesame indicum* L.)	Application of potassium nanofertilizer increased the content of chlorophyll, total carbohydrate, and unsaturated fatty acids, and activities of CAT and POD enzymes	Mahdavi Khorami et al. (2020)
Zn-nanofertilizer	Castor bean (*Ricinus communis* L.)	Enhanced the Chl a, Chl b, and total Chl, number of capsules, the number of seeds, seed and oil yield, protein content, as well as activities of CAT, SOD, and POD enzymes	Rahbari et al. (2021)
Nano-vermicompost	Tomato (*Solanum lycopersicon* L.)	Increased the growth and dry matter content, chlorophyll contents, photosynthesis, and PSII activity	Ahanger et al. (2021)
Nano potassium fertilizer	Peppermint (*Mentha piperita* L.)	Improved the plant height, Chl a and b contents, RWC, fresh and dry yield, proline, and total soluble carbohydrates	Saedi et al. (2020)
Zinc (Zn) and iron (Fe) nanoparticles	Bean (*Phaseolus vulgaris* L.)	Leaf area, protein content, antioxidant enzyme activity, proline, and soluble sugar content were increased by the application of nanofertilizer	Fatollahpour Grangah et al. (2020)
Potassium nano-silica	Maize (*Zea mays* L.)	Application of nano-silica mitigated the negative effects of drought stress by increasing the nutrient absorption under drought stress	Aqaei et al. (2020)
Nano-nutrient solution of biochar	Maize	Enhanced the shoot length, shoot fresh and dry weight, number of leaves and flowers, Chl, carotenoids, and total phenolic contents	Mubashir et al. (2023)

7.3.2 Heavy Metal Stress

Application of zinc oxide NPs (ZnO NPs) exhibited an increase in the growth, Zn content, activity of antioxidative enzymes, yield, and decreased As accumulation in mung bean cultivars under As stress (Rani et al. 2023). ZnO NPs augmented the growth attributes, activity of antioxidative enzymes, i.e., APX, DHAR, MDHAR, and GR, and content of non-enzymatic antioxidants such as AsA and GSH in rice cultivars under Cr toxicity (Prakash et al. 2022). Similarly, an increase in the plant growth parameters, plant biomass, chlorophyll a, b and carotenoid contents, and functioning of antioxidative enzymes such as SOD, CAT, POD, and APX was found in ZnO NPs treated in cotton (*Gossypium hirsutum* L.) seedlings under Cd and Pb stresses (Priyanka et al. 2021). Application of FeO NPs as fertilizer increased the coriander's plant tolerance to Pb and Cd stress by improving the plant height and declining the metal content (Fahad et al. 2022). *Azadirachta indica*–mediated ZnO NPs increased the growth parameters, physiological attributes, soil nutrient status, and the activities of SOD, CAT, and POD enzymes in maize plants (Mehmood et al. 2023). Yusefi-Tanha et al. (2020) found that the use of ZnO NPs as nanofertilizer improved the morphology and antioxidant defense system of soybean (*Glycine* max) under Zn-deficient conditions.

7.3.3 Salinity Stress

Application of nanofertilizer improved the plant height, leaf number, stem diameter, and flowering characteristics in tomato plants under salinity stress (Sajyan et al. 2018). Foliar application of ZnO and Fe_3O_4 nanofertilizer significantly improved the *Moringa peregrina* plant tolerance to salinity stress by increasing the growth parameters (plant height, root length, number of leaves and branches, shoot and root fresh and dry weights), photosynthetic pigments, i.e., chlorophyll, and carotenoids, protein, proline, and total carbohydrate contents, and activities of antioxidative enzymes (POD and SOD) (Soliman et al. 2015). The application of calcium phosphate NPs (CaP-NPs) as a soil fertilizer increased the plant growth attributes, photosynthetic pigments, content of total soluble sugars, proline, and total phenolics as well as activities of antioxidant enzymes in broad bean plants under salinity stress (Nasrallah et al. 2022). El-Shal et al. (2022) evaluated the effect of different nanofertilizers, i.e., K_2SO_4 and ZnO and SiO_2, on two alfalfa (*Medicago sativa* L.) genotypes under salinity stress and found that shoot dry weight, plant height, number of flowers, number of tillers, root length, root fresh and dry weight, RWC, proline, and SOD enzyme activity were increased by nanofertilizers under stressed conditions. Application of nano-nitrogen fertilizer exhibited an increase in values of all macro- and micronutrients and spikes and top dry matter concentrations in wheat grains (El-Ashry et al. 2019). The ZnO NPs, synthesized using *Mentha* extract, significantly mitigated the salinity stress in *Vicia faba* by increasing the plant growth attributes, photosynthetic pigments, i.e., chlorophyll a, b, carotenoids, and total pigments, activities of antioxidants, osmolytes, and secondary metabolites (Mogazy and Hanafy 2022) (Figure 7.2).

7.4 CONCLUSION

The development of nanomaterials has greatly impacted the field of plant and agricultural science. With their unique physical and chemical properties, nanomaterials have opened up new possibilities for designing and developing advanced materials that are more efficient, durable, and versatile. The potential applications of nanomaterials are vast and range from electronics and medicine to energy and environmental engineering. Nanofertilizers have shown potential in improving crop growth, yield,

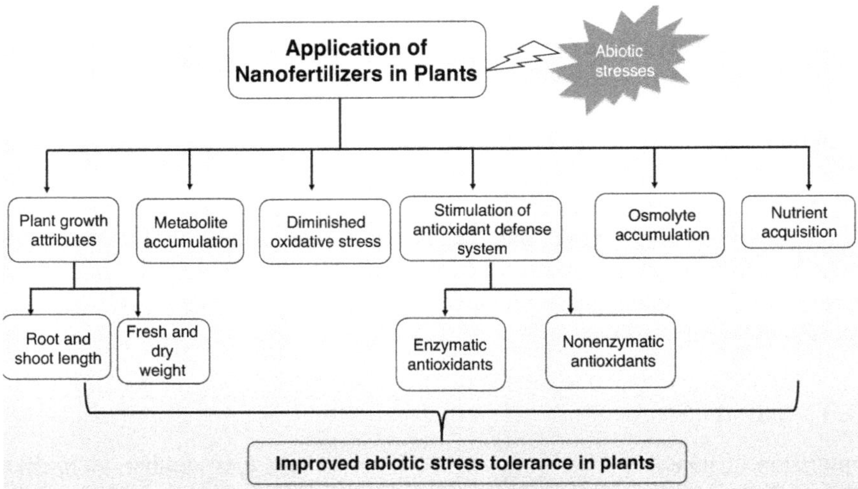

FIGURE 7.2 Diagrammatic representation of the role of nanofertilizers in abiotic stress tolerance in plants.

and quality while minimizing the environmental impact of excess fertilizer. They can also help reduce abiotic stresses and the toxicity of heavy metals in soil. Their use is a promising advancement in agriculture that can increase food production and sustainability. Nanotechnology can revolutionize agriculture by developing pesticides and fertilizers using NPs and nanocapsules. Their use can improve efficiency, reduce chemical usage, and minimize environmental impact. This innovation can address challenges facing the industry, including food security and sustainability. It is exciting to think about the possibilities that the development of nanomaterials will bring in the future.

REFERENCES

Abdel-Aziz, H., Hasaneen, M.N., Omar, A. 2018. Effect of foliar application of nano chitosan NPK fertilizer on the chemical composition of wheat grains. *Egyptian Journal of Botany* 58(1), 87–95.

Adhikari, T., Sarkar, D., Mashayekhi, H., Xing, B. 2016. Growth and enzymatic activity of maize (*Zea mays* L.) plant: solution culture test for copper dioxide nanoparticles. *Journal of Plant Nutrition* 39(1), 99–115. https://doi.org/10.1080/01904167.2015.1109103

Adisa, I.O., Pullagurala, V.L.R., Peralta-Videa, J.R., Dimkpa, C.O., Elmer, W.H., Gardea-Torresdey, J.L., White, J.C. 2019. Recent advances in nano-enabled fertilizers and pesticides: a critical review of mechanisms of action. *Environmental Science: Nano* 6(7), 2002–2030. https://doi.org/10.1039/C9EN00265K

Ahanger, M.A., Qi, M., Huang, Z., Xu, X., Begum, N., Qin, C., Zhang, L. 2021. Improving growth and photosynthetic performance of drought-stressed tomato by application of nano-organic fertilizer involves up-regulation of nitrogen, antioxidant and osmolyte metabolism. *Ecotoxicology and Environmental Safety* 216, 112195. https://doi.org/10.1016/j.ecoenv.2021.112195

Ahmadian, K., Jalilian, J., Pirzad, A. 2021. Nano-fertilizers improved drought tolerance in wheat under deficit irrigation. *Agricultural Water Management* 244, 106544. https://doi.org/10.1016/j.agwat.2020.106544

Ahmed, T., Shahid, M., Noman, M., Niazi, M.B.K., Mahmood, F., Manzoor, I. et al. 2020. Silver nanoparticles synthesized by using *Bacillus cereus* SZT1 ameliorated the damage of bacterial leaf blight pathogen in rice. *Pathogens* 9(3), 160. https://doi.org/10.3390/pathogens9030160

Aqaei, P., Weisany, W., Diyanat, M., Razmi, J., Struik, P.C. 2020. Response of maize (*Zea mays* L.) to potassium nano-silica application under drought stress. *Journal of Plant Nutrition* 43(9), 1205–1216. https://doi.org/10.1080/01904167.2020.1727508

Astaneh, N., Bazrafshan, F., Zare, M., Amiri, B., Bahrani, A. 2021. Nano-fertilizer prevents environmental pollution and improves physiological traits of wheat grown under drought stress conditions. *Scientia Agropecuaria* 12(1), 41–47. https://doi.org/10.17268/sci.agropecu.2021.006

Athawale, V., Paralikar, P., Ingle, A.P., Rai, M. 2018. Biogenically engineered nanoparticles inhibit *Fusarium oxysporum* causing soft-rot of ginger. *IET Nanobiotechnology* 12(8), 1084–1089. https://doi.org/10.1049/iet-nbt.2017.0216

Ayyaz, A., Fang, R., Ma, J., Hannan, F., Huang, Q., Sun, Y., Farooq, M.A. 2022. Calcium nanoparticles (Ca-NPs) improve drought stress tolerance in *Brassica napus* by modulating photosystem II, nutrient acquisition and antioxidant performance. *NanoImpact* 28, 100423. https://doi.org/10.1016/j.impact.2022.100423

Bala, R., Kalia, A., Dhaliwal, S.S. 2019. Evaluation of efficacy of ZnO nanoparticles as remedial zinc nanofertilizer for rice. *Journal of Soil Science and Plant Nutrition*, 19(2), 379–389.

Borel, T., Sabliov, C.M. 2014. Nanodelivery of bioactive components for food applications: types of delivery systems, properties, and their effect on ADME profiles and toxicity of nanoparticles. *Annual Review of Food Science and Technology* 5, 197–213. https://doi.org/10.1146/annurev-food-030713-092354

Cromwell, W.A., Yang, J., Starr, J.L., Jo, Y.K. 2014. Nematicidal effects of silver nanoparticles on root-knot nematode in bermudagrass. *Journal of Nematology* 46(3), 261–266.

Deepa, M., Sudhakar, P., Nagamadhuri, K.V., Balakrishna Reddy, K., Giridhara Krishna, T., Prasad, T.N.V.K.V. 2015. First evidence on phloem transport of nanoscale calcium oxide in groundnut using solution culture technique. *Applied Nanoscience* 5, 545–551. https://doi.org/10.1007/s13204-014-0348-8

Demirer, G.S., Zhang, H., Goh, N.S., Pinals, R.L., Chang, R., Landry, M.P. 2020. Carbon nanocarriers deliver siRNA to intact plant cells for efficient gene knockdown. *Science Advances* 6(26), eaaz0495. https://doi.org/10.1126/sciadv.aaz0495

Dimkpa, C.O., White, J.C., Elmer, W.H., Gardea-Torresdey, J. 2017. Nanoparticle and ionic Zn promote nutrient loading of sorghum grain under low NPK fertilization. *Journal of Agricultural and Food Chemistry* 65(39), 8552–8559. https://doi.org/10.1021/acs.jafc.7b02961

El-Ashry, S.M., El-Faham, S.Y., El-Dok, S. 2019. Wheat plant dry matter and grain nutrient status and its responses to nanofertilizer under salinity condition. *Plant Archives* 19, 2053–2063.

El-Shal, R.M., El-Naggar, A.H., El-Beshbeshy, T.R., Mahmoud, E.K., El-Kader, N.I.A., Missaui, A.M., El-Sharkawy, M.S. 2022. Effect of nano-fertilizers on alfalfa plants grown under different salt stresses in hydroponic system. *Agriculture* 12(8), 1113. https://doi.org/10.3390/agriculture12081113

Fahad, B.A., Agheem, M.H., Memon, S.A., Baloch, A.R., Tunio, A., Qasim, S. 2022. Efficient mitigation of cadmium and lead toxicity in coriander plant utilizing magnetite (Fe_3O_4) nanofertilizer as growth regulator and antimicrobial agent. *International Journal of Environmental Analytical Chemistry* 102(16), 3868–3879. https://doi.org/10.1080/03067319.2020.1776861

Fatollahpour Grangah, M., Rashidi, V., Mirshekari, B., Khalilvand Behrouzyar, E., Farahvash, F. 2020. Effects of nano-fertilizers on physiological and yield characteristics of pinto bean cultivars under water deficit stress. *Journal of Plant Nutrition* 43(19), 2898–2910. https://doi.org/10.1080/01904167.2020.1799001

Gholinezhad, E. 2017. Effect of drought stress and Fe nano-fertilizer on seed yield, morphological traits, essential oil percentage and yield of dill (*Anethum graveolens* L.). *Journal of Essential Oil-Bearing Plants* 20(4), 1006–1017. https://doi.org/10.1080/0972060X.2017.1362999

Giraldo, J.P., Wu, H., Newkirk, G.M., Kruss, S. 2019. Nanobiotechnology approaches for engineering smart plant sensors. *Nature Nanotechnology* 14(6), 541–553. https://doi.org/10.1038/s41565-019-0470-6

Huang, Z., Rajasekaran, P., Ozcan, A., Santra, S. 2018. Antimicrobial magnesium hydroxide nanoparticles as an alternative to Cu biocide for crop protection. *Journal of Agricultural and Food Chemistry* 66(33), 8679–8686. https://doi.org/10.1021/acs.jafc.8b01727

Imada, K., Sakai, S., Kajihara, H., Tanaka, S., Ito, S. 2016. Magnesium oxide nanoparticles induce systemic resistance in tomato against bacterial wilt disease. *Plant Pathology* 65(4), 551–560. https://doi.org/10.1111/ppa.12419

Irfan, S.A., Razali, R., KuShaari, K., Mansor, N., Azeem, B., Versypt, A.N.F. 2018. A review of mathematical modeling and simulation of controlled-release fertilizers. *Journal of Controlled Release* 271, 45–54. https://doi.org/10.1016/j.jconrel.2017.12.020

Irshad, M.A., Nawaz, R., Ur Rehman, M.Z., Adrees, M., Rizwan, M., Ali, S., Tasleem, S. 2021. Synthesis, characterization and advanced sustainable applications of titanium dioxide nanoparticles: a review. *Ecotoxicology and Environmental Safety* 212, 111978. https://doi.org/10.1016/j.ecoenv.2021.111978

Jat, S.K., Bhattacharya, J., Sharma, M.K. 2020. Nanomaterial-based gene delivery: a promising method for plant genome engineering. *Journal of Materials Chemistry B* 8(19), 4165–4175. https://doi.org/10.1039/D0TB00217H

Joshi, A., Kaur, S., Singh, P., Dharamvir, K., Nayyar, H., Verma, G. 2018. Tracking multi-walled carbon nanotubes inside oat (*Avena sativa* L.) plants and assessing their effect on growth, yield, and mammalian (human) cell viability. *Applied Nanoscience* 8, 1399–1414. https://doi.org/10.1007/s13204-018-0801-1

Kah, M., Tufenkji, N., White, J.C. 2019. Nano-enabled strategies to enhance crop nutrition and protection. *Nature Nanotechnology* 14(6), 532–540. https://doi.org/10.1038/s41565-019-0439-5

Kazemi, E., Ganjali, H.R., Mehraban, A., Ghasemi, A. 2022. Yield and biochemical properties of grain sorghum (*Sorghum bicolor* L. Moench) affected by nano-fertilizer under field drought stress. *Cereal Research Communications* 50(3), 397–405. https://doi.org/10.1007/s42976-021-00198-2

Khalifa, N.S., Hasaneen, M.N. 2018. The effect of chitosan–PMAA–NPK nanofertilizer on *Pisum sativum* plants. *3 Biotech* 8(4), 193. https://doi.org/10.1007/s13205-018-1221-3

Khandelwal, N., Barbole, R.S., Banerjee, S.S., Chate, G.P., Biradar, A.V., Khandare, J.J., Giri, A.P. 2016. Budding trends in integrated pest management using advanced micro- and nano-materials: challenges and perspectives. *Journal of Environmental Management* 184, 157–169. https://doi.org/10.1016/j.jenvman.2016.09.064

Kookana, R.S., Boxall, A.B., Reeves, P.T., Ashauer, R., Beulke, S., Chaudhry, Q. et al. 2014. Nanopesticides: guiding principles for regulatory evaluation of environmental risks. *Journal of Agricultural and Food Chemistry* 62(19), 4227–4240. https://doi.org/10.1021/jf500232f

Kumar, S., Bhanjana, G., Sharma, A., Dilbaghi, N., Sidhu, M.C., Kim, K.H. 2017. Development of nanoformulation approaches for the control of weeds. *Science of the Total Environment* 586, 1272–1278. https://doi.org/10.1016/j.scitotenv.2017.02.020

Kumar, S., Nehra, M., Dilbaghi, N., Marrazza, G., Hassan, A.A., Kim, K.H. 2019. Nano-based smart pesticide formulations: emerging opportunities for agriculture. *Journal of Controlled Release* 294, 131–153. https://doi.org/10.1016/j.jconrel.2018.12.012

Li, J., Zafar, S., Javaid, A., Perveen, S., Hasnain, Z., Ihtisham, M., Abbas, M. 2023. Zinc nanoparticles (ZnNPs): high-fidelity amelioration in turnip (*Brassica rapa* L.) production under drought stress. *Sustainability* 15(8), 6512. https://doi.org/10.3390/su15086512

Li, T., Liu, B., Spalding, M.H., Weeks, D.P., Yang, B. 2012. High-efficiency TALEN-based gene editing produces disease-resistant rice. *Nature Biotechnology* 30(5), 390–392. https://doi.org/10.1038/nbt.2199

Liang, Y., Guo, M., Fan, C., Dong, H., Ding, G., Zhang, W., Tang, G., Yang, J., Kong, D., Cao, Y. 2017. Development of novel urease-responsive pendimethalin microcapsules using silica-IPTS-PEI as controlled-release carrier materials. *ACS Sustainable Chemistry & Engineering* 5(6), 4802–4810. https://doi.org/10.1021/acssuschemeng.7b00208

Liu, R., Lal, R. 2014. Synthetic apatite nanoparticles as a phosphorus fertilizer for soybean (Glycine max). *Scientific Reports*, 4(1), 5686.

Liu, R., Lal, R. 2015. Potentials of engineered nanoparticles as fertilizers for increasing agronomic productions. *Science of the Total Environment* 514, 131–139. https://doi.org/10.1016/j.scitotenv.2015.01.104

Liu, Y., Wu, T., White, J.C., Lin, D. 2021. A new strategy using nanoscale zero-valent iron to simultaneously promote remediation and safe crop production in contaminated soil. *Nature Nanotechnology* 16(2), 197–205. https://doi.org/10.1038/s41565-020-00803-1

Lowry, G.V., Avellan, A., Gilbertson, L.M. 2019. Opportunities and challenges for nanotechnology in the agri-tech revolution. *Nature Nanotechnology* 14(6), 517–522. https://doi.org/10.1038/s41565-019-0461-7

Mahdavi Khorami, A., Masoud Sinaki, J., Amini Dehaghi, M., Rezvan, S., Damavandi, A. 2020. Sesame (*Sesamum indicum* L.) biochemical and physiological responses as affected by applying chemical, biological, and nano-fertilizers in field water stress conditions. *Journal of Plant Nutrition* 43(3), 456–475. https://doi.org/10.1080/01904167.2019.1676905

Mehmood, S., Ou, W., Ahmed, W., Bundschuh, J., Rizwan, M., Mahmood, M., Li, W. 2023. ZnO nanoparticles mediated by *Azadirachta indica* as nano fertilizer: improvement in physiological and biochemical indices of *Zea mays* grown in Cr-contaminated soil. *Environmental Pollution* 339, 122755. https://doi.org/10.1016/j.envpol.2023.122755

Mizani, A., Sinaki, J.M., Rezvan, S., Esfahlani, M.A., Damavandi, A. 2022. Study of physiological, growth, and yield responses of sesame to the application of potassium nano-fertilizer, foliar application of fulvic acid and chitosan under drought stress. *Journal of Plant Nutrition* 45(15), 2251–2266. https://doi.org/10.22059/jci.2022.331235.2618

Mody, V.V., Cox, A., Shah, S., Singh, A., Bevins, W., Parihar, H. 2014. Magnetic nanoparticle drug delivery systems for targeting tumor. *Applied Nanoscience* 4, 385–392. https://doi.org/10.1007/s13204-013-0216-y

Mogazy, A.M., Hanafy, R.S. 2022. Foliar spray of biosynthesized zinc oxide nanoparticles alleviate salinity stress effect on *Vicia faba* plants. *Journal of Soil Science and Plant Nutrition* 22(2), 2647–2662. https://doi.org/10.1007/s42729-022-00833-9

Moulick, R.G., Das, S., Debnath, N., Bandyopadhyay, K. 2020. Potential use of nanotechnology in sustainable and "smart" agriculture: advancements made in the last decade. *Plant Biotechnology Reports* 14(5), 505–513. https://doi.org/10.1007/s11816-020-00631-8

Mubashir, A., Nisa, Z.U., Shah, A.A., Kiran, M., Hussain, I., Ali, N., AbdElgawad, H. 2023. Effect of foliar application of nano-nutrients solution on growth and biochemical attributes of tomato (*Solanum lycopersicum*) under drought stress. *Frontiers in Plant Science* 13, 1066790. https://doi.org/10.3389/fpls.2022.1066790

Nasrallah, A.K., Kheder, A.A., Kord, M.A., Fouad, A.S., El-Mogy, M.M., Atia, M.A. 2022. Mitigation of salinity stress effects on broad bean productivity using calcium phosphate nanoparticles application. *Horticulturae* 8(1), 75. https://doi.org/10.3390/horticulturae8010075

Palmqvist, N.M., Seisenbaeva, G.A., Svedlindh, P., Kessler, V.G. 2017. Maghemite nanoparticles act as nanozymes, improving growth and abiotic stress tolerance in *Brassica napus*. *Nanoscale Research Letters* 12, 631. https://doi.org/10.1186/s11671-017-2404-2

Paret, M.L., Vallad, G.E., Averett, D.R., Jones, J.B., Olson, S.M. 2013. Photocatalysis: effect of light-activated nanoscale formulations of TiO_2 on *Xanthomonas perforans* and control of bacterial spot of tomato. *Phytopathology* 103(3), 228–236. https://doi.org/10.1094/phyto-08-12-0183-r

Pradhan, S., Patra, P., Das, S., Chandra, S., Mitra, S., Dey, K.K., Akbar, S., Palit, P., Goswami, A. 2013. Photochemical modulation of biosafe manganese nanoparticles on *Vigna radiata*: a detailed molecular, biochemical, and biophysical study. *Environmental Science & Technology* 47(22), 13122–13131. https://doi.org/10.1021/es402659t

Prakash, V., Rai, P., Sharma, N.C., Singh, V.P., Tripathi, D.K., Sharma, S., Sahi, S. 2022. Application of zinc oxide nanoparticles as fertilizer boosts growth in rice plant and alleviates chromium stress by regulating genes involved in oxidative stress. *Chemosphere* 303, 134554. https://doi.org/10.1016/j.chemosphere.2022.134554

Priyanka, N., Geetha, N., Manish, T., Sahi, S.V., Venkatachalam, P. 2021. Zinc oxide nanocatalyst mediates cadmium and lead toxicity tolerance mechanism by differential regulation of photosynthetic machinery and antioxidant enzyme levels in cotton seedlings. *Toxicology Reports* 8, 295–302. https://doi.org/10.1016/j.toxrep.2021.01.013

Rahbari, A., Sinaki, J.M., Damavandi, A., Rezvan, S. 2021. Castor bean (*Ricinus communis* L.) responses to drought stress and foliar application of Zn-nano fertilizer and humic acid: grain yield, oil content, antioxidant activity, and photosynthetic pigments. *Notulae Botanicae Horti Agrobotanici Cluj-Napoca* 49(4), 12003. https://doi.org/10.15835/nbha49412003

Raliya, R., Biswas, P., Tarafdar, J.C. 2015. TiO_2 nanoparticle biosynthesis and its physiological effect on mung bean (*Vigna radiata* L.). *Biotechnology Reports* 5, 22–26. https://doi.org/10.1016/j.btre.2014.11.002

Rani, S., Kumari, N., Sharma, V. 2023. Biosynthesized zinc oxide nanoparticles as a novel nanofertilizer influence growth, yield, antioxidant enzymes, and arsenic accumulation in mungbean (*Vigna radiata* L.) under arsenic stress. *Journal of Soil Science and Plant Nutrition* 23(2), 2360–2380. https://doi.org/10.1007/s42729-023-01188-5

Rao, K.J., Paria, S. 2013. Use of sulfur nanoparticles as a green pesticide on *Fusarium solani* and *Venturia inaequalis* phytopathogens. *RSC Advances* 3(26), 10471–10478. https://doi.org/10.1039/C3RA40500A

Rossi, L., Fedenia, L.N., Sharifan, H., Ma, X., Lombardini, L. 2019. Effects of foliar application of zinc sulfate and zinc nanoparticles in coffee (*Coffea arabica* L.) plants. *Plant Physiology and Biochemistry* 135, 160–166. https://doi.org/10.1016/j.plaphy.2018.12.005

Saedi, F., Sirousmehr, A., Javadi, T. 2020. Effect of nano-potassium fertilizer on some morpho-physiological characters of peppermint (*Mentha piperita* L.) under drought stress. *Journal of Plant Research (Iranian Journal of Biology)* 33(1), 94–110.

Saffari, M.R., Jafarzadeh Kenarsari, M., Farnia, A., Sasani, S. 2023. Growth analysis and oil quality of canola (*Brassica napus* L.) treated with zinc nanochelate and zinc sulfate under different irrigation regimes. *Gesunde Pflanzen* 75(5), 1615–1624. https://doi.org/10.1007/s10343-022-00825-w

Sajyan, T.K., Naim, L., Sebaaly, Z., Rizkallah, J., Shaban, N., Sassine, Y.N. 2018. Alleviating the adverse effects of salinity stress on tomato crop (*Solanum lycopersicum*) using nano-fertilizer as foliar application. *Acta Horticulturae* 1253, 33–40. https://doi.org/10.17660/ActaHortic.2019.1253.5

Salem, N.M., Albanna, L.S., Awwad, A.M. 2016. Toxic heavy metals accumulation in tomato plant (*Solanum lycopersicum*). *ARPN Journal of Agricultural and Biological Science* 11, 399–404.

Salem, N.M., Albanna, L.S., Awwad, A.M., Ibrahim, Q.M., Abdeen, A.O. 2015. Green synthesis of nano-sized sulfur and its effect on plant growth. *Journal of Agricultural Science* 8(1), 188. https://doi.org/10.5539/jas.v8n1p188

Sassine, Y.N., Alturki, S.M., Germanos, M., Shaban, N., Sattar, M.N., Sajyan, T.K. 2020. Mitigation of salt stress on tomato crop by using foliar spraying or fertigation of various products. *Journal of Plant Nutrition* 43(16), 2493–2507. https://doi.org/10.1080/01904167.2020.1766079

Shah, A.A., Aslam, S., Akbar, M., Ahmad, A., Khan, W.U., Yasin, N.A., Ali, S. 2021. Combined effect of *Bacillus fortis* IAGS 223 and zinc oxide nanoparticles to alleviate cadmium phytotoxicity in *Cucumis melo*. *Plant Physiology and Biochemistry* 158, 1–12. https://doi.org/10.1016/j.plaphy.2020.10.035

Sidhu, A., Barmota, H., Bala, A. 2017. Antifungal evaluation studies of copper sulfide nano-aquaformulations and its impact on seed quality of rice (*Oryza sativa*). *Applied Nanoscience* 7, 681–689. https://doi.org/10.1007/s13204-017-0606-7

Sohrabi, Y., Omer, K.M., Yazdani, M., Sharifi Kalyani, F. 2023. The effect of nano-fertilizer of paulownia on morpho-physiological traits and dry matter yield of basil under different irrigation levels. *Journal of Plant Nutrition* 46(9), 1868–1888. https://doi.org/10.1080/01904167.2022.2105711

Soliman, A.S., El-Feky, S.A., Darwish, E. 2015. Alleviation of salt stress on *Moringa peregrina* using foliar application of nanofertilizers. *Journal of Horticulture and Forestry* 7(2), 36–47. https://doi.org/10.5897/JHF2014.0379

Song, J., Zhao, H., Zhao, G., Xiang, Y., Liu, Y. 2019. Novel semi-IPN nanocomposites with functions of both nutrient slow-release and water retention. 1. Microscopic structure, water absorbency, and degradation performance. *Journal of Agricultural and Food Chemistry* 67(27), 7587–7597. https://doi.org/10.1021/acs.jafc.9b00888

Tang, W., Yan, T., Wang, F., Yang, J., Wu, J., Wang, J., Yue, T., Li, Z. 2019. Rapid fabrication of wearable carbon nanotube/graphite strain sensor for real-time monitoring of plant growth. *Carbon* 147, 295–302. https://doi.org/10.1016/j.carbon.2019.03.038

Taran, N.Y., Gonchar, O.M., Lopatko, K.G., Batsmanova, L.M., Patyka, M.V., Volkogon, M.V. 2014. The effect of colloidal solution of molybdenum nanoparticles on the microbial composition in rhizosphere of *Cicer arietinum* L. *Nanoscale Research Letters* 9, 289. https://doi.org/10.1186/1556-276X-9-289

Wang, A., Kang, F., Wang, Z., Shao, Q., Li, Z., Zhu, G., Lu, J., Li, Y.Y. 2019a. Facile synthesis of nitrogen-rich carbon dots as fertilizers for mung bean sprouts. *Advanced Sustainable Systems* 3(3), 1800132. https://doi.org/10.1002/adsu.201800132

Wang, J.W., Grandio, E.G., Newkirk, G.M., Demirer, G.S., Butrus, S., Giraldo, J.P., Landry, M.P. 2019b. Nanoparticle-mediated genetic engineering of plants. *Molecular Plant* 12(8), 1037–1040. https://doi.org/10.1016/j.molp.2019.06.010

Wang, P., Lombi, E., Zhao, F.J., Kopittke, P.M. 2016. Nanotechnology: a new opportunity in plant sciences. *Trends in Plant Science* 21(8), 699–712. https://doi.org/10.1016/j.tplants.2016.04.005

Wang, Y., Yan, J., Wen, N., Xiong, H., Cai, S., He, Q., Hu, Y., Peng, D., Liu, Z., Liu, Y. 2020. Metal–organic frameworks for stimuli-responsive drug delivery. *Biomaterials* 230, 119619.

Wei, W., Lu, X., Wang, Z., Pérez, B., Liu, J., Wu, C., Dong, M., Feng, F., Mu, H., Guo, Z. 2017. Single-component solid lipid nanocarriers prepared with ultra-long chain amphiphilic lipids. *Journal of Colloid and Interface Science* 505, 392–401.

White, J.C., Gardea-Torresdey, J. 2018. Achieving food security through the very small. *Nature Nanotechnology* 13(8), 627–629.

Yin, J., Wang, Y., Gilbertson, L.M. 2018. Opportunities to advance sustainable design of nano-enabled agriculture identified through a literature review. *Environmental Science: Nano* 5(1), 11–26.

Yusefi-Tanha, E., Fallah, S., Rostamnejadi, A., Pokhrel, L.R. 2020. Zinc oxide nanoparticles (ZnO NPs) as a novel nanofertilizer: influence on seed yield and antioxidant defense system in soil-grown soybean (*Glycine max* cv. Kowsar). *Science of the Total Environment* 738, 140240.

Zhang, H., Zhang, H., Demirer, G.S., González-Grandío, E., Fan, C., Landry, M.P. 2020. Engineering DNA nanostructures for siRNA delivery in plants. *Nature Protocols* 15(9), 3064–3087.

Zhao, L., Lu, L., Wang, A., Zhang, H., Huang, M., Wu, H., Xing, B., Wang, Z., Ji, R. 2020. Nano-biotechnology in agriculture: use of nanomaterials to promote plant growth and stress tolerance. *Journal of Agricultural and Food Chemistry* 68(7), 1935–1947.

Zhong, X., Su, G., Zeng, Q., Li, G., Xu, H., Wu, H., Zhou, X. 2023. Preparation of salicylic acid-functionalized nanopesticides and their applications in enhancing salt stress resistance. *ACS Applied Materials & Interfaces* 15(37), 43282–43293.

8 Nano-Enabled Targeted Delivery of Agrochemicals

Mechanisms, Applications, and Future Prospects

*Ashutosh Sharma, Sonika Sharma, Gitika Kumawat, Priyanka Sharma, Jeenia Thalyari, Varsha Rattan, Pratibha Sharma, and Indu Sharma**

8.1 INTRODUCTION

Around 60% increase in food production is required by the year 2050, and the major challenge in achieving this is crop pests and diseases (Izuafa et al. 2025). Traditional agricultural practices may have ecological hazards, mainly due to the indiscriminate use of considerable quantities of broad-spectrum agrochemicals, of which only a fraction reaches the targets (Izuafa et al. 2025; Santana et al. 2020). For instance, only 30–50% nitrogen and 45% phosphorous applied are absorbed, while the remaining may result in off-target toxicity (Izuafa et al. 2025; Yadav 2025). To ensure global food security in a sustainable manner, more effective nutrient management and more precise plant protection interventions are the prerequisites. Further, to optimize the efficiency of agrochemicals and enhance productivity, strategies include precision nutrient management, integration of organic and inorganic fertilizers, use of biodegradable polymer coatings, and support for rhizosphere microbiota (Carmona et al. 2021; Nakachew et al. 2024). Moreover, continuous monitoring and evaluation of agrochemicals are also essential to ensure the effectiveness and improved decision-making. Recent research emphasizes the potential of nanotechnology to develop new nano-formulations of agrochemicals (like nano-fertilizers, nano-herbicides, nano-biostimulants, nano-pesticides, *etc.*), which enhance nutrient uptake and efficiency and mitigate adverse environmental impacts of pesticides (Chauhan et al. 2023; Preeti et al. 2024; Soni and Sharma 2023; Upasna et al. 2023). Therefore, the use of such nano-agrochemicals enables safer and sustainable food production with reduced chemical dosage and ecological hazards (Shraogi et al. 2025). Nano-sensors also aid in the rapid detection of pathogens and nutrient deficiencies, promoting proactive management.

DOI: 10.1201/9781032659435-8

Nano-herbicides have also been developed for sustainable weed management by enhancing herbicide delivery *via* nanoscale carriers, allowing for controlled release and increased solubility (Kusumavathi et al. 2025; Mohana et al. 2020). Conversely, nano-biostimulants combine biostimulants with nanoparticles (NPs), allowing for controlled nutrient release, reduced stress, improved nutrient uptake, better plant–microbe interactions, and improved nutrient availability in the rhizosphere (Eliaspour et al. 2020; Timmusk et al. 2018). Nano-pesticides utilize nanoscale materials to improve pesticide effectiveness, enhancing adhesion, penetration, and controlled release (Baker et al. 2017; Ganguli 2019; Kumar et al. 2019). They also have reduced application rates while improving the bioavailability and reducing ecological toxicities.

Nano-forms of macronutrients like nitrogen (N), phosphorus, (P) and potassium (K) may facilitate their controlled release with distinct kinetics compared to traditional fertilizers (Carmona et al. 2021). For example, calcium phosphate NPs or nanosized calcium (ortho)phosphate (CaP), including nanocrystalline apatite and amorphous calcium phosphate, demonstrated reduced solubility and dual delivery of P and N, thus conserving non-renewable phosphate resources. Besides, these NPs can also be modified with other nutrients and reveal higher nutrient use efficiency (NUE) in various crops (Carmona et al. 2022). The use of nano-herbicides reduces environmental impact and phytotoxicity risk while ensuring effective weed control at lower doses (Kusumavathi et al. 2025). The economic implications of nanomaterial production deserve careful consideration, especially given the tight profit margins in the agricultural industry, while also promoting an efficient and environmentally friendly manufacturing process (Carmona et al. 2022). However, scaling up the production of nano-fertilizers also remains a challenge, particularly in identifying different formulations that effectively impact crops (Majeed et al. 2024). Nevertheless, continuous supply of raw materials to meet market demand, along with the risk of overuse, raises environmental concerns. This chapter, therefore, compiles the recent advancements in nano-enabled targeted delivery of agrochemicals for sustainable agriculture.

8.2 NANO-ENABLED DELIVERY SYSTEMS

Nanoscale carriers provide a controlled and targeted delivery of agrochemicals, thereby minimizing their ecotoxicity (Izuafa et al. 2025; Nair et al. 2010). The integration of nanotechnology into agriculture has opened new paths for crop protection, nutrient management, and sustainable farming practices.

8.2.1 Nanomaterials and Nanocarriers

Nanomaterials are particles sized in the nanometre (10^{-9}m) range utilized in various sectors, including agriculture (An et al. 2022). They are now being explored not only as pesticide carriers, but also as active agents, plant growth regulators, fertilizers, soil conditioners, biosensors, and environmental remediation tools (Kumar et al. 2019). There are diverse uses of nanomaterials, which include carbon nanotubes (CNTs), cobalt oxide NPs, iron NPs, and silica NPs, each serving distinct purposes due to their unique properties, such as carriers (An et al. 2022; Kour et al. 2021; Singh et al. 2021). Organic materials, like polymers and liposomes, are also employed as nanocarriers, particularly for encapsulating pesticides. Collectively, these are known as nanostructured materials or engineered nanomaterials. Their targeted mode of action minimizes adverse effects on non-target organisms and ecosystems, while maximizing their efficiency (Kumar et al. 2019). The integration of nanomaterials with pesticides enhances their surface area and reactivity, promoting effective interactions in agricultural applications (Ahlawat et al. 2021). Various nanocarriers, like polymer-based formulations, nanodispersions, and other NPs, have been developed using physical and chemical methods depending on the pesticide properties. Polymer nanoproducts have garnered significant interest due to the associated benefits of improved solubility, controlled release, and reduced toxicity. Further, nanocarrier-based herbicides offer better protection of the active ingredient and enable site-specific, sustained availability (Mohana et al. 2020). A wide range of nanocarriers, including polymeric NPs, nano-clays, liposomes, silica NPs, nanoemulsions, and biopolymer-based systems, have been explored for this purpose. They improve leaf adhesion and penetration, facilitate targeted transport within weed tissues, and reduce off-target effects. Furthermore, stimulus-responsive nano-herbicides releasing the active ingredients in response to environmental signals like pH, moisture, or enzymatic activity provide new opportunities for precision weed control.

8.2.2 Active and Passive Targeting

NPs enable agrochemical delivery systems act *via* two major primary processes, namely, passive and active pathways (Figure 8.1). The passive targeting exploits natural pathways of plants, whereas active targeting deals with precision through biorecognition.

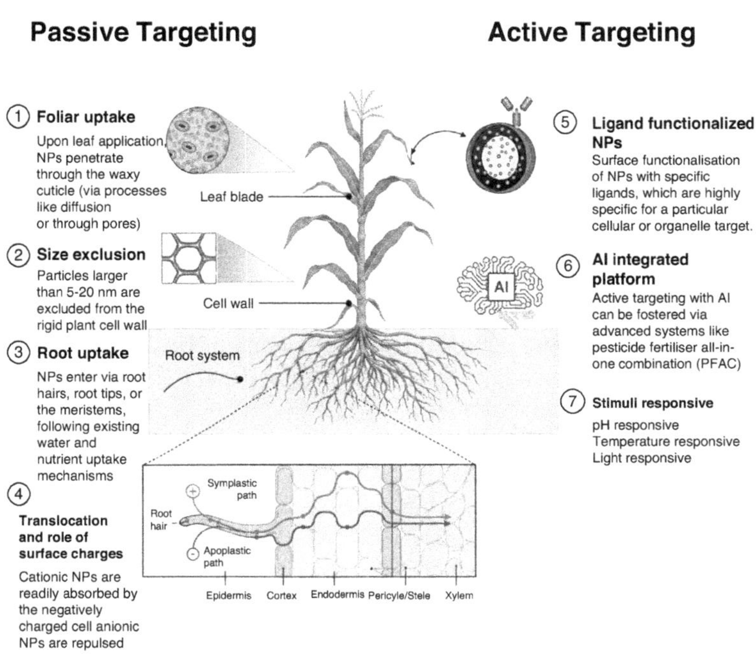

FIGURE 8.1 Nanoparticle-enabled agrochemical delivery systems *via* two primary processes, namely passive and active pathways.

8.2.2.1 Passive Targeting

Passive targeting is the movement of NPs mainly through roots and/or leaves, without the help of any external power or biorecognition motifs. Both the inherent physical and chemical characteristics of NPs and their interaction with the anatomical barriers of plants play a crucial role (Sembada and Lenggoro 2024).

8.2.2.1.1 Entry Routes and Size Exclusion Limits

The 'gateways' via which NPs enter the plant system are characterized by a size exclusion limit (SEL) as follows.

8.2.2.1.1.1 Root Uptake The existing water and nutrient uptake mechanisms are used by NPs to enter *via* root hairs, root tips, or rhizodermis (Sembada and Lenggoro 2024; Yadav 2025). Root uptake provides longer-lasting effects than foliar applications due to their reduced susceptibility to weathering (Sembada and Lenggoro 2024).

8.2.2.1.1.2 Foliar Uptake Upon leaf application, NPs enter either via the waxy cuticle by processes like diffusion or through pores of approximately 2 nm in size, i.e. the stomata, which have larger openings ranging from 3 to 12 μm in width (Sembada and Lenggoro 2024; Yadav 2025).

8.2.2.1.1.3 Physical Barriers Particles larger than 5–20 μm are excluded from the rigid plant cell wall. However, few studies have reported the entry of particles up to 50 nm via poorly understood mechanisms, such as temporary enlargement of pores (Cunningham et al. 2018; Nair et al. 2010).

8.2.2.1.2 Translocation and Role of Surface Charges

Upon internalization, NPs are translocated *via* the apoplastic and symplastic pathways. The apoplastic transport is characterized by the movement of NPs through the extracellular spaces between the cell walls and within the xylem (Sembada and Lenggoro 2024; Yadav 2025), whereas the symplastic transport involves movement through the cytoplasm of adjacent cells, facilitated by plasmodesmata (Yadav 2025). Moreover, the surface charges of NPs also play a critical role in dictating their mobility (Li et al. 2016; Yadav 2025). Cationic NPs are readily absorbed by negatively charged cell walls and mucosal secretions, which leads to high accumulation in roots but low translocation to other tissues. On the other hand, anionic NPs experience repulsive forces by the cell wall, which favours efficient bidirectional translocation from roots to shoots through the vascular system.

8.2.2.2 Active Targeting

Active targeting deals with surface functionalization of NPs with specific ligands such as antibodies, peptides, or signal molecules, which are highly specific for a particular cellular or organelle target. This strategy leads to more efficient localization and enhanced therapeutic efficacy of the encapsulated agrochemicals while minimizing their off-target exposure (Izuafa et al. 2025; Puangpathumanond et al. 2025; Santana et al. 2020).

8.2.2.2.1 Biorecognition Motifs for Cellular and Organelle Targeting

- **Stomata Targeting (SENDS)**: Puangpathumanond et al. (2025) reported a Surface ligand-Engineered NPs for targeted Delivery to Stomata (SENDS) system that makes use of antibodies to target arabinan, which is a pectic polysaccharide found abundantly in the cell walls of stomatal guard cells. This improves defence efficacy by 20-fold compared to non-targeted particles, by facilitating the direct delivery of antimicrobial agents to the primary entry points for many pathogens, such as *Xanthomonas campestris.*
- **Organelle Targeting**: It was suggested that chloroplast guiding peptides (derived from the Rubisco small subunit, RbcS) that mimic natural protein precursors can be coated on NPs for highly specific organelle targeting (Santana et al. 2020). These peptides allow NPs to specifically recognize and bind to the ToC (translocation of the outer membrane of chloroplasts) complex, which further helps in achieving colocalization rates of over 74% within the photosynthetic organelles.
- **Ligand Facilitated Internalization**: Li et al. (2016) suggested that endocytosis processes can further be strengthened by functionalizing surfaces of NPs with ligands like cysteine. Cysteine serves as a signal molecule that helps identify receptors on cell membranes, which significantly increases the internalization of gold NPs in crops like rice and tomato.

8.2.2.2.2 AI-Integrated and Stimuli-Responsive Platforms

Ji et al. (2021) showed how active targeting with AI can be achieved via advanced systems like Pesticide-Fertilizer All-in-one Combination (PFAC). These systems make use of deep learning algorithms to help identify and classify weeds in the field. Upon target identification, external triggers such as near-infrared (NIR) irradiation from drones help induce a phase change in nanocarriers, thus facilitating selective release of herbicides on weed surfaces. Figure 8.1 illustrates the delivery systems for NP-enabled agrochemicals through two primary processes: passive pathways and active pathways.

8.2.3 Controlled Release and Environmental Triggers

Stimuli-responsive release ensures that the active ingredients are released in response to certain environmental stimuli and supplied only when and where they are needed (Izuafa et al. 2025; Yadav 2025; Yin et al. 2023). These smart-carriers respond to environmental triggers like changes in pH, temperature, and light. Response to pH fluctuations has been reported in Chitosan-coated NPs, which can release copper ions specifically in acidic environments that are characteristic of fungal infections or in the alkaline midguts of pest larvae (Izuafa et al. 2025; Yin et al. 2023). Nanogels and composites can be specifically engineered to release cargos in response to temperature changes that represent the pest activity peaks or via photosensitive switches like azobenzene.

8.2.4 Nano-Enabled Delivery Systems Other than Nanoparticles

Despite spherical NPs being the most common, several other nano-enabled systems provide specialized delivery features. Cunningham et al. (2018) showed that nanotubes, particularly CNTs, are capable of delivering DNA and siRNA for gene silencing purposes without the need for an external mechanical force, thus serving as high-aspect-ratio molecular transporters. Yadav (2025) reported that the solubility of hydrophobic pesticides can be improved with the help of nanoemulsions, while liposomes (spherical lipid bilayers) help deliver micronutrients like iron and magnesium by fusing with plant cell membranes. Additionally, nanofibre arrays have been explored for the controlled integration of plasmid DNA into viable cells. Nanocapsules or coated liposomes offer protected environments for sensitive biological ingredients, thus ensuring successful targeting without premature degradation (Nair et al. 2010; Yadav 2025).

8.3 MECHANISMS OF CONTROLLED RELEASE

Nano-enabled delivery systems for agrochemicals aim to increase the efficiency of fertilizers and pesticides by releasing active ingredients in either a sustained or stimulus-responsive manner (Shao et al. 2022). A sustained released pattern helps in achieving higher nutrient uptake by the plant, thereby resulting in a higher NUE (Nawadkar et al. 2024). The release of cargo occurs in three different phases: the lag phase, during which water enters into the nanocarrier to dissolve the solid core; the constant release phase, where a saturated solution builds up and moves outwards due to a constant motive force; and the decay period, where the internal concentration decreases and the release rate slows down (Nawadkar et al. 2024).

8.3.1 Mechanisms of Nanoparticle-Mediated Release

NPs possess properties like high specific surface area, strong adsorption capacity, the ability to undergo surface modifications for enhanced targeted affinity, and the ability to be actively taken up by the cells and thus to achieve targeted delivery (Fincheira et al. 2023; Yin et al. 2023). The release mechanisms from these particles often use physical, chemical, and biological triggers (Figure 8.2).

8.3.1.1 Diffusion and Osmotic Swelling

The hydrostatic pressure in the case of polymeric NPs causes the polymer network to swell, resulting in an expansion in volume and causing the cargo to diffuse through relaxed polymer chains. This process involves two moving fronts – an inward-moving swelling interface and an outward-moving polymer interface. Both these together help release the cargo as the polymer chains relax.

8.3.1.2 Erosion and Degradation

Gradual erosion of the polymer matrix also serves as one of the release mechanisms. This type of release mechanism provides reproducible kinetics while simultaneously protecting sensitive active ingredients (Di Rago et al. 2025). In the

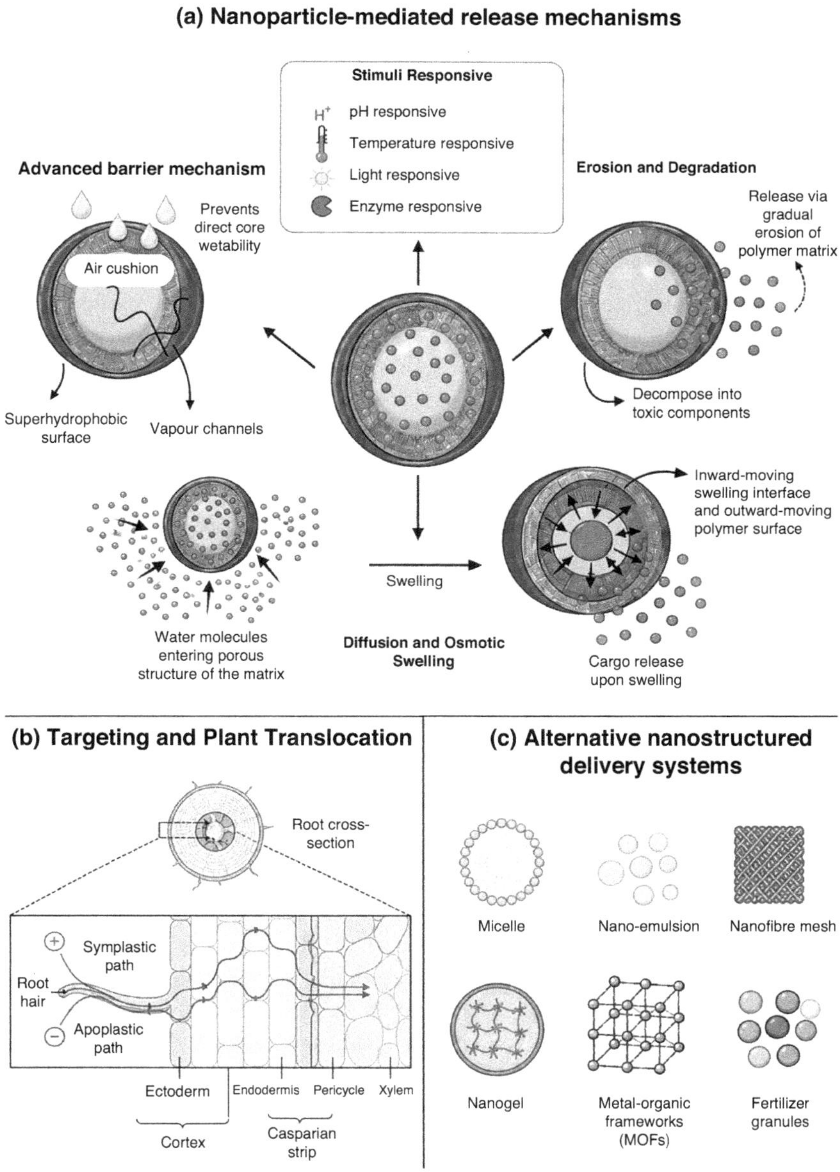

FIGURE 8.2 Illustration of the nano-enabled agrochemical delivery mechanism into plant tissues.

case of biodegradable NPs, like those synthesized from polylactic acid (PLA) or poly(lactic-co-glycolide) (PLGA), release occurs with the help of either hydrolysis by water or degradation by soil microorganisms. These NPs then further decompose into non-toxic components (Nawadkar et al. 2024; Shao et al. 2022).

8.3.1.3 Advanced Barrier Mechanisms

Smart NP designs include a unique multilayer air shield. Here, a superhydrophobic surface roughness creates an 'air cushion', which prevents the core from directly dissolving by water while also simultaneously allowing controlled vapour channels for release. A sustained release profile can be seen in the case of self-healing nanostructures, which use SA-modified particles to automatically repair micro-holes in the coating (Nawadkar et al. 2024; Shao et al. 2022).

8.3.1.4 pH Responsive

Camara et al. (2019) stated that ionizable functional groups like carboxyls (pKa ~4.8) or amines (pKa ~ 6.3) play a crucial role in pH-responsive systems. At specific pH levels, protonation or deprotonation occurs. This causes polymeric NPs to either swell or shrink, thereby opening mass-transfer channels (Camara et al. 2019; Shen et al. 2023). Chitosan-based NPs are protonated at acidic pH, which results in electrostatic repulsion that causes the polymer mesh to open for the release of cargo (Camara et al. 2019; Fincheira et al. 2023). Shen et al. (2023) reported that ZIF-8 NPs are stable at neutral pH, but decompose under acidic conditions, characteristic of fungal infection sites.

8.3.1.5 Temperature Responsive

Some NPs use polymers with a lower critical solution temperature (LCST), such as PNIPAm. Such NPs undergo structural changes at specific temperatures (Camara et al. 2019; Shen et al. 2023). Above the LCST, polymer chains collapse. Depending on whether the polymer is a gatekeeping shell or an internal matrix, either triggering or inhibition will occur (Shen et al. 2023).

8.3.1.6 Light Responsive

Such NPs help in the site-specific and time-controlled release of active ingredients upon light irradiation. These NPs can absorb light of different wavelengths in the UV, visible, and infrared regions (Camara et al. 2019). Release is triggered with the help of UV light by using photolabile moieties like azobenzene or o-nitrobenzyl. These undergo either structural isomerization or photolysis. On the other hand, photothermal materials like polydopamine (PDA) or graphene oxide (GO) lead to temperature-sensitive release by converting NIR light into heat (Camara et al. 2019; Shen et al. 2023).

8.3.1.7 Redox and Enzyme Responsive

Redox systems often rely on disulphide bond cleavage by glutathione. Glutathione is known to serve as an indicator of plant stress or biotic invasion (Camara et al. 2019; Shao et al. 2022; Shen et al. 2023). Enzyme-responsive systems, on the other hand, use nanocarriers composed of starch, pectin, or lipids. These act as substrates for specific enzymes that are produced by target pests (Camara et al. 2019; Shen et al. 2023). For example, starch-coated or cyclodextrin-functionalized NPs can be degraded by enzymes such as alpha-amylase, which are produced by pests. This ensures the release of pesticide only upon ingestion (Di Rago et al. 2025; Shen et al. 2023).

8.3.2 Targeting and Plant Translocation Mechanisms

For any delivery system to be considered effective, it must reach the intended target plant tissue. Yin et al. (2023) suggested that targeting can be enhanced via surface modifications of NPs to improve adhesion to leaves, thus resisting wash off by rainwater. Once applied, NPs enter plants via several pathways such as the apoplastic pathway, symplastic pathway, and foliar entry (Figure 8.2b).

8.3.3 Alternative Nanostructured Delivery Systems

In addition to spherical NPs, several other nanostructured systems provide controlled-release functionalities (Figure 8.2c). Di Rago et al. (2025) described micelles as self-assembled aggregates of amphiphilic molecules, which are highly effective sources for delivering lipophilic pesticides, thus providing improved wettability and anti-splash behaviour on foliage. Nanoemulsions (20–500 nm) facilitate deeper penetration into the crop leafage via enhanced dissolution of hydrophobic compounds in water-based assays (Di Rago et al. 2025; Fincheira et al. 2023). Further, nanofibres are frequently utilized for seed coatings and pheromone delivery. These are often produced via co-axial electrospinning and provide a high-porosity one-dimensional structure that facilitates superior gas circulation as compared to nano-coatings (Di Rago et al. 2025). Fincheira et al. (2023) described another alternative called 'nanogels', which are three-dimensional cross-linked polymer networks combining high water absorption capacity along with nanoscale delivery. This improves the target contact efficiency and UV protection.

8.4 APPLICATIONS OF NANO-ENABLED DELIVERY OF AGROCHEMICALS

Use of nanotechnology in agriculture has recently emerged to improve the efficiency, precision, and sustainability of agrochemical usage (Prasad et al. 2017; Chaud et al. 2021). Nano-enabled delivery systems overcome the inherent limitations of conventional agrochemicals like low solubility, rapid degradation, non-target toxicity, leaching, volatilization, and poor uptake efficiency by plants (An et al. 2022). They also enhance the efficiency and precision of crop inputs by improving stability, bioavailability, and controlled release of active ingredients. Certain nano-formulations have been used for spraying them on the crops at nano-concentrations (Sharma et al. 2022). Their applications include nano-pesticides (fungicides, insecticides, and bactericides), nano-fertilizers, nano-herbicides, and nano-biostimulants (including nano-PGRs), which reduce input doses, minimize environmental losses, and improve crop growth and protection (Table 8.1).

8.4.1 Nano-Pesticides (Fungicides, Insecticides, and Bactericides)

Nano-pesticides represent an advanced approach in pest management by utilizing nanoscale materials to improve the delivery and performance of active ingredients. The small size of NPs allows better adhesion to plant surfaces, enhanced

TABLE 8.1
Applications of Nano-Enabled Delivery of Agrochemicals in Agriculture

Nano-Enabled Input	Type/Examples	Key Functions	Major Advantages	Agricultural Applications
Nano-pesticides	Nano-fungicides, nano-insecticides, nano-bactericides	Targeted delivery, controlled release, enhanced penetration into pest/pathogen tissues	Reduced dose requirement, prolonged efficacy, lower environmental contamination	Management of fungal, insect, and bacterial diseases with higher precision
Nano-fertilizers	Nano-N, nano-P, nano-K, nano-micronutrients (Zn, Fe, Cu)	Slow and synchronized nutrient release, improved nutrient uptake	Higher nutrient use efficiency, reduced leaching and volatilization losses	Enhanced crop growth, yield improvement, soil fertility management
Nano-herbicides	Nano-encapsulated selective and non-selective herbicides	Site-specific weed targeting, controlled herbicide availability	Reduced phytotoxicity, minimized off-target effects	Efficient weed control with lower chemical load
Nano-biostimulants	Nano-organic compounds, nano-extracts	Enhancement of metabolic and physiological processes	Improved plant vigour, stress tolerance	Promotion of growth under abiotic and biotic stress conditions
Nano-PGRs	Nano-GA, nano-auxins, nano-cytokinins, nano-ethylene regulators	Regulation of cell division, elongation, flowering, and senescence	Enhanced bioavailability, reduced degradation	Improved seed germination, flowering, yield, and crop quality
Integrated nano-formulations	Combined nano-fertilizer + pesticide or PGR	Multifunctional delivery systems	Reduced number of applications, cost-effective	Precision and sustainable crop management systems

penetration into target organisms, and controlled or sustained release of pesticides over time (Baker et al. 2017). As a result, nano-pesticides achieve higher efficacy at lower application rates, reduce pesticide residues, and minimize environmental contamination. Nano-pesticides exhibit novel properties attributable to their size, structure, or surface chemistry (Ganguli 2019). Recent developments emphasize the green synthesis of nano-agrochemicals using biological systems such as plants, enzymes, and microorganisms. Biogenic synthesis offers superior control over NP size, morphology, and surface properties while avoiding hazardous chemicals commonly used in physical and chemical synthesis methods. In plant-mediated synthesis, phytochemicals present in plant extracts act as reducing and stabilizing agents (Sharma et al. 2025). These biomolecules donate electrons to metal ions, reducing them into stable NPs under controlled conditions of temperature, pH, and reactant concentration. Furthermore, NPs can be surface-functionalized with polymers or bioactive ligands to enhance stability, solubility, controlled release, and target specificity (An et al. 2022).

Nano-pesticide delivery systems include different mechanisms such as nanospheres, nanocapsules, nanocontainers, nanoemulsions, nanogels, liposomes, and inorganic nanocarriers. Among these, nanocapsules, nanogels, and micelles are particularly effective for controlled-release applications (Kah et al. 2013). Polymer- and clay-based nanocarriers are widely preferred due to their biocompatibility, low cost, and stimulus-responsive behaviour. Natural polymers such as chitosan, cellulose, and polylactide are commonly used for encapsulating active ingredients like avermectin, atrazine, and glyphosate (Wang et al. 2022).

Bacillus thuringiensis (*Bt*) is a widely used microbial insecticide effective against lepidopteran and coleopteran pests and is considered safe for non-target organisms. However, its field performance is often limited by susceptibility to ultraviolet radiation, rainfall, and digestive enzymes, leading to reduced persistence. Nanotechnology has significantly improved *Bt* formulations by enhancing UV resistance and environmental stability. Nanomaterials such as nano-$Mg(OH)_2$ adsorb *Bt* toxins (e.g. Cry11Aa), improving UV protection and facilitating enzymatic activation in the insect midgut. Similarly, immobilization of *Bt* chitinase on nanosilica and incorporation of nano-TiO_2, nano-ZnO, and nano-SiO_2 into microcapsule formulations have resulted in prolonged insecticidal activity (Pan et al. 2017).

Avermectin is a broad-spectrum antiparasitic and bactericidal agent used against nematodes and arthropods. Despite its effectiveness, avermectin suffers from poor solubility, rapid degradation, and short residual activity. Nanosilica-based delivery systems have emerged as efficient nanocarriers that enhance avermectin stability, enable controlled release, and improve field efficacy (Yang et al. 2022).

8.4.2 Nanobotanical Insecticides

Nanotechnology, particularly nanoemulsion systems, has addressed the problems of poor stability and rapid degradation of botanical pesticides by enhancing solubility, chemical stability, and persistence. Essential oil-based nanoemulsions have demonstrated broad-spectrum insecticidal activity. Neem oil nanoemulsions formulated with plant-based adjuvants have shown high efficacy against *Bemisia tabaci*, while tobacco-based nanoemulsions have exhibited superior aphid control compared to crude extracts, underscoring the potential of nanotechnology in botanical pest management (Pavoni et al. 2019).

8.4.3 Nanotechnology in Fungicides

Nanotechnology provides eco-friendly solutions through the development of nano-biofungicides, which exhibit enhanced bioactivity, stability, and targeted delivery (Alghuthaymi et al. 2015). Biologically synthesized silver NPs have shown strong antifungal activity against *Fusarium* spp. Similarly, phytofabricated ZnO NPs, ZnO-based nano-biohybrids, and combinations of beneficial microbes with metal oxide NPs have demonstrated effective disease suppression, including powdery mildew and leaf spot diseases. Chitin- and chitosan-based nanocarriers enable stimulus-responsive controlled release of fungicides. Chitosan-based nanocomposites (e.g. chitosan–copper, chitosan–silver) disrupt fungal membranes and inhibit mycelial development (Liang et al. 2018). Advanced nanocarriers like porous silica, nanosized calcium carbonate, and multifunctional nanocapsules have been used to deliver fungicides like validamycin, thifluzamide, and kasugamycin, improving foliar retention, photostability, and synergistic effects.

8.4.4 Nanotechnology in Bactericides

Nanotechnology has significantly enhanced the performance of bactericides by improving stability, delivery, and antibacterial potency. Encapsulation of essential oils (peppermint, green tea, and lemongrass) in chitosan or polymer-based NPs has preserved phenolic content and increased antibacterial activity against *Staphylococcus aureus* and *Escherichia coli*. Chitosan–metal composites and Cu NP-based nanogels have demonstrated effective control of plant pathogens such as *Pseudomonas syringae* with improved safety profiles (Ma et al. 2022). Pure nanomaterials such as silver, zinc oxide, copper oxide, and magnesium oxide NPs also exhibit strong intrinsic antibacterial activity. Green-synthesized silver NPs show reduced phytotoxicity and long-term antimicrobial effects, making them suitable as nano-pesticides or growth regulators. Shape-dependent antibacterial activity has been observed in CuO NPs, while biosynthesized MgO NPs effectively suppress rice bacterial pathogens. Core–shell ZnO@ZIF-8 nanospheres loaded with berberine have shown synergistic antibacterial effects through ROS generation, DNA damage, and membrane disruption (Liang et al. 2022).

8.4.5 Nano-Fertilizers

Nano-fertilizers are nanoscale nutrient carriers that are more efficient than traditional fertilizer options (Abhiram 2023; Sharma et al. 2022). They consist of nutrient formulations encapsulated or coated with nanomaterials, allowing for controlled nutrient release and gradual dispersion in the soil (Jithendar et al. 2024). Nanomaterials used for coating nutrient formulations commonly include silica, iron, titanium dioxide, zinc oxide, aluminium oxide, gold nanorods, cerium oxide, ZnCdSe/ZnS and InP/

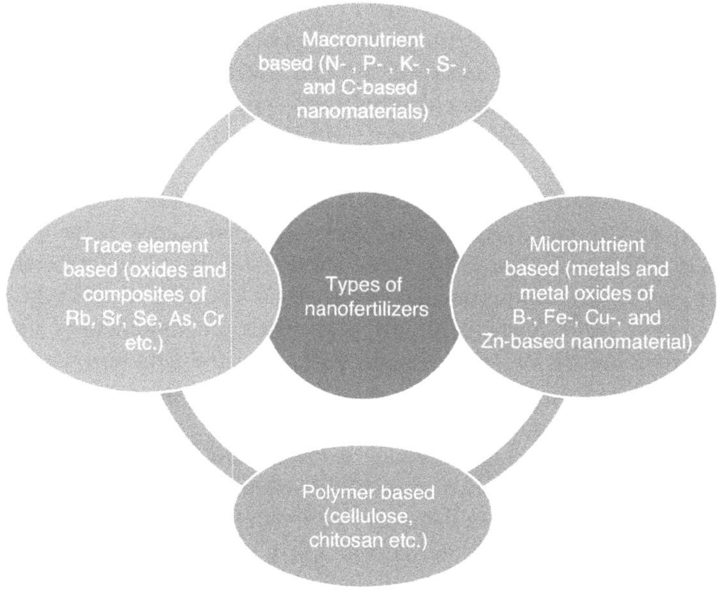

FIGURE 8.3 Different types of nano-fertilizers.

ZnS core–shell structures, and Mn/ZnSe quantum dots. The performance of these nano-fertilizers in enhancing plant growth is strongly influenced by factors such as the size, composition, concentration, and chemical behaviour of these nanomaterials, as well as specific crop species. Upon application, nutrients are gradually released into the soil when the NP suspensions in the nano-fertilizers interact with water (Singh et al. 2021a). Nano-fertilizers can be categorized according to their mode of action, nutrient composition, and physical consistency. These categories include controlled-release nano-fertilizers, targeted-delivery nano-fertilizers, plant growth-promoting nano-fertilizers, and formulations aiming to reduce water and nutrient losses. They also encompass inorganic, organic, and hybrid nano-fertilizers; nutrient-loaded nano-fertilizers; and consistency-based types such as surface-coated, synthetic polymer-coated, biologically coated, and nanocarrier-based nano-fertilizers (Yadav et al. 2023). The major types of nano-fertilizers are macronutrient-based nano-fertilizers, micronutrient-based nano-fertilizers, polymer-based nano-fertilizers, zeolite-based nano-fertilizers (Sharma et al. 2022a), carbon-based nano-fertilizers, and trace-element-based nano-fertilizers (Figure 8.3).

8.4.6 Nano-Herbicides

Nano-herbicides have the ability to improve the efficiency of herbicide delivery (Srimathi et al. 2025), reduce dosage requirements, and minimize off-target toxicity, thus making them promising alternatives to conventional agrochemicals. Nano-enabled herbicide delivery systems have emerged as an advanced strategy for improving weed management while minimizing adverse environmental impacts

(Kusumavathi et al. 2025). The controlled delivery of nano-herbicide formulations minimizes losses due to spray drift, photodegradation, volatilization, and leaching, thereby reducing the contamination of soil and water resources. The gradual and site-specific release of herbicides enhances weed control efficacy at lower doses and reduces the risk of phytotoxicity to crops, contributing to more efficient and sustainable weed management practices. Nano-herbicides can be engineered from organic, inorganic, or hybrid nanomaterials, each offering unique physicochemical properties such as high surface area, tunable size, and controlled-release behaviour. Several studies have demonstrated superior weed suppression by nano-formulated herbicides compared to their conventional counterparts (Oliveira et al. 2015; Pontes et al. 2021; Sousa et al. 2018; Xiang et al. 2017). However, a comprehensive understanding of their mechanisms of action, uptake pathways, and long-term environmental risks remains limited, emphasizing the need for detailed risk assessment and plant–nanomaterial interaction studies.

Conventional herbicides are known to pose ecological risks by affecting non-target plant species and associated wildlife (Brühl et al. 2013; Kalia and Gosal 2011). In this context, nano-enabled and biologically derived herbicide systems have gained attention as safer alternatives. Early developments in bioherbicides, particularly mycoherbicides, laid the foundation for environmentally compatible weed control, while recent advances have focused on nano-based carriers for both chemical and bio-based herbicides. For instance, chitosan and alginate NPs have been successfully used to deliver paraquat, improving efficacy while potentially reducing environmental exposure (Maruyama et al. 2016). The enhanced performance of nano-herbicides has been attributed to improved interaction between nanocarriers and plant tissues, influencing penetration through leaves, roots, and seeds, as well as intracellular transport and target site activity. Studies have shown stronger growth inhibition of weeds such as cogon grass using diuron loaded onto chitosan-based NPs compared to conventional formulations (Chen et al. 2018). Despite these advances, challenges remain in achieving precise subcellular targeting and understanding the fate of nano-herbicides in target and non-target organisms. The development of smart nano-herbicides incorporating biorecognition elements has been suggested as a promising direction to overcome these limitations (Santana et al. 2020). Overall, nano-herbicides support the principles of precision agriculture by enabling site-specific application. Through integration with nano-sensors or remote sensing technologies, weed infestations can be detected early, and nano-herbicides can be applied only to the affected areas (El-Shal et al. 2022; Nongbet et al. 2022).

8.4.7 Nano-Biostimulants (Including Nano-PGRs)

The combination of biostimulants with NPs leads to the development of nano-biostimulants. This approach typically involves encapsulating a biostimulant within an appropriate nanomaterial. Nano-biostimulants help alleviate both biotic and abiotic stresses in plants. They also enhance nutrient uptake efficiency from the soil surface, thereby reducing the dependence on chemical fertilizers (Eliaspour et al. 2020). Furthermore, NPs significantly influence plant–microbe interactions directly by improving nutrient availability in the rhizosphere and indirectly by stimulating microbial community (Timmusk et al. 2018). Biostimulants can be regarded as fertilizer-like agents because of their multiple beneficial effects on crop growth and productivity. They enhance plant stress resistance and promote development by improving nutrient uptake efficiency (du Jardin 2015).

8.5 BENEFITS OF NANO-ENABLED TARGETED DELIVERY

Nano-enabled delivery systems are more target specific and have resulted in the reduction in the dose of agrochemicals, which could be an environmental contaminant, thereby leading to their lesser bio-accumulation in the ecosystem (An et al. 2022; Fincheira et al. 2023). This reduction in the dosage is also linked to the reduction in farm expenses by making crop-production and crop-protection interventions involving the usage of such agrochemicals more cost-effective (Marella et al. 2021; Ullah et al. 2024). Besides decreasing the cost, such nano-formulations of agrochemicals also improve their bioavailability to plants, which is linked to the increased uptake of nano-formulations by the plant owing to their physicochemical characteristics like increased surface area and enhanced solubility, aiding in better tissue penetration (Yadav 2025). One more reason to improve the bioavailability of agrochemicals in their nano-formulations is that they are less prone to degradation and are more targeted with prolonged efficacy, especially in the case of nano-pesticides (Fincheira et al. 2023; Wang et al. 2022). On the environmental front, agrochemical nano-formulations have lower detrimental effects than their conventional counterparts, which may minimize their exposure to non-target species, including beneficial ones (PGPR, biocontrol agents, wildlife, and even humans) and have also minimized ecotoxicological hazards (Chaud et al. 2021; Mathew et al. 2025). Recently, some technological interventions have led towards the development of macrocycle-based supramolecular and stimuli-responsive nano-agrochemicals that can release agrochemicals in response to certain environmental signals, aligning delivery with the plant's requirements (Song et al. 2025).

8.6 CHALLENGES IN WIDESPREAD ADOPTION OF NANO-AGROCHEMICALS

Nanotechnology is transforming sustainable agriculture by improving agrochemical precision, enhancing soil health, and facilitating real-time monitoring with nano-sensors (Shraogi et al. 2025). Various challenges, such as toxicity and environmental concerns related to synthetic nanomaterials, remain, highlighting the need for further research and regulations. This chapter explored the role of nanotechnology in addressing agricultural challenges and its potential for sustainable food production and security. The production of nano-based agrochemicals involves high costs and raises several biosafety concerns among the public. This situation necessitates careful evaluation of safety, production costs, and public perceptions. Other safety issues include the potential impact on non-target organisms and the possible alteration of microbial diversity in the soil. Current assessments are mainly conducted in laboratory settings and lack real-world context. It is suggested that combining nutrient cycling and crop productivity models with nano-informatics could enhance the design of nanoscale agrochemicals (An et al. 2022). Furthermore, extensive field testing is essential to understand both the effectiveness and the environmental risks associated with these substances. More research is needed to weigh the benefits against the emerging risks and large-scale applications of nano-agrochemicals.

8.7 FUTURE PROSPECTS AND EMERGING TRENDS

Future research on nano-agrochemicals will mainly focus on improving their performance and decreasing their cost of production (An et al. 2022). Further, there is a need to set standards for their quality and safety, develop controlled-release systems responsive to environmental changes, and create guidelines for large-scale applications. For their widespread adoption among the farming community, the appropriate knowledge about their usage may be disseminated through agricultural extension workers. Recently, the nano-based delivery technology has entered the field of food production too; therefore, more safety and toxicological evaluations on humans and livestock are also needed. Further, for their large-scale adoption, the scaling up of their production technology is needed so that their use becomes more economical to farmers. Moreover, long-term environmental risk assessments that study their effect on environment are needed. Since the industrial production of nano-agrochemicals is influenced by policy factors, support from various governments is needed in the initial stage of their popularization, especially in agrarian economies. Further, drafting and implementing global supportive policies and public education will be essential for the advancements and implications of nano-agrochemicals for sustainable agriculture.

NOTE:

* Corresponding author.

REFERENCES

Abhiram, G. 2023. Contributions of nano-nitrogen fertilizers to sustainable development goals: a comprehensive review. *Nitrogen* 4(4), 397–415.

Ahlawat, S., Malik, K., Kumari, N., Malik, K., Sandeep, A. 2021. Nano-enabled pesticides in agriculture: budding opportunities and challenges. *Journal of Nanoscience and Nanotechnology* 21(6), 3337–3350.

Alghuthaymi, M.A., Almoammar, H., Rai, M., Said-Galiev, E., Abd-Elsalam, K.A. 2015. Myconanoparticles: synthesis and their role in phytopathogens management. *Biotechnology & Biotechnological Equipment* 29(2), 221–236.

An, C., Sun, C., Li, N., Huang, B., Jiang, J., Shen, Y. et al. 2022. Nanomaterials and nanotechnology for the delivery of agrochemicals: strategies towards sustainable agriculture. *Journal of Nanobiotechnology* 20(1), 11.

Baker, S., Volova, T., Prudnikova, S.V., Satish, S., Prasad, N. 2017. Nanoagroparticles emerging trends and future prospect in modern agriculture system. *Environmental Toxicology and Pharmacology* 53, 10–17.

Brühl, C.A., Schmidt, T., Pieper, S., Alscher, A. 2013. Terrestrial pesticide exposure of amphibians: an underestimated cause of global decline? *Scientific Reports* 3(1), 1135.

Camara, M.C., Campos, E.V.R., Monteiro, R.A., do Espirito Santo Pereira, A., de Freitas Proença, P.L., Fraceto, L.F. 2019. Development of stimuli-responsive nano-based pesticides: emerging opportunities for agriculture. *Journal of Nanobiotechnology* 17(1), 100.

Carmona, F.J., Dal Sasso, G., Ramírez-Rodríguez, G.B., Pii, Y., Delgado-López, J.M., Guagliardi, A., Masciocchi, N. 2021. Urea-functionalized amorphous calcium phosphate nanofertilizers: optimizing the synthetic strategy towards environmental sustainability and manufacturing costs. *Scientific Reports* 11(1), 3419.

Carmona, F.J., Guagliardi, A., Masciocchi, N. 2022. Nanosized calcium phosphates as novel macronutrient nano-fertilizers. *Nanomaterials* 12(15), 2709.

Chaud, M., Souto, E.B., Zielinska, A., Severino, P., Batain, F., Oliveira-Junior, J., Alves, T. 2021. Nanopesticides in agriculture: benefits and challenge in agricultural productivity, toxicological risks to human health and environment. *Toxics* 9(6), 131.

Chauhan, A., Pallvi, R.P., Ludarmani, S.A. 2023. Effect of pre-soaking of potato tubers in nano-urea and nano-zinc on growth, quality and yield. *Journal of Pharmacy Innovation* 12(7), 980–995.

Chen, C., Zhang, G., Dai, Z., Xiang, Y., Liu, B., Bian, P., Zheng, K., Wu, Z., Cai, D. 2018. Fabrication of light-responsively controlled-release herbicide using a nanocomposite. *Chemical Engineering Journal* 349, 101–110.

Cunningham, F.J., Goh, N.S., Demirer, G.S., Matos, J.L., Landry, M.P. 2018. Nanoparticle-mediated delivery towards advancing plant genetic engineering. *Trends in Biotechnology* 36(9), 882–897.

Di Rago, S., Morandi, F., Pizzetti, F., Rossi, F. 2025. Advances in controlled release systems for sustainable crop production: a review of nano-, micro-, and macro-formulations. *Advanced Agrochem* 4(3), 188–206.

Du Jardin, P. 2015. Plant biostimulants: definition, concept, main categories and regulation. *Scientia Horticulturae* 196, 3–14.

Eliaspour, S., Seyed Sharifi, R., Shirkhani, A., Farzaneh, S. 2020. Effects of biofertilizers and iron nano-oxide on maize yield and physiological properties under optimal irrigation and drought stress conditions. *Food Science & Nutrition* 8(11), 5985–5998.

El-Shal, R.M., El-Naggar, A.H., El-Beshbeshy, T.R., Mahmoud, E.K., El-Kader, N.I.A., Missaui, A.M., Du, D., Ghoneim, A.M., El-Sharkawy, M.S. 2022. Effect of nano-fertilizers on alfalfa plants grown under different salt stresses in hydroponic system. *Agriculture* 12(8), 1113.

Fincheira, P., Hoffmann, N., Tortella, G., Ruiz, A., Cornejo, P., Diez, M.C., Seabra, A.B., Benavides-Mendoza, A., Rubilar, O. 2023. Eco-efficient systems based on nanocarriers for the controlled release of fertilizers and pesticides: toward smart agriculture. *Nanomaterials* 13(13), 1978.

Ganguli, P. 2019. Patenting issues in the development of nanobiopesticides. In: Koul, O. (ed), *Nano-Biopesticides Today and Future Perspectives*, pp. 367–395. Academic Press, New York.

Izuafa, A., Chimbekujwo, K.I., Raji, R.O., Oyewole, O.A., Oyewale, R.O., Abioye, O.P. 2025. Application of nanoparticles for targeted management of pests, pathogens and disease of plants. *Plant Nano Biology* 13, 100177.

Ji, Y., Ma, S., Lv, S., Wang, Y., Lu, S., Liu, M. 2021. Nanomaterials for targeted delivery of agrochemicals by an all-in-one combination strategy and deep learning. *ACS Applied Materials & Interfaces* 13(36), 43374–43386.

Jithendar, B., Kumar, R., Rana, N. 2024. Revolutionizing crop nutrition: exploring nano fertilizers in agriculture. *International Journal of Plant & Soil Science* 36(6), 327–339.

Kah, M., Beulke, S., Tiede, K., Hofmann, T. 2013. Nanopesticides: state of knowledge, environmental fate, and exposure modeling. *Critical Reviews in Environmental Science and Technology* 43(16), 1823–1867.

Kalia, A., Gosal, S.K. 2011. Effect of pesticide application on soil microorganisms. *Archives of Agronomy and Soil Science* 57(6), 569–596.

Kour, J., Khanna, K., Sharma, P., Arora, P., Dhiman, S., Kaur, R., Sharma, A., Ohri, P., Bhardwaj, R. 2021. Variability, behaviour and impact of nanoparticles in the environment. In: Singh, V.P., Singh, S., Tripathi, D.K., Prasad, S.M., Chauhan, D.K. (eds), *Plant Responses to Nanomaterials: Recent Interventions, and Physiological and Biochemical Responses*, pp. 315–328. Springer International Publishing, Cham.

Kumar, S., Nehra, M., Dilbaghi, N., Marrazza, G., Hassan, A.A., Kim, K.H. 2019. Nano-based smart pesticide formulations: emerging opportunities for agriculture. *Journal of Controlled Release* 294, 131–153.

Kusumavathi, K., Rautaray, S.K., Sarkar, S., Dash, S., Sahoo, T.R., Swain, S.K., Sethi, D. 2025. Nano-herbicides a sustainable strategy for weed control. *Plant Nano Biology* 11, 100132.

Li, H., Ye, X., Guo, X., Geng, Z., Wang, G. 2016. Effects of surface ligands on the uptake and transport of gold nanoparticles in rice and tomato. *Journal of Hazardous Materials* 314, 188–196.

Liang, R., Li, X., Yuan, W., Jin, S., Hou, S., Wang, M., Wang, H. 2018. Antifungal activity of nanochitin whisker against crown rot diseases of wheat. *Journal of Agricultural and Food Chemistry* 66(38), 9907–9913.

Liang, W., Cheng, J., Zhang, J., Xiong, Q., Jin, M., Zhao, J. 2022. pH-Responsive on-demand alkaloids release from core–shell ZnO@ ZIF-8 nanosphere for synergistic control of bacterial wilt disease. *ACS Nano* 16(2), 2762–2773.

Ma, X., Zhu, X., Qu, S., Cai, L., Ma, G., Fan, G., Sun, X. 2022. Fabrication of copper nanoparticle composite nanogel for high-efficiency management of *Pseudomonas syringae* pv. *tabaci* on tobacco. *Pest Management Science* 78(5), 2074–2085.

Majeed, Z., Manzoor, E., Javed, M.R. 2024. Calcium phosphates nanofertilizers: synthesis and biosafety. In: Abd-Elsalam, K.A. (ed), *Nanofertilizer Synthesis*, pp. 129–143. Elsevier, Amsterdam, Netherlands.

Marella, S., Kumar, A.N., Tollamadugu, N.P. 2021. Nanotechnology-based innovative technologies for high agricultural productivity: opportunities, challenges, and future perspectives. *Recent Developments in Applied Microbiology and Biochemistry* 2021, 211–220.

Maruyama, C.R., Guilger, M., Pascoli, M., Bileshy-José, N., Abhilash, P.C., Fraceto, L.F., De Lima, R. 2016. Nanoparticles based on chitosan as carriers for the combined herbicides imazapic and imazapyr. *Scientific Reports* 6(1), 19768.

Mathew, J.T., Hussaini, J., Inobeme, A., Adetunji, C.O., Tanko, M.S. 2025. Nanocarriers for delivering pesticides. In: Chen, J. (ed), *Nanocarriers in Plant Science and Agriculture*, pp. 273–296. IGI Global Scientific Publishing, Hershey, PA.

Mohana, N.C., Mithun, P.R., Rao, H.Y., Mahendra, C., Satish, S. 2020. Nanoparticle applications in sustainable agriculture, poultry, and food: trends and perspective. In: Susai, R., Anita, M. (eds), *Nanotoxicity*, pp. 341–353. Elsevier, Amsterdam, Netherlands.

Nair, R., Varghese, S.H., Nair, B.G., Maekawa, T., Yoshida, Y., Kumar, D.S. 2010. Nanoparticulate material delivery to plants. *Plant Science* 179(3), 154–163.

Nakachew, K., Yigermal, H., Assefa, F., Gelaye, Y., Ali, S. 2024. Review on enhancing the efficiency of fertilizer utilization: strategies for optimal nutrient management. *Open Agriculture* 9(1), 20220356.

Nawadkar, A., Chaskar, J., Tiwari, A., Chaskar, A. 2024. Controlled release of nanofertilizers: a promising approach for efficient delivery of nutrients in plants. *Journal of Crop Improvement* 38(3), 189–218.

Nongbet, A., Mishra, A.K., Mohanta, Y.K., Mahanta, S., Ray, M.K., Khan, M., Baek, K.H., Chakrabartty, I. 2022. Nanofertilizers: a smart and sustainable attribute to modern agriculture. *Plants* 11(19), 2587.

Oliveira, H.C., Stolf-Moreira, R., Martinez, C.B.R., Grillo, R., de Jesus, M.B., Fraceto, L.F. 2015. Nanoencapsulation enhances the post-emergence herbicidal activity of atrazine against mustard plants. *PLoS One* 10(7), e0132971.

Pan, X., Xu, Z., Li, L., Shao, E., Chen, S., Huang, T. et al. 2017. Adsorption of insecticidal crystal protein Cry11Aa onto nano-Mg $(OH)_2$: effects on bioactivity and anti-ultraviolet ability. *Journal of Agricultural and Food Chemistry* 65(43), 9428–9434.

Pavoni, L., Pavela, R., Cespi, M., Bonacucina, G., Maggi, F., Zeni, V., Canale, A., Lucchi, A., Bruschi, F., Benelli, G. 2019. Green micro-and nanoemulsions for managing parasites, vectors and pests. *Nanomaterials* 9(9), 1285.

Pontes, M.S., Antunes, D.R., Oliveira, I.P., Forini, M.M., Santos, J.S., Arruda, G.J., Caires, A.R., Santiago, E.F., Grillo, R. 2021. Chitosan/tripolyphosphate nanoformulation carrying paraquat: insights on its enhanced herbicidal activity. *Environmental Science: Nano* 8(5), 1336–1351.

Prasad, R., Bhattacharyya, A., Nguyen, Q.D. 2017. Nanotechnology in sustainable agriculture: recent developments, challenges, and perspectives. *Frontiers in Microbiology* 8, 1014.

Preeti, S.N., Maurya, V., Sharma, S., Kumar, R., Sharma, A. 2024. Effect of seed priming by nano-urea and nano-zinc on growth and yield of rat-tail radish (*Raphanus sativus* var. *caudatus*). *International Journal of Research in Agronomy* 7(4), 79–90.

Puangpathumanond, S., Chee, H.L., Sevencan, C., Yang, X., Lau, O.S., Lew, T.T.S. 2025. Stomata-targeted nanocarriers enhance plant defense against pathogen colonization. *Nature Communications* 16(1), 4816.

Santana, I., Wu, H., Hu, P., Giraldo, J.P. 2020. Targeted delivery of nanomaterials with chemical cargoes in plants enabled by a biorecognition motif. *Nature Communications* 11(1), 2045.

Sembada, A.A., Lenggoro, I.W. 2024. Transport of nanoparticles into plants and their detection methods. *Nanomaterials* 14(2), 131.

Shao, C., Zhao, H., Wang, P. 2022. Recent development in functional nanomaterials for sustainable and smart agricultural chemical technologies. *Nano Convergence* 9(1), 11.

Sharma, B., Lakra, U., Sharma, R., Sharma, S.R. 2022. A comprehensive review on nanopesticides and nanofertilizers-A boon for agriculture. *Nano-Enabled Agrochemicals in Agriculture* 2022, 273–290.

Sharma, D., Sharma, A., Gautam, H.R. 2025. Green synthesized nanoparticles for disease management in vegetable crops: a review. *Plant Nano Biology* 13, 100179. https://doi.org/10.1016/j.plana.2025.100179

Sharma, V., Javed, B., Byrne, H., Curtin, J., Tian, F. 2022a. Zeolites as carriers of nano-fertilizers: from structures and principles to prospects and challenges. *Applied Nano* 3(3), 163–186.

Shen, M., Liu, S., Jiang, C., Zhang, T., Chen, W. 2023. Recent advances in stimuli-response mechanisms of nano-enabled controlled-release fertilizers and pesticides. *Eco-Environment & Health* 2(3), 161–175.

Shraogi, N., Singh, A., Verma, R., Chakraborty, C., Mishra, V., Kushwaha, S., Ghosh, D., Patnaik, S. 2025. Recent trends in nano-enabled solutions for sustainable agriculture and global food security. In: Thakur, S. (ed), *Application of Nanotechnology for Sustainable Agriculture and Food Security Volume 2: Innovations in Pre-and Post-Harvest Crop Management*, pp. 125–154. American Chemical Society, Washington, DC.

Singh, A.D., Sharma, I., Khanna, K., Handa, N., Kumar, P., Ibrahim, M., Ohri, P., Bhardwaj, R. 2021. Impact of cobalt oxide nanoparticles on the morpho-physiological and biochemical response in plants. In: Singh, V.P., Singh, S., Tripathi, D.K., Prasad, S.M., Chauhan, D.K. (eds), *Plant Responses to Nanomaterials: Recent Interventions, and Physiological and Biochemical Responses*, pp. 249–267. Springer International Publishing, Cham.

Singh, H., Sharma, A., Bhardwaj, S.K., Arya, S.K., Bhardwaj, N., Khatri, M. 2021a. Recent advances in the applications of nano-agrochemicals for sustainable agricultural development. *Environmental Science: Processes & Impacts* 23(2), 213–239.

Song, Y.H., Bian, Q., Wang, F., Liu, J., Yang, Y.H., Zhang, Y.M., Liu, Y. 2025. Water-soluble stimuli-responsive supramolecular nanoagrochemicals based on macrocycle compounds. *Coordination Chemistry Reviews* 524, 216299.

Soni, S., Sharma, I. 2023. Effect of indole-3-acetic acid and nano-urea on agronomic properties of radish crops. *International Journal of Plant & Soil Science* 35(20), 269–276.

Sousa, G.F., Gomes, D.G., Campos, E.V., Oliveira, J.L., Fraceto, L.F., Stolf-Moreira, R., Oliveira, H.C. 2018. Post-emergence herbicidal activity of nanoatrazine against susceptible weeds. *Frontiers in Environmental Science* 6, 12.

Srimathi, K., Vanathi, J., Vignesh, S. 2025. Nanoherbicides: a smart approach to sustainable weed management. *Agricultural sciences* 11, 59–80.

Timmusk, S., Seisenbaeva, G., Behers, L. 2018. Titania (TiO2) nanoparticles enhance the performance of growth-promoting rhizobacteria. *Scientific Reports* 8(1), 617.

Ullah, Q., Fatima, F., Memon, S.U.R., Khan, M.A., Ismail, A., Rasool, M., Ali, H., Qasim, M., Ullah, U. 2024. Nanotechnological innovations in agrochemicals: enhancing efficacy and environmental stewardship in pesticide and herbicide applications. *Middle East Research Journal of Agriculture and Food Science* 4(05), 164–178.

Upasna, K.A., Rattan, P., Ludarmani, S.A. 2023. Effect of seed priming by nano-urea and nano-zinc on the growth, yield and quality of radish (*Raphanus sativus* L.). *Journal of Pharmaceutical Innovation* 12(10), 604–617.

Wang, D., Saleh, N.B., Byro, A., Zepp, R., Sahle-Demessie, E., Luxton, T.P. et al. 2022. Nano-enabled pesticides for sustainable agriculture and global food security. *Nature Nanotechnology* 17(4), 347–360.

Xiang, Y., Zhang, G., Chi, Y., Cai, D., Wu, Z. 2017. Fabrication of a controllable nanopesticide system with magnetic collectability. *Chemical Engineering Journal* 328, 320–330.

Yadav, A., Yadav, K., Abd-Elsalam, K.A. 2023. Nanofertilizers: types, delivery and advantages in agricultural sustainability. *Agrochemicals* 2(2), 296–336.

Yadav, M. 2025. Nanoparticle-facilitated targeted nutrient delivery in plants: breakthroughs and mechanistic insights. *Plant Nano Biology* 12, 100156.

Yang, Y., Jiang, Q., Peng, M., Zhou, Z., Du, X., Yin, M., Shen, J., Yan, S. 2022. A star polyamine-based nanocarrier delivery system for enhanced avermectin contact and stomach toxicity against green peach aphids. *Nanomaterials* 12(9), 1445.

Yin, J., Su, X., Yan, S., Shen, J. 2023. Multifunctional nanoparticles and nanopesticides in agricultural application. *Nanomaterials* 13(7), 1255.

9 Nano-Enabled Strategies for Smart Agriculture

Upma Arora, Gurneet Kaur,
and Priyanka Sharma

9.1 INTRODUCTION: SMART AGRICULTURE

Recently, the agricultural sector is getting renewed and reshaped due to technological innovations especially the digital innovations have led to the transformation of conventional farming systems to the advanced concepts like smart farming. The main aim of using digital technologies coupled with artificial intelligence in modern agricultural domain is to increase the production quantity and quality thus boosting the food security in the world. The current scenario demands the digitalization of agriculture on economic, legal and ecological fronts to achieve the goal of sustainable development. The strategical minimization of the input by judicial use of natural resources with minimal environmental impact and maximization of the output in terms of enhanced productivity of quality crops is what smart agriculture is all about (Garg et al. 2023). Smart agricultural practices employ the smart agricultural devices to generate digital data which are then assessed, and the information is used to make data-based decisions that consequently lead to the precise application and implementation in the fields to obtain the desired output in terms of improved productivity both in terms of quality and quantity.

The various stages of crop production are continuously monitored by satellites, precision sensors, and drones and utilize the digital systems like the Internet of Things (IoT), big data technologies, and artificial intelligence, to generate, process, and apply the data for soil and climate assessment, earth observation, weather forecast, crop modelling, weed management, yield monitoring, etc. The growth and development of cyber-physical systems have gained an irrefutable relevance for agriculture as it implies an information-driven approach for intelligent solutions in farm-management-systems. The modern farming management involves the use of revolutionary digital tools like the IoT and cloud computing to monitor and optimize the farming processes. Further the use of robots and artificial intelligence equips the modern day farmers to access real-time data regarding their day-to-day agricultural operations involving soil, weather, terrain, plants, livestock, etc. for optimal utilization of resources (Ozdogan et al. 2017).

9.2 DIGITAL TECHNOLOGY AND AGRICULTURE

Over a time, span of around a thousand decades, the agri-technology has gradually evolved from usage of simple hand tools to the utilization of mechanical equipment in commercial farming now-a-days. However, in the most recent times the rapid

DOI: 10.1201/9781032659435-9

development of digital technologies and artificial intelligence has paved the way for the transformation of mechanical engineered farming to 'intelligent' or smart farming.

9.2.1 Tools and Techniques

The most notable smart farming tools and techniques that are revolutionizing the agriculture are given as follows:

Sensors: The electro-technical devices used to measure physical quantities like temperature, light intensity, humidity, soil nutrients, acidity/alkalinity, water, atmospheric pressure, speed and direction of wind, etc. and convert these measurements into a signal which is read and interpreted by an instrument.

Location Systems: Drones, Global Positioning System (GPS), satellite navigation, etc. that carry out the supervision and monitoring of farms, gather multispectral images and other information about the condition of the land, irrigation requirements, plant growth, presence of weeds, pathogens, and pests.

Communication Systems: Mobile phones, tablets, etc. based on mobile network connection or internet connection, LoraWan.

Data Analytics: Analytics solutions like Geographic Information System (GIS) to integrate the large amount of data obtained from drones and other measuring instruments and analysing it to process through data pipelines for optimal solutions.

Software: Software solutions that are specialized and target farm specific aspects and applications.

Robotics: Autonomous tractors, microphones, processors, robotic fertilizer, weed and pest management.

Blockchain: Blockchain monitors the crops from field to suppliers and thus trace the supply chain that can help in optimal supply management.

9.2.2 Internet of Things (IoT)

The IoT integrates the physical world and digital systems. IoT includes the use of various digital tools like commodity sensors, wireless sensor networks, embedded systems to monitor various aspects of agricultural farms including almost everything from assessment of soil parameters, measuring solar radiations, temperature, to plant growth parameters as leaf moisture and stem diameter etc. The information thus gathered is analysed using real-time analytics to make intelligent management decisions and practical implementation in the fields is carried out with the help of control systems and automation. All the controls are exercised remotely through network infrastructure. A complex network system is formed by connecting the sensors and machines; the farmers with the help of their phone or tablet can monitor their farms without visiting the fields and thus can take strategic decisions for managing their farm.

Smart farming hence focuses on optimizing the human labour while managing the farms using modern digital technologies to boost the crop yield and also to ameliorate the agriproducts quality.

9.3 E-AGRICULTURAL PRACTICES

The combined application of data and IoT-driven analytics technologies like big data analytics and robotics is bringing about a revolution in modern agriculture that can be termed as agriculture 4.0 which employs various smart farming practices that include:

Soil Quality Monitoring: For obtaining good yield of healthy crops, the soil quality plays a major role. By understanding the soil conditions accurately using IoT smart farming sensors in the soil, the farmers can optimize applications in order to restore the soil where required precisely through smart farming techniques.

Climate Monitoring and Weather Forecasting: By the use of smart agricultural devices including weather stations and cloud computing, the farmers are able to strategize their farming processes like sowing of seeds, irrigation, harvesting etc. according to climate conditions.

Livestock Monitoring: The tracking devices monitor location of the livestock that lead to early detection of tampering and thus helps in quick stolen livestock recovery. The livestock health monitoring helps in early detection of the sick animals that reduces the cost of treatment and the labour dependence as well.

Sustainable Crop Management: All time accessibility of crop farming specific data including soil quality, weather conditions, and crop health enables the farmer to efficiently monitor the crop growth and take preventive measures towards any unwanted outcomes.

Crop Yield Optimization: The immediate processing of the data regarding the weather conditions, soil and crop health on the field can help in preventing the crop damage due to pests or pathogens and devise a practical plan for harvesting and storage of agricultural produce that leads to maximizing the farm productivity.

Farm Security: Rural crime in agriculture sector is quite detrimental to farmer's income and profits. By the use of smart technologies like motion sensors, GPS tags, tractor guidance GPS Tags, pressure pads, electric fence theft monitoring, perimeter protection by light beam sensors and closed-circuit television (CCTV) monitoring, the stolen assets can be tracked down and a swift action can be taken to reduce the theft and increase the asset recovery rates.

Greenhouse Automation: The accurate real-time information on temperature, lighting, humidity, and soil conditions of the greenhouse eliminates the need of manual intervention. Rather, the farmers with the help of smart weather stations automatically adjust and control the greenhouse environment to match the requisite parameters. This highly reduces the production loss, energy loss, and labour costs.

9.4 SMART PRECISION FARMING

Precision farming means precise measurement of field conditions and variations if any and using IoT-based approaches to apply controlled and accurate farming techniques. The overall management of a successfully operating smart farm requires a cloud computing-based single dashboard that receives the data from all the IoT monitoring devices and sensors in the field, analyses and processes the received information, provides the 'intelligent' solutions helping the farmers in decision making and streamlining their actions by implementing the precise automated operational measures with the help of robotics and automated machines.

The use of technologies like nanosensors, GPS, remote sensors, unmanned aerial vehicles, wireless sensor networks, and multimedia devices in farming practices save time, reduce the inputs and augment the outputs, making precision farming supported by IoT the best approach to attain sustainable development goals (El-Ramady et al. 2023). The crop health and productivity are highly dependent on various abiotic and biotic factors like temperature, humidity, light intensity, soil conditions, presence of pests, pathogens, and weeds etc. Networks of nanosensors like infrared nanosensor films mounted on leaves that detect the onset of plant water stress and miniaturized tensiometer nanosensors that trigger irrigation only when needed enable optimized field-scale water management. The photosynthetically active radiations (PAR) are detected using quantum sensors and another parameter i.e., solar radiation flux density is measured using a pyranometer. The time and amount of precipitation is measured using a rain gauge. Wind speed is measured using anemometers. The DS18B20 is an example of a digital thermometer which is used to monitor the soil temperature remotely (Brick et al. 2022). The point data of test results for chemical and physical characteristics of soil are geostatistically interpolated, and the digitally recorded data from the sensors are analysed through GIS to mark out the management zones and the action needed. Plant phenotyping is done by remote sensing and computer vision-aided imaging. Geospatial mapping is of much importance in monitoring, predicting/forecasting and thus controlling the variations in soil health parameters, crop quality, and pest infestations. Based on collected data, the programmable machines attached to sprayers like variable rate spreader equipped with optical sensors and direct nozzle injection systems help in precise irrigation and delivering the agrochemicals at the right time in right intensities. GIS along with satellite assisted remote sensing provides an early warning system for the occurrence of a natural disaster, its risks and damage assessment, and decision support. Automatic guidance systems have revolutionized the farm scouting thereby reducing the farmer's stress. Automated harvesters make use of robotic technology for harvesting the crops (Garg et al. 2023).

The smart farming revolution is anticipated to bring a significant drop in use of pesticides and fertilizers in the fields whereas a substantial increase in overall efficiency and productivity can be expected. Therefore, smart farming seems to be capable of fulfilling the dream of sustainable agricultural development with precise resource-efficient approach forming the basis of qualitative as well as quantitative improvement of agricultural produce. However, while implementing the smart farming solutions, few key points need to be considered as summarized in Figure 9.1 and given below:

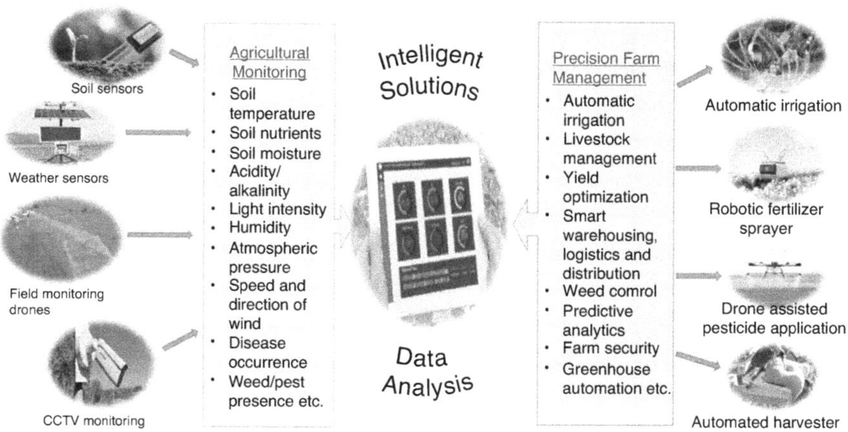

FIGURE 9.1 Working of a smart farm.

Hardware and its Maintenance: The selection of hardware components like sensors and monitors depend on the type of information required and most importantly they must collect data accurately. Further, the sensors and other hardware components used in the field should be durable, able to withstand the variations in environmental conditions and easy to maintain.

Data Analytics: Powerful data analytics and software platforms should be employed to process and analyse the data in order to obtain actionable insights.

Mobility and Connectivity: Mobility i.e., access to information from anywhere and at any time is a key component of smart farming. This can be ensured by high degree of wireless range so that the devices can get effectively interconnected and further can efficiently communicate with the central server and reporting dashboard. The network must be able to withstand bad weather conditions for non-disruptive operations.

Data Delivery and Data Security: Safe and timely delivery of data from field-based sensors and GPS tags on machinery significantly influences the decision making process. Working with large sets of data means more security risks. Smart farming devices and applications must be encrypted and secure to avoid the potential risks of data theft and hacking attacks.

9.5 NANO-ENABLED STRATEGIES FOR AUTOMATED/ SMART AGRICULTURE

The surface area to volume ratio of nanoparticles (NPs) is quite high due to their very small size (in the range of 1–100 nm) that attributes to their high sorption capacity, reactivity, and biocompatibility (Deng et al. 2023). The engineered nanoparticles (ENPs) act as "smart delivery system" in modern day agriculture as they follow controlled-release kinetic mechanisms for delivery to targeted sites (Jain et al. 2021). Recent developments in nano-enabled agriculture include the use of NPs in almost all farming processes ranging from nanodetectors, nanosensors, nanofertilizers, nanopesticides,

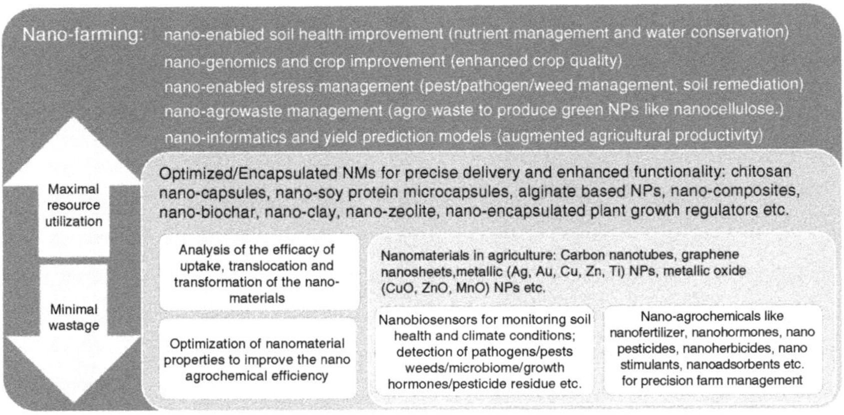

FIGURE 9.2 Nano-enabled smart farming strategies for achieving sustainability in agriculture.

nanoherbicides, nanoadditives, nanocleansers, nanodegraders, and nanodelivery systems exhibit a great promise for enhancing crop production and lessening the environmental footprint (Xin et al. 2020; Asif et al. 2022), as depicted in Figure 9.2.

9.5.1 Nanosensors for Farm Monitoring

Nanosensors very precisely/accurately monitor the nutrient status, moisture content, temperature and pH of soil, pesticides, heavy metals, and occurrence of pathogens and disease in plants etc. (Sharma and Kumar 2024) through detecting presence/absence of certain biochemicals in cell organelles, variations in physical parameters like temperature of pants, analysing some phenomenon etc. (Harini et al. 2023). A typical nanobiosensor is composed of three parts:

i. **Bioreceptor**: a probe, the sensitized biological element like antibodies and enzymes that interacts with analytes, accepts the analyte's chemical signal, and then transmits the signal to the transducer
ii. **Transducer**: an interface that measures and converts physical signals received from the bioreceptor into the electrical signals
iii. **Detector**: the device (microprocessor) implied in processing these signals further amplifying for analyses and presentation as real-time data (Mandal et al. 2020).

Nanobiosensors are designed using a wide range of nanomaterials (NMs) like metal NPs (gold, silver, copper) or nonmetallic carbon-based NPs (graphite, graphene, carbon nanotubes). Chromophore based nanobiosensors detect the analyte's presence/absence through the type and intensity of emitted fluorescence. The sensors employing carbon nanotubes and nanosensors like quantum dots (QDs) work by converting the chemical signals directly to optical signals. Volatile substances are most precisely detected by nanoscale carbon structures. Carbon dots (CDs) are better than conventional semiconductors QDs and very suitable for detection of microorganism as they are highly biocompatible (Miguel-Rojas and Pérez-de-Luque 2023). For example, an infrared plant temperature sensor has been developed to detect the presence of green bug infestation in winter wheat that detects the changes in plants' temperature due to infestation (Michels et al. 2000). Few examples of nanobiosensors are listed in Table 9.1.

TABLE 9.1
Some Nanosensors and their Applications

Nanosensors	Application	References
Single-walled CNTs based nanosensor	Monitoring of plant health and detection of H_2O_2, an indicator of the onset of biotic or abiotic stresses in *Arabidopsis thaliana*	Wu et al. (2020)
Reduced graphene quantum dots (rGQDs) and MWCNTs based nanosensor	Rapid detection of diazinon (DZ)	Talari et al. (2021)
Graphene quantum dots, chitosan, and nickel molybdate nanocomposites based nanosensors	Sensitive and selective determination of DZ	Ghiasi et al. (2021)
Gold nanoparticles (AuNPs) based nanobiosensor	Detection of urea, urease activity, and urease inhibition	Deng et al. (2016)
Gold nanoparticles (AuNPs) functionalized with single-stranded oligonucleotides	Detection of the genomic DNA of *Ralstonia solanacearum*, causal agent of potato bacterial wilt	Khaledian et al. (2017)
Gold NPs based nanosensors	Detection of the plant bacterial pathogen *Pantoea stewartii*	Zhao et al. (2014)
Silver nanorods	Detection of the plant pathogen *Phytophthora ramorum*	Yüksel et al. (2015)
Quantum dot (QD)-based complex nanobiosensor	Detection of the plant pathogen Citrus Tristeza virus	Safarnejad et al. (2017)
Sulphur-doped graphene QDS based nanosensor	Ultrasensitive detection of carbamate insecticides like carbofuran	Nair et al. (2020)
Molybdenum disulphide (MoS_2), 2D layered transition-metal dichalcogenide nanosensor	Highly diversified environmental monitoring	Sha and Bhattacharyya (2020)
Acetylcholinesterase (AChE) biosensor, based on AuNPs-MoS_2-reduced graphene oxide/polyimide flexible film (rGO/PI)	Detection of the insecticide paraoxon	Jia et al. (2020)
Enzyme, optical, and electrochemical sensors	Detection of pesticide mixtures like methyl parathion and chlorpyrifos	Viswanathan et al. (2009)
DNA biosensor using zinc oxide NPs/chitosan nanocomposites	Detection of *Trichoderma harzianum*, a biological control agent against various plant pathogens	Siddiquee et al. (2014)

9.5.2 Nano-Enabled Soil Health Improvement

Soil health is the most important factor for crop productivity. The soil health is assessed in terms of its key indicators including the soil temperature, soil pH, soil moisture holding capacity, nutrient composition, soil salinity, presence/absence of contaminants, and soil microflora etc. The NMs that are used to enhance the soil health include the following:

Nanoadsorbents that adsorb various hazardous contaminants (including heavy metals, organo-chemicals, and compounds with biocidal properties) and *nanocatalysts* that accelerate chemical reactions through nucleation effect and catalytic effect and improve the mechanical properties like shear strength, compressibility, permeability, aggregate number, and bearing capacity leading to enhancement of overall stability of soil are derived from the nanoadditive substances such as nanoclay, nanosilica, nano-aqueous adhesives, and nano-alumina (Awasthi et al. 2019; Huang et al. 2022; Gholami et al. 2016; Zhao et al. 2022)

Nanofilters like activated carbon nanofibers or zinc oxide and copper NP-based nanofilters can purify and disinfect the air by reducing CO_2, N_2O and CH_4 emissions through increased carbon sequestration, denitrification and inhibiting methanogenesis respectively (Oertel et al. 2016; Sharma and Kumar 2024).

Hydrophilic nanomaterials like nanoclay minerals enhance moisture retention in soil and reduce irrigation needs and also increase the available water content and reduce the drainage and evaporation losses in sandy soils (Wei et al. 2018). Cellulose nanofibers and polyacrylamide nanogels have been reported to improve capillary storage leading to increase in moisture retention and the available water to plants in light soils while minimizing the water losses (Zohuriaan-Mehr and Kabiri 2008; Li et al. 2014). Such ammendments can help in improving the overall soil health thus sustaining the agricultural productivity in the face of growing resource scarcity (Kansotia et al. 2024)

9.5.3 Nanofertilizers and Precise Plant Nutrition

Nanofertilizers are the nanoscale materials used in fields for plant nutrition to enhance plant growth and yield and are generally classified as macronutrient and micronutrient nanofertilizers, on the basis of nutrient type; controlled release, targeted delivery, and plant growth stimulating NMs on the basis of their action (Marchiol et al. 2020). While working on maize, Adnan et al. (2020) reported that when compared to urea, the nano-nitrogen fertilizer exhibited improved nitrogen use efficacy leading to enhanced growth and productivity. Another study conducted by Jakhar et al. (2022) using chitosan-coated zinc oxide NPs under saline conditions showed increased zinc content and activity of antioxidants in wheat thereby ameliorating the grain yield. Enhanced tomato growth in hydroponic conditions by nano-phosphorus (Zhu et al. 2023); and NP induced plant immune response to various biotic and abiotic stresses (Seleiman et al. 2020) have been reported in the literature. Nanosols and multi-walled carbon nanotubes have been reported to enhance the availability of nitrogen and phosphorus and their uptake by stimulating the soil enzymes urease, phosphatase, and dehydrogenase (Rajput et al. 2021). Nano-encapsulation of these nanocarriers further improve their solubility, mobility, stability, and bioavailability. The polymers used for encapsulation of nutrients prevent their degradation and leaching. Further the slow and controlled release of nutrients in the soil facilitates their efficient uptake by plant tissues (Noman et al. 2023a). The utilization of nanofertilizers is financially advantageous as it reduces the nutrient loss by decreasing the draining and volatilization along with enhancing the quality as well as the quantity of the agricultural produce making it a promising way to attain sustainability in agriculture (Naidu et al. 2022).

9.5.4 Nano-Genomics and Crop Improvement

Compared to the traditional plant breeding practices which required extensive manual labour for crop improvement and diversifying the gene pool, NP-mediated genetic engineering techniques offer a versatile platform to introduce the desirable genetic traits into crops (Cunningham et al. 2018). The highly specific genome editing methods like transcription activator-like effector nucleases (TALENs), zinc-finger nucleases (ZFNs), and clustered regularly interspaced short palindromic repeats (CRISPR) have revolutionized the whole scenario of crop improvement (Yaqoob et al. 2023). Although CRISPR/Cas method has exhibited the most promising results but major constraint is the effective delivery of the desired genes or transcripts (deoxyribonucleic acid (DNA)/ribonucleic acid (RNA)) into plant cells which is typically done through physical methods like gene gun/cationic delivery or through biological tools like Agrobacterium/viral infection (Sharma and Lew 2022). RNA, DNA complex with NPs like carbon nanocarriers have been found to be very effective for delivering genetic materials into the plant genome (Demirer et al. 2020). The CRISPR/Cas9 plasmids encapsulated with poly (lactic-co-glycolic acid) NPs and the lipid nanoparticle (LNP) Cas9 (Cas9 encapsulated with lipid NPs) delivery system are highly stable and ensure target-specific delivery without damaging tissues as they have high editing efficiency and don't show any off-target mutations (Qiu et al. 2021; Santana et al. 2020). NPs used as small-interfering RNAs (siRNAs) carriers that mediate RNA interference (RNAi) responses are a powerful tool for targeted gene silencing which in turn modulates the expression of specific genes responsible for disease susceptibility or insect resistance etc. (Noman et al. 2023b). However, the interruption of important enzymatic activities and potential reactive oxygen species (ROS) generation are the major risk factors for which the direct contact of plant cells with NPs should be avoided/minimized. Moreover, the toxic impacts and safety of different NPs should be assessed properly before using them for nano-genome editing (Raza et al. 2023).

9.5.5 Nano-Enabled Plant Disease Management

Highly sensitive and rapid nanobiosensors like QD nanoprobes, gold NP-based assays, graphene, carbon nanotube electrodes, piezoelectric nanobiosensors, and paper-based nano-enabled nucleic acid assays offer on-site monitoring of plant health and sensitive and early detection of plant pathogens that facilitates prompt disease management. Further, nanobiosensors like carbon nanotube gas sensors and fluorescent silica NP tags are highly sensitive for on-site detection of insects and pests that helps in precision targeting and reducing pesticide use (Kansotia et al. 2024).

NPs in the form of nanocomposites, nanopesticides, and nanoherbicides are very effective in disease management and countering phytopathogens. NMs help in disease management in two ways i.e., by enhancing the crop resilience and by crop protection.

9.5.5.1 Nano-Enabled Crop Resilience

Metallic NPs like silver/copper NPs, nanocomposites such as chitosan-coated iron/magnesium NPs have been reported to show antimicrobial properties (Ahmed et al. 2020) and prevent the development of drug-resistant microbial strains (Malandrakis et al. 2019; Hamid and Saleem 2022) thus helping in efficient disease management. Sulphur NPs and silica NPs trigger the synthesis of antimicrobial compounds like phenolic compounds, pathogenesis-related (PR) proteins and hydrolytic enzymes such as chitinase, pectinase and β-1,3-glucanase thereby augmenting the plant defense (Fu et al. 2020). Also, the copper NPs and silver NPs have been shown to function as immune modulators by stimulating salicylic acid (SA) and jasmonic acid (JA) mediated signalling and defense responses. Modulation of phytohormone signalling pathways by NPs like nitrogen-doped carbon bolsters plant immunity against a broad spectrum of pathogens and activates systemic acquired resistance (SAR) that is characterized by enhanced resistance to subsequent pathogen attacks (Turgut and Bezirganoğlu 2022). This equips the crops with an enduring and comprehensive shield against pathogens, reduces disease incidence and lowers the use of chemical pesticides thereby contributing to sustainable and environmentally friendly agriculture (Noman et al. 2023a).

9.5.5.2 Nano-Enabled Crop Protection

Chemical protectants and antimicrobial agents like pesticides, bactericides and fungicides have been inevitably used in contemporary agriculture for protecting the crops from phytopathogens (Athanassiou et al. 2018). Recently, the nanopesticides have been put to use as target-specific control of pests and developing intelligent nanosystems for sustainable crop productivity (Nuruzzaman et al. 2016). By encapsulating or binding the chemical protectants with NPs, e.g., mesoporous silica offers a precise and controlled delivery for prochloraz. Azoxystrobin-loaded silica NPs, nanoscale fungicides like fenhexamid and polyhexamethylene biguanide, ensure target-specific delivery with minimal off-target effects. Furthermore, pH-sensitive carbendazim- loaded mesoporous selenium like NPs have been found to be very specific and release their load only in response to certain environmental conditions (Mosa et al. 2022). NPs-mediated crop protection strategies provide an effective solution to environmental problems like accumulation of non-biodegradable components in the food chain and offer long term utility .

9.5.6 Nano-Enabled Weed Management

Nanoherbicides, the nano-enabled weed-controlling substances include three fundamental categories: (i) organic materials – organic nanocarrier encapsulated active ingredients; (ii) inorganic materials – formed of clay minerals, metals and metal oxides; and (iii) hybrid NMs having both organic and inorganic components (Forini et al. 2020; Soltani et al. 2022). Nanoherbicides enhance the adsorption of the herbicide and its biological activity by ensuring better adhesion to the waxy leaf surface. Nano-encapsulation results in the development of novel nanoherbicides that utilize some specific biorecognition motifs to precisely target the subcellular compartments. These can further be improved using strategic design and machine learning (Santana et al. 2020). Polycaprolactone (PCL)-nano-encapsulated herbicides (ametryn, atrazine, metribuzin and simazine) are considered to be most suitable as they are biodegradable, have good biocompatibility, can be prepared from inexpensive materials and have low toxicity thus making them more eco-friendly products (Zargar et al. 2023).

9.5.7 Post Harvest Nano-Encapsulation of Agriproducts

Nano-encapsulation based production of nanostructured food ingredients improves the texture, solubility, colour, flavour, and stability of foods during processing and storage and also increases the bioavailability of the food nutrients (Pateiro et al. 2021). Nanofillers, e.g., chitosan-coated NPs and nanocellulose have been used to reinforce the protein/polymer-based plastic films. titanium dioxide (TiO_2), ZnO, and SiO_2 are amongst the most often used inorganic nanofillers to reinforce the protein films. TiO_2 is chemically inert with low toxicity and possesses antimicrobial activity (Lei et al. 2014). ZnO is thermally resistant and exhibits antibacterial activity. Also, approved by US FDA as GRAS (generally recognized as safe), ZnO is used for packaging of food. Hydrophobic substances including essential oils, vegetable oils, and waxes are used as nanoemulsions in the milk proteins or carbohydrates-based films to refine the water vapor barrier properties. NPs combined with essential oils like oregano essential oil possess effective antimicrobial particularly antiviral action and further coating them with cellulose nanocrystals (CNCs) result in multilayer films that ascertain their biodegradability in food packaging. Additionally, including antioxidants like omega-3 oils, beta carotenes and vitamin D in the lipid phase can provide nutraceutical benefits to the food by preventing the oxidation due to air (Ur Rahim et al. 2021). However, food safety controls and risk assessments must be rigorously done before commercialization. Strict policies and guidelines need to be formulated so as to provide effective regulatory systems for the food industry to follow for utilizing the NPs in a safe and ethical manner. Therefore, rigorous toxicological screening methodologies under stringent government guidelines and legislations are essential for the lawful nano-encapsulation of agriproducts (Neme et al. 2021).

9.5.8 Nano-Enabled Stress Smart Agriculture

Global climate change is a cause of concern for agriculture sector as environmental factors significantly influence plant growth and development, hence the need of crop plants that are adapted to grow under current hostile conditions with elevated levels of heat, cold, drought and salinity stress (Cooper and Messina 2023). In recent years, several metal NPs, such as zinc/iron/silver/gold NPs; metal oxide NPs like cerium oxide (CeO_2)/zinc oxide (ZnO)/copper oxide (CuO)/TiO_2/silica (SiO_2) NPs; and carbon-based NPs (e.g., carbon nanotubes etc.), have been investigated for determining their impact on alleviating the antagonistic effects of environmental stress factors. There are various reports that show that seed nano-priming can help in breaking the seed dormancy as the nutrient and water uptake capacity of plants increases (Acharya et al. 2020); can activate the hydrogen peroxide (H_2O_2) signalling and ameliorate the antioxidant immune system by reducing malondialdehyde (MDA) content (Li et al. 2022) and enhancing antioxidative enzymes activities like those of catalase (CAT), superoxide dismutase (SOD), ascorbate peroxidase (APX), and peroxidase (POD) (Salam et al. 2022); and can enhance proline and other osmoprotectant levels (Waqas Mazhar et al. 2022). Also, the addition of TiO-NPs has been shown to improve the iron, carbohydrate and protein contents in the wheat plants (Amin et al. 2023) and addition of graphene oxide-NPs to soil has been reported to increase membrane integrity, photosynthetic components, oxidant and non-oxidant metabolites in sweet basil (Ganjavi et al. 2021). Foliar application of salicylic acid and nano-Fe_2O_3 NPs has been found to promote the enzymatic activities of H+-ATPase and H+- PPase thereby relieving salinity toxicity and enhancing plant growth (Ghassemi-Golezani and Abdoli 2021). Therefore, the application of NPs by addition to soil, foliar treatment or nanopriming regulates the molecular machinery, modulates the signalling molecules like H_2O_2, upregulates the oxidative stress-related genes expression and enzymatic activities thereby enabling the plants to tackle with various environmental stresses (Raza et al. 2023).

9.5.9 Nano-Enabled Remediation

The accumulation of mismanaged agro-wastes and heavy metal contamination of agricultural soils due to various anthropogenic activities are detrimental to environmental safety and quality of the crops. Therefore, nanomanagement of agro-wastes employing nanotechnology-based biorefining of agro-wastes is being recently developed as a green approach to ascertain sustainable development in agriculture sector. The strategies include production of lignin-based NMs by biorefining of the agricultural refuse to be used as fertilizers/soil conditioners/soil modifiers; and transforming farm-waste into nano-enabled energy applications (El-Ramady et al. 2023). NMs are commonly used for remediating contaminated soils as mobilizing agents because their large specific surface area and higher reactivity enables them to enhance the removal of soil contaminants by increasing their mobility and bioavailability. On the other hand, NMs can also function as immobilizing agents as they bind the contaminants within the soil matrix thus decreasing their mobility and bioavailability (Sharma and Kumar 2024). Various NMs like single/multi-walled carbon nanotubes, graphene, TiO_2 NPs, manganese oxide (MnO) NPs, iron oxide (FeO) NPs, iron sulphide (FeS) NPs, magnesium oxide (MgO) NPs, and nanozerovalent iron (nZVI) etc. find a widespread application in degradation and detoxification of inorganic as well as organic contaminants from the soil, water, and air by implying diverse mechanisms like adsorption, immobilization, formation of immobile complexes, reduction, ionic exchange, reductive dehalogenation and nucleophile substitution etc. (Ur Rahim et al. 2021).

9.5.10 Nanoinformatics and Prediction Models for Increased Productivity

Nanoinformatics is expected to play a vital role in improving agricultural production by integrating the experimental data on physicochemical properties of NMs as affected by different soil conditions/climate change conditions and different plant species and developing safer-by-design nano-agrochemicals (Zhang et al. 2021). The increased productivity as anticipated from development of nano-enabled agriculture requires rapid and accurate prediction models for translocation, accumulation and interactions of nano-agrochemicals as well as yield of the crops. Few models like the soil-water-organism nanomaterial fate and speciation in the environment (NanoFASE) model are used for predictive analysis as they help in providing a foresight into the probable fate of specific NMs in particular environmental conditions. Machine learning models can project future trends as they are data-driven approaches with an ability to reveal complex relationships. By compiling high-spatial- resolution laboratory and field data on soil, (acidic or alkaline), climate (temperature and relative humidity), regional crop planting timing and expected crop growth, uptake of NPs, photosynthesis, and oxidative stress, yield prediction models can be developed and region-specific potential of nano-enabled agriculture can be predicted (Deng et al. 2023).

9.6 CONCLUSIONS

The advent of smart agriculture technology making the use of integration of nanotechnology and the digital technologies has revolutionized the modern agriculture and has provided an efficient solution to the anticipated global food security problem. Smart nanotechnologies follow an interdisciplinary approach of enhanced NP-mediated delivery and utilization of nutrients to crops and ensure food safety through nanosensors. NPs in the form of nanofertilizers, nanopesticides, nanoadditives, and nanoweedicides etc. are used for obtaining higher yield and for crop protection. NPs also assist in controlled and regulated gene delivery during genetic crop improvement (Xiaoping et al. 2020). Coupling the nanotechnology applications with the technologies like IoT, location systems and artificial intelligence optimizes the complex farming systems. Nano-enabled smart agriculture thus holds great promise for delivering in future a more efficient, resilient and sustainable agricultural system.

However, the implementation of smart farming comes with its own challenges like cyber security, data storage, operational technological issues, power supplies etc. Further, the recent reports show that nano-enabled products alter the crop quality by varying the composition of mineral nutrients like calcium (Ca), magnesium (Mg), and phosphorus (P) and the contents of some proteins and sugars like albumin, globulin, and prolamin (Gomez et al. 2021).

The lack of detailed knowledge regarding the interactions between NMs and soil/plants, alteration of plant microbiome, effects on biogeochemical cycling, environmental safety and human health risks of NM application etc. also hinder the acceptance of nano-enabled agriculture and its implementation on a commercial scale. Despite these challenges, farming 'smarter' is gradually becoming necessary for sustainable agriculture development. The diminishing biodiversity and earth's resources urgently demand the reduction of environmental impacts and minimizing the environmental constraints. The optimization and preservation of human and natural resources achieved through the technological innovations have improved the overall agricultural outputs. Thus a combined system of, sensors, pressure pads, tags, monitoring devices and automated management tools creates a digital ecosystem that can lead to simple and smart digital transformation of the cotemporary farms.

REFERENCES

Acharya, P., Jayaprakasha, G.K., Crosby, K.M., Jifon, J.L., Patil, B.S. 2020. Nanoparticle-mediated seed priming improves germination, growth, yield, and quality of watermelons (*Citrullus lanatus*) at multi-locations in Texas. *Scientific Reports* 10, 5037. https://doi.org/10.1038/s41598-020-61696-7

Adnan, M., Shah, Z., Fahad, S., Arif, M., Alam, M., Khan, I.A. et al. 2020. Nano-nitrogen vs urea: growth performance of maize under water stress conditions. *Chemosphere* 239, 124785. https://doi.org/10.1016/j.chemosphere.2019.124785

Ahmed, T., Shahid, M., Noman, M., Niazi, M.B.K., Mahmood, F., Manzoor, I., Zhang, Y., Li, B., Yang, Y., Yan, C. 2020. Silver nanoparticles synthesized using *Bacillus cereus* SZT1 ameliorated the damage of bacterial leaf blight pathogen in rice. *Pathogens* 9, 160. https://doi.org/10.3390/pathogens9030160

Amin, A., Manzoor, M., Ramay, M.W., Hassan, A., Hina, K., Syed, A. et al. 2023. Metallic nanoparticles photodegraded antibiotics and co-application improved wheat growth and nutritional quality through stress alleviation. *Chemosphere* 323, 138189. https://doi.org/10.1016/j.chemosphere.2023.138189

Asif, M., Islam, S., Malik, M.A., Dar, Z.M., Masood, A., Shafi, S., Rashid, B., Sidique, S. 2022. Nano pesticides application in agriculture and their impact on environment. In: Asaduzzaman, Md., Afroz, M. (eds), *Agricultural Development in Asia – Potential Use of Nano-Materials and Nano-Technology*. IntechOpen, New York. https://doi.org/10.5772/intechopen.100690

Athanassiou, C., Kavallieratos, N., Benelli, G., Losic, D., Usha Rani, P., Desneux, N. 2018. Nanoparticles for pest control: current status and future perspectives. *Journal of Pest Science* 91, 1–15. https://doi.org/10.1007/s10340-017-0898-0

Awasthi, A., Jadhao, P., Kumari, K. 2019. Clay nano-adsorbent: structures, applications and mechanism for water treatment. *SN Applied Sciences* 1, 1076. https://doi.org/10.1007/s42452-019-0858-9

Brick, E.S.R., Holland, J., Anagnostou, D.E., Brown, K., Desmulliez, M.P. 2022. A review of agroforestry, precision agriculture, and precision livestock farming—the case for a data-driven agroforestry strategy. *Frontiers in Sensors* 3, 998928. https://doi.org/10.3389/fsens.2022.998928

Cooper, M., Messina, C.D. 2023. Breeding crops for drought-affected environments and improved climate resilience. *The Plant Cell* 35, 162–186. https://doi.org/10.1093/plcell/koac260

Cunningham, F.J., Goh, N.S., Demirer, G.S., Matos, J.L., Landry, M.P. 2018. Nanoparticle-mediated delivery towards advancing plant genetic engineering. *Trends in Biotechnology* 36, 882–897. https://doi.org/10.1016/j.tibtech.2018.03.009

Demirer, G.S., Zhang, H., Goh, N.S., Pinals, R.L., Chang, R., Landry, M.P. 2020. Carbon nanocarriers deliver siRNA to intact plant cells for efficient gene knockdown. *Science Advances* 6, eaaz0495. https://doi.org/10.1126/sciadv.aaz0495

Deng, H., Hong, G., Lin, F., Liu, A., Xia, X., Chen, W. 2016. Colorimetric detection of urea, urease, and urease inhibitor based on the peroxidase-like activity of gold nanoparticles. *Analytica Chimica Acta* 915, 74–80. https://doi.org/10.1016/j.aca.2016.02.008

Deng, P., Gao, Y., Mu, L., Hu, X., Yu, F., Jia, Y., Wang, Z., Xing, B. 2023. Development potential of nanoenabled agriculture projected using machine learning. *Proceedings of the National Academy of Sciences of the United States of America* 120, e2301885120. https://doi.org/10.1073/pnas.2301885120

El-Ramady, H., Abdalla, N., Sári, D., Ferroudj, A., Muthu, A., Prokisch, J., Fawzy, Z.F., Brevik, E.C., Solberg, S.Ø. 2023. Nanofarming: promising solutions for the future of the global agricultural industry. *Agronomy* 13, 1600. https://doi.org/10.3390/agronomy13061600

Forini, M.M.L., Antunes, D.R., Cavalcante, L.A.F., Pontes, M.S., Biscalchim, É.R., Sanches, A.O., Santiago, E.F., Fraceto, L.F., Grillo, R. 2020. Fabrication and characterization of a novel herbicide delivery system with magnetic collectability and its phytotoxic effect on photosystem II of aquatic macrophyte. *Journal of Agricultural and Food Chemistry* 68, 11105–11113. https://doi.org/10.1021/acs.jafc.0c03645

Fu, L., Wang, Z., Dhankher, O.P., Xing, B. 2020. Nanotechnology as a new sustainable approach for controlling crop diseases and increasing agricultural production. *Journal of Experimental Botany* 71, 507–519. https://doi.org/10.1093/jxb/erz457

Ganjavi, A.S., Oraei, M., Gohari, G., Akbari, A., Faramarzi, A. 2021. Glycine betaine functionalized graphene oxide as a new engineering nanoparticle lessens salt stress impacts in sweet basil (*Ocimum basilicum* L.). *Plant Physiology and Biochemistry* 162, 14–26. https://doi.org/10.1016/j.plaphy.2021.02.014

Garg, S., Rumjit, N.P., Roy, S. 2023. Smart agriculture and nanotechnology: technology, challenges, and new perspective. *Advanced Agrochemistry* 3, 115–125. https://doi.org/10.1016/j.aac.2023.11.001

Ghassemi-Golezani, K., Abdoli, S. 2021. Improving ATPase and PPase activities, nutrient uptake and growth of salt-stressed ajowan plants by salicylic acid and iron-oxide nanoparticles. *Plant Cell Reports* 40, 559–573. https://doi.org/10.1007/s00299-020-02652-7

Ghiasi, T., Ahmadi, S., Ahmadi, E., Olyai, M.R.T.B., Khodadadi, Z. 2021. Novel electrochemical sensor based on modified glassy carbon electrode with graphene quantum dots, chitosan and nickel molybdate nanocomposites for diazinon and optimal design by the Taguchi method. *Microchemical Journal* 160, 105628. https://doi.org/10.1016/j.microc.2020.105628

Gholami, M., Khodadadi Darban, A., Khormali, F., Ayoubi, S., Zarei, M., Gorji, M. 2016. Review of nano additives in stabilization of soil. *International Journal of Engineering Science Research Technology* 5, 308–314.

Gomez, A., Narayan, M., Zhao, L., Jia, X., Bernal, R.A., Lopez-Moreno, M.L., Peralta-Videa, J.R. 2021. Effects of nano-enabled agricultural strategies on food quality: current knowledge and future research needs. *Journal of Hazardous Materials* 401, 123385. https://doi.org/10.1016/j.jhazmat.2020.123385

Hamid, A., Saleem, S. 2022. Role of nanoparticles in management of plant pathogens and scope in plant transgenics for imparting disease resistance. *Plant Protection Science* 58, 173–184. https://doi.org/10.17221/37/2020-PPS

Harini, K., Girigoswami, K., Pallavi, P., Thirumalai, A., Gowtham, P., Girigoswami, A. 2023. Smart farming approach using nanotechnology: an inevitable role in the application of pesticides. *Journal of Plant Protection Research* 63, 137–158. https://doi.org/10.24425/jppr.2023.145758

Huang, W., Du, J., Lai, H., Zhang, Q., Zhou, C., Liu, Z. 2022. Mechanical properties and mechanism of soil treated with nano-aqueous adhesive (NAA). *Scientific Reports* 12, 14711. https://doi.org/10.1038/s41598-022-19108-5

Jain, D., Sanadhya, S., Saheewala, H., Joshi, A., Bhojiya, A.A., Mohanty, S.R. 2021. The role of nanofertilizers in smart agriculture: an effective approach to increase nutrient use efficiency. In: Kumar Srivastava, D., Kumar Thakur, A., Kumar, P. (eds), *Agricultural Biotechnology: Latest Research and Trends*. Springer, Singapore. https://doi.org/10.1007/978-981-16-2339-4_21

Jakhar, A.M., Aziz, I., Kaleri, A.R., Hasnain, M., Haider, G., Ma, J., Abideen, Z. 2022. Nano-fertilizers: a sustainable technology for improving crop nutrition and food security. *NanoImpact* 27, 101986. https://doi.org/10.1016/j.impact.2022.100411

Jia, L., Zhou, Y., Wu, K., Feng, Q., Wang, C., He, P. 2020. Acetylcholinesterase modified AuNPs–MoS_2–rGO/PI flexible film biosensor: towards efficient fabrication and application in paraoxon detection. *Bioelectrochemistry* 131, 107392. https://doi.org/10.1016/j.bioelechem.2019.107392

Kansotia, K., Naresh, R., Sharma, Y., Sekhar, M., Sachan, P., Baral, K., Pandey, S.K. 2024. Nanotechnology-driven solutions: transforming agriculture for a sustainable and productive future. *Journal of Scientific Research and Reports* 30, 32–51.

Khaledian, S., Nikkhah, M., Shams-Bakhsh, M., Hoseinzadeh, S. 2017. A sensitive biosensor based on gold nanoparticles to detect *Ralstonia solanacearum* in soil. *Journal of General Plant Pathology* 83, 231–239. https://doi.org/10.1007/s10327-017-0721-z

Lei, Q., Huang, Z.Y., Pan, J.Z., Bao, J.Q., Xun, Q.N., Jiang, S., Zhang, Y.T. 2014. Research on antimicrobial activity and packaging performance of degradable protein films. *Advanced Materials Research* 915, 947–953. https://doi.org/10.4028/www.scientific.net/AMR.915-916.947

Li, W., Huang, M., Wu, Y., Liu, Y. 2014. Effects of polyacrylamide on soil organic matter content and aggregate stability of Ultisols in subtropical China. *Soil and Tillage Research* 144, 126–133. https://doi.org/10.1016/j.still.2014.07.002

Li, Y., Liu, M., Yang, X., Zhang, Y., Hui, H., Zhang, D. et al. 2022. Multi-walled carbon nanotubes enhanced the antioxidative system and alleviated salt stress in grape seedlings. *Scientia Horticulturae* 293, 110698. https://doi.org/10.1016/j.scienta.2021.110698

Malandrakis, A.A., Kavroulakis, N., Chrysikopoulos, C.V. 2019. Use of copper, silver and zinc nanoparticles against foliar and soil-borne plant pathogens. *Science of the Total Environment* 670, 292–299. https://doi.org/10.1016/j.scitotenv.2019.03.210

Mandal, N., Adhikary, S., Rakshit, R. 2020. Nanobiosensors: Recent developments in soil health assessment. In: Rakshit, A., Ghosh, S., Chakraborty, S., Philip, V., Datta, A. (eds), *Soil Analysis: Recent Trends and Applications*, pp. 285–304. Springer, Singapore.

Marchiol, L., Iafisco, M., Fellet, G., Adamiano, A. 2020. Nanotechnology support the next agricultural revolution: perspectives to enhancement of nutrient use efficiency. *Advances in Agronomy* 161, 27–116. https://doi.org/10.1016/bs.agron.2019.12.001

Mazhar, W.M., Ishtiaq, M., Hussain, I., Parveen, A., Bhatti, K.H., Azeem, M. et al. 2022. Seed nano-priming with zinc oxide nanoparticles in rice mitigates drought and enhances agronomic profile. *PLoS One* 17, e0264967. https://doi.org/10.1371/journal.pone.0264967

Michels, G.J., Piccinni, G., Rush, C.M., Fritts, D.A. 2000. Using infrared transducers to sense greenbug infestation in winter wheat. In: *Proceedings of the 5th International Conference on Precision Agriculture*. American Society of Agronomy, Precision Agriculture Center, University of Minnesota, St. Paul, MN.

Miguel-Rojas, C., Pérez-de-Luque, A. 2023. Nanobiosensors and nanoformulations in agriculture: new advances and challenges for sustainable agriculture. *Emerging Topics in Life Sciences* 7, 229–238. https://doi.org/10.1042/ETLS20230070

Mosa, M.A., El-Abeid, S.E., Khalifa, M., Elsharouny, T., El-Baz, S.M., Ahmed, A.Y. 2022. Smart pH-responsive system based on hybrid mesoporous silica nanoparticles for delivery of fungicide to control *Fusarium* crown and root rot in tomato. *Journal of Plant Pathology* 104, 979–992. https://doi.org/10.1007/s42161-022-01122-1

Naidu, G., Challa, Y.S.R., Vasudha, D., Rao, K.V.P. 2022. Recent advances in nano-enabled fertilizers towards sustainable agriculture and environment: a mini review. In: *21st Century Nanostructured Materials – Physics, Chemistry, Classification, and Emerging Applications in Industry, Biomedicine, and Agriculture*. IntechOpen, New York. https://doi.org/10.5772/intechopen.103053

Nair, R.V., Thomas, R.T., Mohamed, A., Pillai, S. 2020. Fluorescent turn-off sensor based on sulphur-doped graphene quantum dots in colloidal and film forms for the ultrasensitive detection of carbamate pesticides. *Microchemical Journal* 157, 104971. https://doi.org/10.1016/j.microc.2020.104971

Neme, K., Nafady, A., Uddin, S., Tola, Y.B. 2021. Application of nanotechnology in agriculture, postharvest loss reduction and food processing: food security implication and challenges. *Heliyon* 7, e08539. https://doi.org/10.1016/j.heliyon.2021.e08539

Noman, M., Ahmed, T., Wang, J., Ijaz, M., Shahid, M., Islam, M.S., Azizullah, M.I., Li, D., Song, F. 2023a. Nano-enabled crop resilience against pathogens: potential, mechanisms and strategies. *Crop Health* 1, 15. https://doi.org/10.1007/s44297-023-00015-8

Noman, M., Ahmed, T., White, J.C., Nazir, M.M., Li, D., Song, F. 2023b. *Bacillus altitudinis*-stabilized multifarious copper nanoparticles prevent bacterial fruit blotch in watermelon (*Citrullus lanatus* L.): direct pathogen inhibition, in planta particles accumulation, and host stomatal immunity modulation. *Small* 19, 2207136. https://doi.org/10.1002/smll.202207136

Nuruzzaman, M., Rahman, M.M., Liu, Y., Naidu, R. 2016. Nanoencapsulation, nano-guard for pesticides: a new window for safe application. *Journal of Agricultural and Food Chemistry* 64, 1447–1483. https://doi.org/10.1021/acs.jafc.5b05214

Oertel, C., Matschullat, J., Zurba, K., Zimmermann, F., Erasmi, S. 2016. Greenhouse gas emissions from soils—a review. *Geochemistry* 76, 327–352. https://doi.org/10.1016/j.chemer.2016.04.002

Ozdogan, B., Gacar, A., Aktas, H. 2017. Digital agriculture practices in the context of agriculture 4.0. *Journal of Economics, Finance and Accounting* 4, 186–193. https://doi.org/10.17261/Pressacademia.2017.448

Pateiro, M., Gómez, B., Munekata, P.E.S., Barba, F.J., Putnik, P., Kovačević, D.B., Lorenzo, J.M. 2021. Nanoencapsulation of promising bioactive compounds to improve their absorption, stability, functionality and the appearance of the final food products. *Molecules* 26, 1547. https://doi.org/10.3390/molecules26061547

Qiu, M., Glass, Z., Chen, J., Haas, M., Jin, X., Zhao, X. et al. 2021. Lipid nanoparticle-mediated codelivery of Cas9 mRNA and single-guide RNA achieves liver-specific in vivo genome editing of *Angptl3*. *Proceedings of the National Academy of Sciences of the United States of America* 118, e2020401118. https://doi.org/10.1073/pnas.2020401118

Rajput, V.D., Singh, A., Minkina, T., Rawat, S., Mandzhieva, S., Sushkova, S. et al. 2021. Nano-enabled products: challenges and opportunities for sustainable agriculture. *Plants* 10, 2727. https://doi.org/10.3390/plants10122727

Raza, A., Charagh, S., Salehi, H., Abbas, S., Saeed, F., Poinern, G., Siddique, K.H.M., Varshney, R.K. 2023. Nano-enabled stress-smart agriculture: can nanotechnology deliver drought- and salinity-smart crops? *Journal of Sustainable Agriculture and Environment* 2, 189–214. https://doi.org/10.1002/sae2.12061

Safarnejad, M.R., Samiee, F., Tabatabie, M., Mohsenifar, A. 2017. Development of quantum dot-based nanobiosensors against citrus tristeza virus (CTV). *Sensors & Transducers* 213, 54–60.

Salam, A., Khan, A.R., Liu, L., Yang, S., Azhar, W., Ulhassan, Z. et al. 2022. Seed priming with zinc oxide nanoparticles downplayed ultrastructural damage and improved photosynthetic apparatus in maize under cobalt stress. *Journal of Hazardous Materials* 423, 127021.

Santana, I., Wu, H., Hu, P., Giraldo, J.P. 2020. Targeted delivery of nanomaterials with chemical cargoes in plants enabled by a biorecognition motif. *Nature Communications* 11, 2045. https://doi.org/10.1038/s41467-020-15731-w

Seleiman, M.F., El-Sayed, M., El-Sayed, S., El-Sayed, A. 2020. Nanofertilizer as an eco-friendly innovation for enhancing plant growth, yield components, and biochemical aspects of flax cultivars grown under newly reclaimed sandy soil conditions. *Scientific Reports* 10, 1–15. https://doi.org/10.56093/ijas.v95i11.141995

Sha, R., Bhattacharyya, T.K. 2020. MoS_2-based nanosensors in biomedical and environmental monitoring applications. *Electrochimica Acta* 349, 136370. https://doi.org/10.1016/j.electacta.2020.136370

Sharma, P., Lew, T.T.S. 2022. Principles of nanoparticle design for genome editing in plants. *Frontiers in Genome Editing* 4, 4. https://doi.org/10.3389/fgeed.2022.846624

Sharma, R., Kumar, V. 2024. Nano enabled agriculture for sustainable soil. *Waste Management Bulletin* 2, 152–161.

Siddiquee, S., Rovina, K., Yusof, N.A., Rodrigues, K.F., Suryani, S. 2014. Nanoparticle-enhanced electrochemical biosensor with DNA immobilization and hybridization of *Trichoderma harzianum* gene. *Sensing and Biosensing Research* 2, 16–22. https://doi.org/10.1016/j.sbsr.2014.06.002

Soltani Nejad, M., Samandari Najafabadi, N., Aghighi, S., Pakina, E., Zargar, M. 2022. Evaluation of *Phoma* sp. biomass as an endophytic fungus for synthesis of extracellular gold nanoparticles with antibacterial and antifungal properties. *Molecules* 27, 1181. https://doi.org/10.3390/molecules27041181

Talari, F.F., Bozorg, A., Faridbod, F., Vossoughi, M. 2021. A novel sensitive aptamer-based nanosensor using rGQDs and MWCNTs for rapid detection of diazinon pesticide. *Journal of Environmental Chemical Engineering* 9, 104878. https://doi.org/10.1016/j.jece.2020.104878

Turgut, B.A., Bezirganoğlu, İ. 2022. Foliar application of CaO nanoparticles and salicylic acid on *Medicago sativa* seedlings enhances tolerance against *Fusarium oxysporum*. *Physiological and Molecular Plant Pathology* 122, 101926. https://doi.org/10.1016/j.pmpp.2022.101926

Ur Rahim, H., Qaswar, M., Uddin, M., Giannini, C., Herrera, M.L., Rea, G. 2021. Nano-enable materials promoting sustainability and resilience in modern agriculture. *Nanomaterials* 11, 2068. https://doi.org/10.3390/nano11082068

Viswanathan, S., Radecka, H., Radecki, J. 2009. Electrochemical biosensor for pesticides based on acetylcholinesterase immobilized on polyaniline deposited on vertically assembled carbon nanotubes wrapped with ssDNA. *Biosensors and Bioelectronics* 24, 2772–2777. https://doi.org/10.1016/j.bios.2009.01.044

Wei, Z., Zou, R., Zhang, L.R., Zhang, J., Cai, Y.N., Liu, N., Pan, G.X. 2018. Potential of silicate nanoparticles applied to a highly weathered tropical soil to increase rice yields and nutrient availability. *Science of the Total Environment* 615, 1415–1424. https://doi.org/10.1016/j.scitotenv.2017.09.312

Wu, H., Nibler, R., Morris, V., Herrmann, N., Hu, P., Jeon, S.-J., Kruss, S., Giraldo, J.P. 2020. Monitoring plant health with near-infrared fluorescent H_2O_2 nanosensors. *Nano Letters* 20, 2432–2442. https://doi.org/10.1021/acs.nanolett.9b05159

Xin, X., Judy, X., Jonathan, D., Sumerlin, B.B., He, Z. 2020. Nano-enabled agriculture: from nanoparticles to smart nanodelivery systems. *Environmental Chemistry* 17, 413–421. https://doi.org/10.1071/EN19254

Yaqoob, H., Tariq, A., Bhat, B.A., Bhat, K.A., Nehvi, I.B., Raza, A. et al. 2023. Integrating genomics and genome editing for orphan crop improvement: a bridge between orphan crops and modern agriculture system. *GM Crops & Food* 14, 1–20. https://doi.org/10.1080/21645698.2022.2146952

Yüksel, S., Schwenkbier, L., Pollok, S., Weber, K., Cialla-May, D., Popp, J. 2015. Label-free detection of *Phytophthora ramorum* using surface-enhanced Raman spectroscopy. *Analyst* 140, 7254–7262. https://doi.org/10.1039/C5AN01156F

Zargar, M., Bayat, M., Saquee, F.S., Diakite, S., Ramzanovich, N.M., Akhmadovich, K.A.S. 2023. New advances in nano-enabled weed management using poly(ε-caprolactone)-based nano-herbicides: a review. *Agriculture* 13, 2031. https://doi.org/10.3390/agriculture13102031

Zhang, P., Guo, Z., Ullah, S., Melagraki, G., Afantitis, A., Lynch, I. 2021. Nanotechnology and artificial intelligence to enable sustainable and precision agriculture. *Nature Plants* 7, 864–876. https://doi.org/10.1038/s41477-021-00946-6

Zhao, W., Adeel, M., Zhang, P., Zhou, P., Huang, L., Zhao, Y. et al. 2022. A critical review on surface-modified nano-catalyst application for the photocatalytic degradation of volatile organic compounds. *Environmental Science: Nano* 9, 61–80. https://doi.org/10.1039/D1EN00955A

Zhao, Y., Liu, L., Kong, D., Kuang, H., Wang, L., Xu, C. 2014. Dual amplified electrochemical immunosensor for highly sensitive detection of *Pantoea stewartii* subsp. *stewartii*. *ACS Applied Materials & Interfaces* 6, 21178–21183. https://doi.org/10.1021/am506104r

Zhu, G., Sun, Y., Shakoor, N., Zhao, W., Wang, Q., Wang, Q. et al. 2023. Phosphorus-based nanomaterials as a potential phosphate fertilizer for sustainable agricultural development. *Plant Physiology and Biochemistry* 197, 108172. https://doi.org/10.1016/j.plaphy.2023.108172

Zohuriaan-Mehr, M.J., Kabiri, K. 2008. Superabsorbent polymer materials: a review. *Iranian Polymer Journal* 17, 451–477.

10 Insights into Toxicity of Nano-Fertilizers and Nano-Pesticides

Apurva Gosavi, Sarvesh Jadhav, Aagam Bamb, Amol Kahandal, Anant B. Kanagare, and Chandrakant K. Tagad

10.1 INTRODUCTION

Nanotechnology is increasingly recognized as a transformative technology across multiple sectors, including agriculture, where it offers innovative approaches to enhance productivity, sustainability, and overall efficiency. As the global population steadily increases, the specter of food insecurity looms large. The intensifying pressure to fulfill this burgeoning demand necessitates innovative solutions that optimize agricultural practices while minimizing environmental impact. In this pursuit, nanotechnology has emerged as a beacon of hope, offering a plethora of tools with the potential to revolutionize the agricultural landscape. Among these advancements, nano-agrochemicals, encompassing both nano-pesticides and nano-fertilizers, have ignited significant interest due to their unique properties and promising applications.

Nanotechnology introduces new possibilities in agriculture by enabling the development of nano-pesticides and nano-fertilizers that improve pest control and nutrient management. These nano-based formulations offer the potential to minimize chemical consumption, optimize nutrient supply, reduce environmental impact, and reshape traditional agricultural approaches. The integration of nano-pesticides, designed to precisely target and control pest spread, has great potential to effectively control agricultural pests while minimizing chemical residues in soil and water. Similarly, nano-fertilizers provide a tailored nutrient delivery system, ensuring optimal plant uptake while reducing nutrient runoff and leaching, thereby promoting soil health and fertility, and environmental sustainability. However, alongside these promising advances, it is important to comprehensively assess the potential risks and impacts associated with the deployment of nanomaterials (NMs) in agricultural ecosystems.

The narrative surrounding nano-agrochemicals, however, is not without its complexities. While their potential benefits are undeniable, concerns regarding their potential toxicity to various organisms and ecosystems necessitate a comprehensive understanding of their behavior and potential risks. The nanoscale dimensions of these materials endow them with unique physicochemical properties that can translate into unforeseen biological interactions. Unraveling the intricate mechanisms

 DOI: 10.1201/9781032659435-10

governing their uptake, translocation, and potential accumulation within plants, soil biota, and even humans is paramount for ensuring their responsible development and application. This chapter delves into the intricate world of nano-agrochemicals, dissecting their mechanisms of action, potential pathways of exposure, and emerging insights into their toxicological profiles, thereby serving as a springboard for informed discourse on the multifaceted implications of nano-agrochemicals in sustainable agriculture.

As the agricultural industry explores the application of nanotechnology, it is imperative to further study the complexities of nanotoxicology, the mechanisms of nanoparticle (NP) uptake by plants, and its broader ecological ramifications. Understanding the interactions between NMs and soil microbiomes, aquatic ecosystems and human health is essential for making informed decisions and developing responsible applications of nanotechnology in agriculture. This chapter situates nanotechnology within the agricultural context, emphasizing its capacity to enhance productivity while underscoring the need for robust risk assessment, governance, and regulatory mechanisms.

10.2 NANO-PESTICIDES

Pesticides are substances designed to control unwanted organisms, commonly referred to as pests. These can include insects, weeds, fungi, bacteria, and even rodents. They play a crucial role in agriculture by protecting crops from yield losses and ensuring food security. However, conventional pesticides often raise concerns due to their potential negative impacts on human health and the environment. Nano-pesticides emerge as a novel class of pest control agents engineered at the nanoscale (1–100 nm) encapsulating the active ingredients. Nano-pesticides offer several potential advantages like their smaller size, and targeted delivery systems allow them to deliver active ingredients more efficiently to pests, potentially requiring lower application rates and reducing overall exposure. This also contributes to their reduced environmental impact, as their controlled-release profile kinetics and specific targeting through entrapment of organic pesticides or making emulsions/dispersions can increase soil health and sustainability (Kutawa et al. 2021).

Additionally, NPs possess unique physical and chemical properties at the nanoscale that enable novel modes of pest control not achievable with conventional methods. These novel mechanisms could prove crucial in overcoming existing pest resistance and developing more sustainable pest management strategies. These key differences between conventional pesticides and nano-pesticides are summarized in Table 10.1 (Kannan et al. 2023).

The diverse nature of nano-pesticides necessitates categorization based on several aspects. Functionally, they can target specific pests like insects, fungi, or weeds or act as broad-spectrum antimicrobials. Active ingredients range from organic compounds, such as essential oils, to inorganic materials, such as silver nanoparticles (Ag NPs), each encapsulated in polymers for enhanced efficacy and controlled release. Nano-pesticides can be broadly categorized based on their function, active ingredient, delivery mechanisms, and application method, as described in Table 10.2.

TABLE 10.1
Key Differences between Conventional Pesticides and Nano-Pesticides

Feature	Conventional Pesticides	Nano-Pesticides
Size and mode of action	Direct contact/ingestion	Cellular-/molecular-level interaction
Delivery and specificity	Broad-spectrum, can harm non-target organisms	Targeted delivery, reduced harm to non-target organisms
Environmental impact	Persistent, can contaminate soil and water	Potentially biodegradable, reduced environmental persistence
Efficacy	Variable, may require high application rates	Potentially increased efficacy with lower application rates
Novelty	Established technology	Emerging technology with novel modes of action
Degradation in soil or plant	Slower	Faster
Regulatory guidelines	Available	Partially available (Under development)

TABLE 10.2
Classifications of Nano-Pesticides based on Function, Active Ingredient, Delivery Mechanism, and Applications

Function	Nano-insecticides	Target insects and other arthropods, either through direct contact or ingestion (Manna et al. 2023)
	Nanofungicide	Combat fungal infections in plants and crops (Kutawa et al. 2021)
	Nanoherbicides	Control unwanted plant growth (Zargar et al. 2023)
	Nematicides	Target nematodes, microscopic worms that can damage plant roots (Zinovieva et al. 2023)
Active ingredient	Inorganic nanomaterials	Metal oxides (ZnO, TiO_2), metal nanoparticles (silver, copper), and carbon nanotubes fall under this category. They can act directly as pesticides or serve as carriers for other active ingredients (Abbassy et al. 2017)
	Organic nanomaterials	Polymers, chitosan, and dendrimers are examples. They can improve the solubility of existing pesticides, facilitate targeted delivery, and enhance bioavailability (Zheng et al. 2022)
	Biological nanomaterials	Viruses, bacteria, and plant-derived nanoparticles are being explored for their potential as natural pest control agents (Djiwanti and Kaushik 2019)

(Continued)

TABLE 10.2 (*Continued*)
Classifications of Nano-Pesticides based on Function, Active Ingredient, Delivery Mechanism, and Applications

Delivery mechanisms	Emulsions	Active ingredients are encapsulated within the oil in water emulsion for efficient delivery (Mustafa and Hussein 2020)
	Dispersions	A tiny liquid suspension of active ingredients to target pests (Lu et al. 2021)
	Hydrogels	Hydrophilic gels are used to encapsulate and release pest control ingredients precisely and sustainably (Zhang et al. 2020b)
Application	Foliar spray	Nanoparticles are applied directly to plant leaves for insect and fungal control (Hong et al. 2021)
	Seed treatment	Seeds are coated with nanoparticles for protection against soil-borne pests and diseases (Shelar et al. 2023)
	Soil application	Nanoparticles are incorporated into soil to target pests like nematodes and soil-borne pathogens (Khan and Rizvi 2017)

Although nano-pesticides show significant promise, much more investigation is required to clearly determine their safety, effectiveness, and effects on the environment. Long-term evaluations are especially important to identify possible risks and to support their safe and responsible use. In addition, improving delivery methods, enhancing target specificity, and conducting comprehensive life-cycle assessments are necessary to increase their advantages while reducing potential negative impacts.

10.3 NANO-FERTILIZERS

Fertilizers play a vital role in agriculture by replenishing essential soil nutrients, fostering plant growth, and significantly increasing crop productivity. While research suggests that fertilizers contribute to higher crop yields, an inherent contradiction arises that some fertilizers, if misapplied or used excessively, can unintentionally hinder plant growth, disrupting the delicate nutrient balance in the soil. This imbalance may lead to toxicities or nutrient shortages, impeding the plants from reaching their full potential. Moreover, careless fertilizer use can result in soil degradation, compromising fertility and the soil's ability to retain vital nutrients and moisture necessary for robust plant development.

Recognizing these challenges, the incorporation of nanotechnologies emerges as a focal point in contemporary fertilizer research. This emphasis, particularly in optimizing crop fertilization, aims to minimize nutrient losses. The main objective is to leverage highly engineered nanostructures for the systematic release of nutrients in a regulated and tailored manner to meet the specific nutritional needs of various crops. This precision-driven synchronization not only provides a more focused and scientifically precise method for increasing crop yield but also strives to minimize environmental impact, signifying a major advancement in modifying agricultural practices (El-Saadony et al. 2021).

The groundbreaking potential of these nanostructures is underscored by their ability to usher in a resource-efficient, ecologically sound, and scientifically validated future. The utilization of nano-fertilizers, encapsulating nutrients and employing composites, not only improves plant uptake but also mitigates the negative environmental effects of runoff. This innovative combination not only enhances crop yield but also redefines sustainable pesticides, crucial for addressing food scarcity and safeguarding ecosystems (Taware et al. 2024).

The growing economic interest in fertilizers based on synthetic zeolite reflects a profound awareness of their significance in agriculture. Research efforts focusing on nutrient optimization, incorporating combinations of N, P, K, and micronutrients, demonstrate the importance of balanced fertilization. Studies showcasing the beneficial effects of peaty mixtures on maize development and the addition of organic layers to clay reveal novel approaches to nutrient regulation. Nano-fertilizers and composites shield nutrients, enhancing their effectiveness and slowing down soil erosion. In contrast to conventional fertilizers, nano-fertilizers exhibit longer nutrient availability. Recent research unveiling their release patterns, including the release of nitrogen for up to 1,000 hours, holds transformative potential for sustainable agriculture by boosting crop yield, improving food output, and mitigating the detrimental impacts of conventional fertilizers on agroecosystems (El-Saadony et al. 2021)

However, in addition to their possible advantages, nano-pesticides and nano-fertilizers present several issues and difficulties. These include the necessity for thorough safety evaluations and legal frameworks to guarantee their responsible use in agriculture, as well as their possible environmental impact, including the possibility of NPs leaking into soil and water systems. In order to maintain safe and sustainable agriculture practices, ongoing research and development efforts are concentrated on comprehending the behavior of nano-pesticides and nano-fertilizers, improving their performance, and addressing potential dangers.

10.4 NANOTOXICOLOGY AND UPTAKE OF NANOPARTICLES FROM ROOTS AND LEAVES

In the constantly evolving field of nanotechnology, NPs arise as pioneers of innovation, offering unparalleled opportunities for progress in agriculture, medicine, and beyond. NPs show both positive and negative effects. They may show a good response on plants, but they might hurt other components of the ecosystem. This dualistic nature of these minuscule entities necessitates a comprehensive examination of their ecological impact. Although nano-pesticides show significant promise, further investigation is needed to assess their safety, effectiveness, and environmental impacts fully. Long-term evaluations are especially important for identifying potential risks and supporting their safe and responsible use. In addition, improving delivery methods, enhancing target specificity, and conducting comprehensive life-cycle assessments are necessary to maximize their benefits while minimizing potential negative impacts.

The emergence of nano-biopesticides offers high precision in pest management; however, it requires careful examination for environmental sustainability. Concurrently, the introduction of nano-biofertilizers shows the potential to improve agricultural efficiency. However, a comprehensive assessment of the potential impacts on non-target plant tissues, soil quality, and associated microbial populations is essential. For instance, zinc oxide NPs have been reported to enhance the performance of *Mangifera indica* Linn. by increasing fruit yield (number and weight per tree), improving fruit size, total soluble solids, sugar content, and reducing acidity (Elsheery et al. 2020). In contrast, studies on *Glycine max* have demonstrated adverse effects, including suppressed root growth, decreased cell viability and biomass, elevated superoxide production, reduced foliage mass, and changes in gene expression patterns following ZnO NP exposure (Hossain et al. 2016). This investigation highlights the significance of achieving a delicate balance between harnessing the potential advantages of NPs and ensuring responsible environmental stewardship. It is crucial to acknowledge both the positive and negative aspects of NPs within the complex network of ecosystems.

The primary routes through which NPs enter plants are the roots and foliage. Roots, composed of structures such as the epidermis, cortex, endodermis, and stele, are especially important for the absorption and movement of water, nutrients, and NPs within the plant system. NP uptake occurs through the root epidermis via apoplastic and symplastic pathways, with subsequent translocation via the xylem and phloem. Foliar uptake occurs through leaf epidermis channels and stomatal pathways. Despite the cuticle's protective role, it allows selective NP diffusion, while stomatal features contribute to variability in NP uptake. Understanding these structures and pathways is vital for unravelling the complexities of NP–plant interactions in both environmental and agricultural contexts (Nobel 2009).

10.5 ADVERSE IMPACTS OF NANO-FERTILIZERS AND NANO-PESTICIDES

Nano-pesticides are designed to improve pest management by enabling controlled release and precise targeting of active ingredients. Despite this potential, a major limitation remains, as only a very small fraction, approximately 0.1%, actually reaches the intended pathogen, while the vast majority disperses into the surrounding environment. This inefficiency can result in unintended environmental impacts, including biodiversity loss, contamination of soil and water resources, and the accelerated development of pest and disease resistance. Considering these implications, there is a pressing need for comprehensive research to understand the negative effects and interactions of nano-pesticides before widespread use in plant production. Therefore, careful consideration of potential environmental effects is crucial before adopting nano-pesticides.

The toxicity effects of nano-fertilizers, designed to enhance nutrient delivery to plants and increase nutrient utilization efficiency, are a subject of growing concern. While nano-fertilizers hold promise for boosting crop yields and reducing environmental impact, there are reservations about their potential negative impacts on soil, animals, and human health. Figure 10.1 depicts the possible adverse impacts of nano-fertilizers and nano-pesticides.

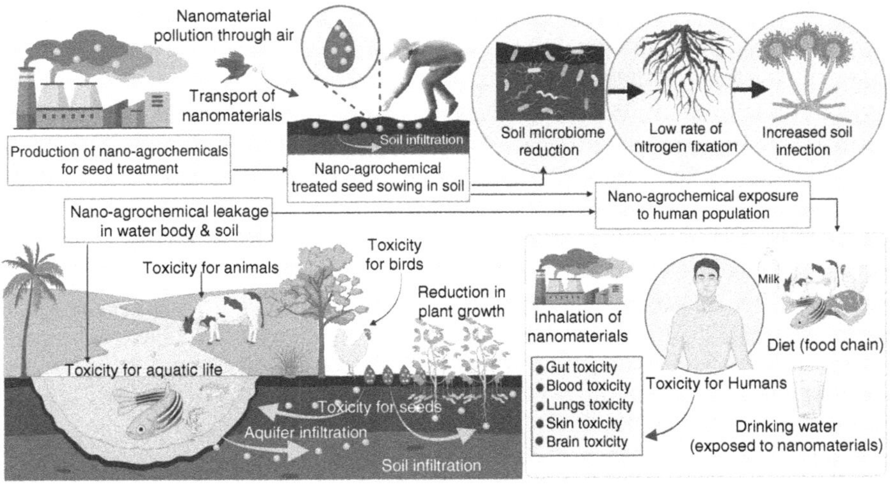

FIGURE 10.1 Adverse impacts of nano-fertilizers and nano-pesticides (Shelar et al. 2023).

10.5.1 Toxicity-Causing Factors and Uptake Mechanism

The potential toxicity of NMs to living organisms is not yet fully understood. Research suggests that their adverse effects are primarily associated with four key mechanisms. First, NPs are capable of quickly crossing cell membranes, which may disrupt photosynthetic activity and interfere with normal cellular functions (Abdel-Aziz et al. 2019). Second, certain NMs can release toxic metal ions, such as those derived from nano-silver, nano-copper, and nano-zinc (Bratovcic 2023). Third, oxidative stress plays a major role, as NMs can rapidly generate reactive oxygen species (ROS). These highly reactive molecules can impair mitochondrial antioxidant systems, damage functional proteins, disturb normal physiological processes, and ultimately lead to cell death (Rajput et al. 2020). Finally, NMs may exhibit a biological amplification effect, whereby they accumulate along food chains, resulting in higher concentrations in organisms at upper trophic levels through biomagnification.

The uptake of NPs by cells is governed by multiple factors, including their surface chemistry, size, shape, stiffness, and chemical composition. These characteristics play a critical role in determining NP stability within biological environments and influence how NPs interact with cells and are subsequently absorbed (Sousa De Almeida et al. 2021). Cellular membranes select specific endocytic pathways based largely on NP size, shape, and surface properties. Parameters such as targeting efficiency, circulation time, saturation levels, and uptake mechanisms are strongly size dependent, while the type of NP can further influence the internalization process. Advances in understanding endocytosis, intracellular transport, and cell–NP interactions have expanded significantly in recent years. In general, larger NPs and aggregated forms are internalized through macropinocytosis or phagocytosis, whereas smaller NPs are predominantly taken up via clathrin- or caveolae-mediated endocytosis. Smaller NPs can also enter and exit cells more readily. Spherical NPs tend to be internalized more efficiently than cylindrical ones, and positively charged NPs are absorbed more rapidly compared to neutral or negatively charged counterparts (Sousa De Almeida et al. 2021).

Toxicity of NPs is affected not only by intrinsic properties such as particle size, shape, surface area, and surface charge, but also by external environmental factors including pH, ionic strength, water hardness, and the presence of organic matter. The detrimental effects of NPs are primarily associated with the production of ROS, genetic damage, and disruption of cellular membranes. The risks posed by nonbiodegradable substances build up and remain in soil, plants, and animals. Because of their large surface area, NPs respond to several environmental factors. They are extremely reactive and dynamic, and the environment may undergo a variety of physical, chemical, or biological changes. Understanding the environmental risks posed by NMs is extremely difficult due to uncertainties surrounding ecotoxicity and environmental conditions.

10.5.2 Phytotoxicity

Plants possess a range of physiological, biochemical, and molecular defense mechanisms that help mitigate the harmful effects of NMs applied via nano-fertilizers and nano-pesticides. However, when present at high concentrations, NMs can interfere with essential plant functions. Key processes such as chlorophyll synthesis, photosynthetic efficiency, transpiration, electron transport, and the activity of the oxygen-evolving complex may be disrupted. These negative effects occur because NMs can bind to DNA within the photosynthetic system, damage DNA strands, alter molecular structures, and suppress gene expression related to chlorophyll formation. The toxicity of NMs is closely linked to their concentration, surface properties, size, shape, and overall chemical characteristics. Numerous studies have reported that excessive NM exposure poses significant risks to plant health, with toxicity increasing at higher doses (Garg et al. 2023). Importantly, NM-induced phytotoxic effects can be altered or mitigated through changes at the molecular, biochemical, and physiological levels.

Many physiological, biochemical, and molecular barriers mitigate the negative effects of NMs used in the production of nano-fertilizers and nano-pesticides. Numerous plant processes, such as the amount of chlorophyll, photosynthetic rates, transpiration, the oxygen-evolving complex, and electron transport activities, may be negatively impacted by elevated NM levels as they adhere to the DNA of the photosystem's structure and the chlorophyll production causing strand nicking and distortion, leading to the reduced expression of genes. NM content, surface charge, size, shape, and phytotoxicity are all correlated. Many scientific studies have demonstrated that NMs are harmful to plants, becoming particularly hazardous at higher concentrations. Modifications can be made to the molecular, biochemical, and physiological levels of NM phytotoxicity.

Studies have shown that exposure to high levels of 25 nm Ag NPs poses significant toxicity to *Oryza sativa* root cells, leading to damage of the cell wall and disruption of vacuolar structure (Zhang et al. 2021). While Ag NP concentrations up to 30 μg/mL were insufficient to penetrate the root cells, higher concentrations resulted in cellular disruption and pronounced toxic effects. The researchers found that although doses of 60 g/mL restrict root growth, levels of 30 g/mL encourage it. The results show that while Ag NP penetration is required to cause damage to plants, plants may benefit from Ag NPs at lesser penetration levels found in nature (Yan and Chen 2019).

The way NPs interact with plants, as well as their method of application, is strongly affected by the physicochemical properties of the NPs and the physiological characteristics of the plant. After entry through roots or leaves, NPs must pass through multiple tissues and structural barriers before reaching the vascular system. In most cases, the plant cell wall serves as a primary obstacle, limiting the movement of NPs into internal plant tissues. To improve NP uptake, plant cells can either create new pores in the cell wall or increase the width of existing ones. Additionally, NPs can enter a cell by using ion channel mechanisms or transport carrier proteins to pass the membrane.

The application of Ag NPs has been reported to markedly reduce root elongation in tomato plants (Song et al. 2013) and to impair seed germination and early seedling development in rice (Thuesombat et al. 2014). Similarly, exposure to TiO_2 NPs in maize seedlings resulted in reduced root hydraulic conductivity, lower transpiration rates, and restricted leaf growth (Asli and Neumann 2009). In addition, Cvjetko et al. (2018) documented the toxic effects of Ag NPs on *Nicotiana tabacum*, highlighting their potential risk to plant health (Cvjetko et al. 2018).

10.5.3 Effects on Soil and Microbial Population

The use of NMs in agriculture is not always beneficial and can have a variety of negative effects on plants, soil, water, and, most importantly, humans because of the long food chain and the ease with which NPs can travel. To fully comprehend the agroecological toxicity of NPs, research must be done on how they behave in soil, plants, and water at different rates and sizes. The harmful effects of nano-fertilizers on soil ecosystems are an important component of continuing agricultural nanotechnology research and concern. While nano-fertilizers have the potential to improve nutrient delivery to plants and increase nutrient use efficiency, their possible negative impacts on soil health must be considered. Because of their small size and distinct physicochemical features, NPs may interact with soil components, potentially affecting soil structure, microbial populations, and nutrient dynamics. The possibility of nano-fertilizers building up in the soil over time and having unforeseen effects is one of the main worries. The delicate equilibrium of the soil ecosystem may be upset by NPs, which may affect the variety and activity of soil microbes. Microbial community changes could have an impact on the processes involved in the cycling of nutrients, which could have a domino effect on the fertility of the soil. Furthermore, the continuous presence of NPs in the soil could cause plants to bioaccumulate them, which could have an effect on the food chain and endanger consumers.

The complicated interaction between bacteria and NPs in the soil affects the latter's fate in the environment in many ways. Beneficial bacteria in soil help with the breakdown of organic matter, nutrient recycling, disease prevention, and enhanced growth. Reportedly, the use of nano-formulations might negatively affect the soil's microbial population, which could eventually deteriorate the soil's quality and agricultural viability. It is commonly known that NMs have a significant impact on the variety, abundance, and vital microbial processes of the soil microbiome, including mineralization, nitrogen fixation, and activities that promote plant growth. Higher dosages of NPs in soil have detrimental consequences that mostly impact the enzymatic activities of soil microorganisms. Important nutrient recycling processes such as ammonification, denitrification, nitrogen fixation, phosphate solubilization, and activities that promote plant growth can be inhibited by NPs. These processes are essential for preserving soil fertility and the environment. According to Zhang et al., soil microorganisms involved in the cycles of phosphorus, carbon, and nitrogen were negatively impacted by Ag NPs, which also altered the pH of the soil (Zhang et al. 2020a).

Depending on dosage and length of exposure, it has been discovered that other metallic metal NPs, like zinc and titanium, negatively impact the soil microbiota (Ameen et al. 2021). Rhizobia and mycorrhizal fungi decreased in a study by Rahman et al. (2020) due to the negative effects of platinum NPs stabilized with poly(vinylpyrrolidone) on the microbiota of pea roots. Numerous studies have reported that NMs decrease the activity of enzymes. Reduced rhizospheric bacteria can also be indicated by measuring urease activity and how it decreases in response to NMs (Lars Michael Skjolding et al. 2016). *P. putida KT2440*, a helpful environmental microbe, was toxic to Ag, CuO, and ZnO NPs, indicating that the application of NPs (B1 mg Ag/L and 10 mg Cu, Zn/L) may have a detrimental effect on soil microbial processes (Pandey et al. 2018; Gajjar et al. 2009).

10.5.4 Effects on the Aquatic Ecosystem

The growing use of NMs and agriproducts derived from them is appealing to a wide range of businesses. Agrochemicals based on NMs may leak into water bodies and become incorporated into the food chain through a process known as bioaccumulation. Water bodies may become contaminated by NPs found in industrial plant effluents. If an NM is dropped into water, it may infiltrate the body of an aquatic organism and disrupt its feeding and functioning. Copper-based nano-pesticides (CuO, Cu, and $Cu(OH)_2$) are currently receiving a lot of interest. According to Carley et al. (2020), they had little influence on the objective, the terrestrial agroecosystem, but had a significant impact on the non-target wetland (aquatic community), particularly on protists.

Luo et al. (2020) reported that the environmental behavior of nano-TiO_2 in aquatic systems is governed by processes such as aggregation, sedimentation, movement through water and sediments, and its interactions with both biotic and abiotic components of the ecosystem. Furthermore, ultraviolet (UV) radiation plays an important role in influencing the effects of metal oxide NPs, particularly those with photocatalytic properties, including ZnO and TiO_2 NPs. ROS are formed in these situations. Subsequently, the exposure of these natural killers to light in surface waters produces reactive intermediates, which in turn modify the ecotoxicological potential of coexisting organic micropollutants, such as pesticides, by catalytic degradation (Lüderwald et al. 2020). Fish are frequently employed to evaluate the robustness and condition of aquatic habitats. TiO_2 NPs are important for managing aquatic health and biodiversity, as they are released into the aquatic environment from various sources and can cause cytogenetic and hematological changes in African catfish, *Clarias gariepinus* (Ogunsuyi et al. 2020).

10.5.5 Bioaccumulation and Trophic Transfer

Evaluating the uptake, retention, and elimination of NMs is essential for understanding their potential trophic transfer and biomagnification, which are key concerns in ecotoxicology (Petersen et al. 2019). Tangaa et al. (2016) identified four major factors that influence the trophic movement of metal NPs: transformations of NPs in the environment, their uptake and accumulation in prey organisms, their internal distribution and fate within the prey, and the digestive physiology of the predator (Tangaa et al. 2016). The possibility of substantial bioaccumulation and biomagnification of Ag NPs in fish and benthic invertebrates, such as prawns and shellfish, raises concerns about the safety of aquatic food resources. Nevertheless, the extent of accumulation varies among NP types. For instance, Ag NPs have been shown to biomagnify within aquatic food webs and exhibit higher bioaccumulation potential than TiO_2 NPs (Wang et al. 2019).

Skjolding et al. (2014) demonstrated that ZnO NPs can be transferred across trophic levels from *Daphnia magna* to zebrafish (*Danio rerio*). Similarly, Nemati et al. (2019) reported that CuO NPs are capable of trophic transfer, as confirmed through a 21-day dietary exposure experiment in which *Amatitlania nigrofasciata* larvae consumed *Artemia salina nauplii* previously exposed to CuO NPs. Successful trophic transfer of NMs depends primarily on their stability and surface properties. Yang et al. (2018) further showed that the adsorption of arsenic (As) onto nano-TiO_2 present in contaminated algal food enhanced the transfer of As along the food chain to *A. salina*. In another study, uptake and translocation analyses revealed that approximately 15% of Ce^{4+} from CeO_2 NPs was reduced to Ce^{3+} in cucumber plants, which were subsequently transported through the phloem. These findings highlight the importance of evaluating the potential health risks posed by NMs entering food chains, as their accumulation and transformation may threaten human health.

10.5.6 Human Toxicity

Overuse of nano-fertilizers in farming techniques can lead to unwelcome and lasting environmental problems. The reactivity and unpredictable nature of NPs have raised concerns about worker safety during manufacturing and field application. Also, it must be noted that the last recipient of the NPs is a mammal, including humans, that inhales, eats, or absorbs contaminated food through the skin.

NMs can enter the bloodstream through ingestion or inhalation, cross biological membranes, and reach organs. Biochemical and genotoxic effects could arise from this. The shape of NPs plays a significant role in determining their toxicity. For example, rod-shaped ZnO NPs have been shown to induce greater toxicity in human lung epithelial cells than spherical ZnO NPs. Likewise, gold NPs with identical functions can exhibit varying toxic effects depending on their morphology; spherical gold NPs tend to be more toxic than rod-shaped forms, potentially due to the faster release of the attached functional molecules (Egbuna et al. 2021). If a particle has the same composition and crystalline structure, its toxicity increases with its size. According to Ettlinger et al. (2022), tiny NPs can induce fibrosis, inflammation, and cell proliferation, thereby leading to tumor formation.

Numerous forms of NPs enhance the expression of viral receptors, leading to macrophage hyperactivity and severe inflammation (Ettlinger et al. 2022). Certain NPs, including SiO_2 and TiO_2, have been reported to reduce the expression of viral and bacterial receptors, which may weaken the body's ability to defend against these pathogens (Mahon et al. 2020). Ag NPs are known to exhibit cytotoxic effects in human lung tissues and have been associated with an increased risk of lung cancer. In addition, Ag NP exposure has been linked to multiple forms of toxicity, including genotoxic effects, size-dependent developmental toxicity, cellular damage, and inflammatory responses (Parveen et al. 2021).

10.6 CONCLUSION

The advancement of nano-fertilizers and nano-pesticides has emerged as a promising approach to sustainable agriculture and global food security. These technologies offer notable advantages over conventional agrochemicals and may help address the increasing demand for food as the global population is projected to exceed nine billion by 2050, while simultaneously reducing environmental impacts. Compared with traditional fertilizers and pesticides, nano-agrochemicals provide benefits such as improved nutrient-use efficiency, reduced water and nutrients losses, targeted delivery, controlled release, and enhanced crop growth and productivity. Despite these advantages, a comprehensive evaluation of the agroecological risks associated with nano-agrochemicals is essential to ensure their long-term safety for soil organisms, crops, human health, and entire ecosystems. Such assessments require a detailed understanding of interactions of NPs with soil physicochemical properties to prevent the disruption of soil microbial communities; investigation of NP bioaccumulation potential; identification of effective yet non-toxic concentration ranges; evaluation of trophic transfer within food webs; and the development of robust risk assessment and management strategies. In addition, establishing regulatory frameworks, standardized guidelines, and farmer awareness programs is crucial to prevent the misuse or overapplication of nano-based agrochemicals.

REFERENCES

Abbassy, M.A., Abdel-Rasoul, M.A., Nassar, A.M.K., Soliman, B.S.M. 2017. Nematicidal activity of silver nanoparticles of botanical products against root-knot nematode, *Meloidogyne incognita*. *Archives of Phytopathology and Plant Protection* 50 (17–18), 909–926. https://doi.org/10.1080/03235408.2017.1405608

Abdel-Aziz, H.M.M., Hasaneen, M.N.A., Omer, A.M. 2019. Impact of engineered nanomaterials either alone or loaded with NPK on growth and productivity of French bean plants: seed priming vs foliar application. *South African Journal of Botany* 125, 102–108. https://doi.org/10.1016/J.SAJB.2019.07.005

Ameen, F., Alsamhary, K., Alabdullatif, J.A., ALNadhari, S. 2021. A review on metal-based nanoparticles and their toxicity to beneficial soil bacteria and fungi. *Ecotoxicology and Environmental Safety* 213, 112027. https://doi.org/10.1016/J.ECOENV.2021.112027

Asli, S., Neumann, P.M. 2009. Colloidal suspensions of clay or titanium dioxide nanoparticles can inhibit leaf growth and transpiration via physical effects on root water transport. *Plant, Cell & Environment* 32(5), 577–584. https://doi.org/10.1111/J.1365-3040.2009.01952.X

Bratovcic, A. 2023. Different applications of nanomaterials and their impact on the environment. *SSRG International Journal of Material Science and Engineering* 5, 1–7. https://www.internationaljournalssrg.org

Carley, L.N., Panchagavi, R., Song, X., Davenport, S., Bergemann, C.M., McCumber, A.W. et al. (2020). Long-term effects of copper nanopesticides on soil and sediment community diversity in two outdoor mesocosm experiments. *Environmental Science & Technology*, 54(14), 8878–8889.

Cvjetko, P., Zovko, M., Štefanić, P., Petra, B., Renata, T., Mirta, D. et al. 2018. Phytotoxic effects of silver nanoparticles in tobacco plants. *Environmental Science and Pollution Research* 25(6), 5590–5602. https://doi.org/10.1007/S11356-017-0928-8

Djiwanti, S.R., Kaushik, S. 2019. Nanopesticide: future application of nanomaterials in plant protection. In: Prasad, R. (eds), *Nanotechnology in the Life Sciences*, pp. 255–298. Springer, Cham. https://doi.org/10.1007/978-3-030-16379-2_10

Egbuna, C., Parmar, V.K., Jeevanandam, J., Ezzat, S.M., Patrick-Iwuanyanwu, K.C., Adetunji, C.O. et al. 2021. Toxicity of nanoparticles in biomedical application: nanotoxicology. *Journal of Toxicology* 2021, 9954443. https://doi.org/10.1155/2021/9954443

El-Saadony, M.T., ALmoshadak, A.S., Shafi, M.E., Albaqami, N.M., Saad, A.M., El-Tahan, A.M. et al. 2021. Vital roles of sustainable nano-fertilizers in improving plant quality and quantity—an updated review. *Saudi Journal of Biological Sciences* 28(12), 7349–7359. https://doi.org/10.1016/J.SJBS.2021.08.032

Elsheery, N.I., Helaly, M.N., El-Hoseiny, H.M., Alam-Eldein, S.M. 2020. Zinc oxide and silicone nanoparticles to improve the resistance mechanism and annual productivity of salt-stressed mango trees. *Agronomy* 10(4), 558. https://doi.org/10.3390/AGRONOMY10040558

Ettlinger, R., Lächelt, U., Gref, R., Horcajada, P., Lammers, T., Serre, C., Couvreur, P., Morris, R.E., Wuttke, S. 2022. Toxicity of metal–organic framework nanoparticles: From essential analyses to potential applications. *Chemical Society Reviews* 51(2), 464–484. https://doi.org/10.1039/D1CS00918D

Gajjar, P., Pettee, B., Britt, D.W., Huang, W., Johnson, W.P., Anderson, A.J. 2009. Antimicrobial activities of commercial nanoparticles against an environmental soil microbe, *Pseudomonas putida* KT2440. *Journal of Biological Engineering* 3(1), 1–13. https://doi.org/10.1186/1754-1611-3-9

Garg, D., Sridhar, K., Inbaraj, S., Chawla, B., Tripathi, P., Sharma, M. et al. 2023. Nano-biofertilizer formulations for agriculture: a systematic review on recent advances and prospective applications. *Bioengineering* 10(9), 1010. https://doi.org/10.3390/BIOENGINEERING10091010

Hong, J., Wang, C., Wagner, D.C., Gardea-Torresdey, J.L., He, F., Rico, C.M. 2021. Foliar application of nanoparticles: mechanisms of absorption, transfer, and multiple impacts. *Environmental Science: Nano* 8(5), 1196–1210. https://doi.org/10.1039/D0EN01129K

Hossain, Z., Mustafa, G., Sakata, K., Komatsu, S. 2016. Insights into the proteomic response of soybean towards Al_2O_3, ZnO, and Ag nanoparticles stress. *Journal of Hazardous Materials* 304, 291–305. https://doi.org/10.1016/J.JHAZMAT.2015.10.071

Kannan, M., Bojan, N., Swaminathan, J., Zicarelli, G., Hemalatha, D., Zhang, Y., Ramesh, M., Faggio, C. 2023. Nanopesticides in agricultural pest management and their environmental risks: a review. *International Journal of Environmental Science and Technology* 20(9), 10507–10532. https://doi.org/10.1007/S13762-023-04795-Y

Khan, M.R., Rizvi, T.F. 2017. *Application of Nanofertilizer and Nanopesticides for Improvements in Crop Production and Protection*, pp. 405–427. Springer, Cham. https://doi.org/10.1007/978-3-319-46835-8_15

Kutawa, A.B., Ahmad, K., Ali, A., Hussein, M.Z., Wahab, M.A.A., Adamu, A., Ismaila, A.A., Gunasena, M.T., Rahman, M.Z., Hossain, M.I. 2021. Trends in nanotechnology and its potentialities to control plant pathogenic fungi: a review. *Biology* 10(9), 881. https://doi.org/10.3390/BIOLOGY10090881

Lu, Z., Zhang, C., Gao, Y., Wang, W., Lin, J., Du, F. 2021. Simple, effective, and energy-efficient strategy to construct a stable pesticide nanodispersion. *ACS Agricultural Science and Technology* 1(4), 329–337. https://doi.org/10.1021/ACSAGSCITECH.1C00018

Lüderwald, S., Meyer, F., Gerstle, V., Friedrichs, L., Rolfing, K., Schreiner, V.C., Bakanov, N., Schulz, R., Bundschuh, M. 2020. Reduction of pesticide toxicity under field-relevant conditions? The interaction of titanium dioxide nanoparticles, ultraviolet, and natural organic matter. *Environmental Toxicology and Chemistry* 39(11), 2237–2246. https://doi.org/10.1002/ETC.4851

Luo, Z., Li, Z., Xie, Z., Sokolova, I.M., Song, L., Peijnenburg, W.J.G.M., Hu, M., Wang, Y. 2020. Rethinking nano-TiO_2 safety: overview of toxic effects in humans and aquatic animals. *Small* 16(36), 2002019. https://doi.org/10.1002/SMLL.202002019

Mahon, O.R., Browe, D.C., Gonzalez-Fernandez, T., Pitacco, P., Whelan, I.T., Von Euw, S. et al. 2020. Nano-particle mediated M2 macrophage polarization enhances bone formation and MSC osteogenesis in an IL-10 dependent manner. *Biomaterials* 239, 119833. https://doi.org/10.1016/J.BIOMATERIALS.2020.119833

Manna, S., Roy, S., Dolai, A., Ravula, A.R., Perumal, V., Das, A. 2023. Current and future prospects of "all-organic" nanoinsecticides for agricultural insect pest management. *Frontiers in Nanotechnology* 4, 1082128. https://doi.org/10.3389/FNANO.2022.1082128

Mustafa, I.F., Hussein, M.Z. 2020. Synthesis and technology of nanoemulsion-based pesticide formulation. *Nanomaterials* 10(8), 1608. https://doi.org/10.3390/NANO10081608

Nemati, T., Sarkheil, M., Johari, S.A. 2019. Trophic transfer of CuO nanoparticles from *Artemia salina* nauplii to convict cichlid (*Amatitlania nigrofasciata*) larvae: uptake, accumulation and elimination. *Environmental Science and Pollution Research* 26(10), 9610–9618. https://doi.org/10.1007/S11356-019-04263-6

Nobel, P.S. 2009. *Physicochemical and Environmental Plant Physiology*, 4th Edition, pp. 1–582. Elsevier, Amsterdam, Netherlands. https://doi.org/10.1016/B978-0-12-374143-1.X0001-4

Ogunsuyi, O.M., Adegoye, E.O., Ogunsuyi, O.I., Alabi, O.A., Alimba, C.G., Bakare, A.A. 2020. Titanium dioxide nanoparticles-induced cytogenotoxicity and alterations in haematological indices of *Clarias gariepinus*. *Toxicologic Pathology* 48(6), 807–815. https://doi.org/10.1177/0748233720948682

Pandey, S., Giri, K., Kumar, R., Mishra, G., Rishi, R.R. 2018. Nanopesticides: opportunities in crop protection and associated environmental risks. *Proceedings of the National Academy of Sciences India Section B* 88(4), 1287–1308. https://doi.org/10.1007/S40011-016-0791-2

Parveen, R., Maiti, P.K., Murmu, N., Datta, A. 2021. Preparation of serum capped silver nanoparticles for selective killing of microbial cells sparing host cells. *Scientific Reports* 11(1), 1–12. https://doi.org/10.1038/S41598-021-91031-7

Petersen, E.J., Mortimer, M., Burgess, R.M., Handy, R., Hanna, S., Ho, K.T. et al. 2019. Strategies for robust and accurate experimental approaches to quantify nanomaterial bioaccumulation across a broad range of organisms. *Environmental Science: Nano* 6(6), 1619–1656. https://doi.org/10.1039/C8EN01378K

Rahman, M.S., Chakraborty, A., Mazumdar, S., Nandi, N.C., Bhuiyan, M.N.I., Alauddin, S.M., Khan, I.A., Hossain, M.J. 2020. Effects of poly(vinylpyrrolidone) protected platinum nanoparticles on seed germination and growth performance of *Pisum sativum*. *Nano-Structures & Nano-Objects* 21, 100408. https://doi.org/10.1016/J.NANOSO.2019.100408

Rajput, V., Minkina, T., Mazarji, M., Shende, S., Sushkova, S., Mandzhieva, S., Burachevskaya, M., Chaplygin, V., Singh, A., Jatav, H. 2020. Accumulation of nanoparticles in the soil–plant systems and their effects on human health. *Annals of Agricultural Sciences* 65(2), 137–143. https://doi.org/10.1016/J.AOAS.2020.08.001

Shelar, A., Nile, S.H., Singh, A.V., Rothenstein, D., Bill, J., Xiao, J., Chaskar, M., Kai, G., Patil, R. 2023. Recent advances in nano-enabled seed treatment strategies for sustainable agriculture: challenges, risk assessment, and future perspectives. *Nano-Micro Letters* 15(1), 1–37. https://doi.org/10.1007/S40820-023-01025-5

Skjolding, L.M., Sørensen, S.N., Hartmann, N.B., Hjorth, R., Hansen, S.F., Baun, A. 2016. Aquatic ecotoxicity testing of nanoparticles—The quest to disclose nanoparticle effects. *Angewandte Chemie International Edition* 55(49), 15224–15239. https://doi.org/10.1002/ANIE.201604964

Skjolding, L.M., Winther-Nielsen, M., Baun, A. 2014. Trophic transfer of differently functionalized zinc oxide nanoparticles from crustaceans (*Daphnia magna*) to zebrafish (*Danio rerio*). *Aquatic Toxicology* 157, 101–108. https://doi.org/10.1016/J.AQUATOX.2014.10.005

Song, U., Jun, H., Waldman, B., Roh, J., Kim, Y., Yi, J., Lee, E.J. 2013. Functional analyses of nanoparticle toxicity: a comparative study of the effects of TiO_2 and Ag on tomatoes (*Lycopersicon esculentum*). *Ecotoxicology and Environmental Safety* 93, 60–67. https://doi.org/10.1016/J.ECOENV.2013.03.033

Sousa De Almeida, M., Susnik, E., Drasler, B., Taladriz-Blanco, P., Petri-Fink, A., Rothen-Rutishauser, B. 2021. Understanding nanoparticle endocytosis to improve targeting strategies in nanomedicine. *Chemical Society Reviews* 50(9), 5397–5434. https://doi.org/10.1039/D0CS01127D

Tangaa, S.R., Selck, H., Winther-Nielsen, M., Khan, F.R. 2016. Trophic transfer of metal-based nanoparticles in aquatic environments: a review and recommendations for future research focus. *Environmental Science: Nano* 3(5), 966–981. https://doi.org/10.1039/C5EN00280J

Taware, A.S., Rathod, P.B., Katariya, A.P., Tagad, C.K., Wagh, P.S., Sonar, J.P., Deshmukh, S.U., Kanagare, A.B. 2024. Technological advancement in the development of nano fertilizers for sustainable agriculture. *Journal of Soil Science and Plant Nutrition* 24, 1–17. https://doi.org/10.1007/S42729-024-01674-4

Thuesombat, P., Hannongbua, S., Akasit, S., Chadchawan, S. 2014. Effect of silver nanoparticles on rice (*Oryza sativa* L. cv. KDML 105) seed germination and seedling growth. *Ecotoxicology and Environmental Safety* 104(1), 302–309. https://doi.org/10.1016/J.ECOENV.2014.03.022

Wang, X., Zhong, W., Xiao, B., Liu, Q., Yang, L., Covaci, A., Zhu, L. 2019. Bioavailability and biomagnification of organophosphate esters in the food web of Taihu Lake, China. *Environment International* 125, 25–32. https://doi.org/10.1016/J.ENVINT.2019.01.018

Yan, A., Chen, Z. 2019. Impacts of silver nanoparticles on plants: a focus on the phytotoxicity and underlying mechanism. *International Journal of Molecular Sciences* 20(5), 1003. https://doi.org/10.3390/IJMS20051003

Yang, F., Zeng, L., Luo, Z., Wang, Z., Huang, F., Wang, Q., Drobne, D., Yan, C. 2018. Complex role of titanium dioxide nanoparticles in the trophic transfer of arsenic. *Aquatic Toxicology* 198, 231–239. https://doi.org/10.1016/J.AQUATOX.2018.03.009

Zargar, M., Bayat, M., Saquee, F.S., Diakite, S., Ramzanovich, N.M., Akhmadovich, K.A.S. 2023. New advances in nano-enabled weed management using poly(ε-caprolactone)-based nanoherbicides: a review. *Agriculture* 13(10), 2031. https://doi.org/10.3390/AGRICULTURE13102031

Zhang, H., Huang, M., Zhang, W., Gardea-Torresdey, J.L., White, J.C., Ji, R., Zhao, L. 2020a. Silver nanoparticles alter soil microbial community compositions and metabolite profiles. *Environmental Science and Technology* 54(6), 3334–3342. https://doi.org/10.1021/ACS.EST.9B07562

Zhang, Q., Wu, K., Qian, H., Ramachandran, B., Jiang, F. 2021. Advances of characterization and evaluation methods for the compatibility and assembly structure stability of food soft matter. *Trends in Food Science & Technology* 112, 753–763. https://doi.org/10.1016/J.TIFS.2021.04.034

Zhang, Z.J., Shang, X.F., Yang, L., Shi, Y.B., Liu, Y.Q., Li, J.C., Yang, G.Z., Yang, C.J. 2020b. Engineering of pegylated camptothecin into nanomicelles and supramolecular hydrogels for pesticide combination control. *Frontiers in Chemistry* 7, 491237. https://doi.org/10.3389/FCHEM.2019.00922

Zheng, Q., Qin, D., Wang, R., Yan, W., Zhao, W., Shen, S., Huang, S., Cheng, D., Zhao, C., Zhang, Z. 2022. Novel application of biodegradable chitosan in agriculture: using green nanopesticides to control *Solenopsis invicta. International Journal of Biological Macromolecules* 220, 193–203. https://doi.org/10.1016/J.IJBIOMAC.2022.08.066

Zinovieva, S.V., Udalova, Z.V., Khasanova, O.S. 2023. Nanomaterials in plant protection against parasitic nematodes. *Biology Bulletin Reviews* 13(6), 703–722. https://doi.org/10.1134/S2079086423060178

11 Advancements in the Application of Nanomaterials (NMs) in Sustainable Agriculture

Isha Madaan, Shamsher Singh, Gurvarinder Kaur, Akanksha Sharma, Simranpreet Kaur, Manjot Kaur, Jahanvi, Jetina, Nivedita Sharma, Tarandeep Kaur, Renu Bhardwaj, and Geetika Sirhindi

11.1 INTRODUCTION

Nanotechnology has emerged as an advanced and promising approach in sustainable agriculture, providing novel strategies to improve crop yield, reduce environmental burden, and ensure more efficient use of agricultural resources. The integration of nanomaterials (NMs) in agriculture is promoting an ideal reposition towards more sustainable and effective farming practices. There are multifaceted applications of NMs in the agricultural sector. Nanotechnology has opened up new possibilities in sustainable agriculture by providing tools to address various challenges such as nutrient deficiencies, pest and disease management, soil degradation, and water scarcity. The application of nanoparticles (NPs) in agriculture has shown promising results in enhancing crop yields, improving nutrient uptake efficiency, and reducing the effect of conventional farming practices on environment. One of the major key advantages of NMs is their ability to deliver nutrients and agrochemicals to plants in a controlled manner, which helps in reducing the extensive use of fertilizers and pesticides. NPs can encapsulate nutrients and protect them from leaching or volatilization, ensuring their delivery in readily available forms to the target plant systems. This targeted delivery improves nutrient use efficiency as well as eliminates the probable chances of environmental pollution (Kaphle et al. 2018).

In addition to this, NPs are being utilized to develop highly effective, eco-friendly novel pesticides and fungicides in comparison to traditional chemical-based products. These nanopesticides can target specific pests or pathogens, reducing the requirement of broad-spectrum chemicals that can harm advantageous organisms and pollute the environment. It helps in not only improving soil quality but also reducing the risk of food contamination (Okey-Onyesolu et al. 2021).

 DOI: 10.1201/9781032659435-11

Furthermore, nanotechnology is playing a transformative role in precision agriculture by facilitating continuous and real-time assessment of crop performance and surrounding environmental factors. Such advancements support efficient resource management, minimize losses, and promote environmentally sustainable farming systems. Despite the promising applications of NMs in agriculture, several challenges remain unresolved, particularly regarding their potential risks to human health and ecological systems, as well as associated regulatory and ethical concerns. NMs possess significant potential to advance sustainable agriculture through novel approaches that boost crop yield, enhance soil quality, and mitigate adverse environmental effects (Wang et al. 2024). Ongoing research and technological advancements are therefore crucial to ensure the safe, efficient, and responsible application of NMs for the long-term sustainability of agriculture and food production.

11.2 TYPES AND BIOSYNTHESIS OF NANOMATERIALS

The size range of the nanometre scale of atoms and molecules is the pivotal characteristic of NMs (Yu et al. 2020). This uniqueness of NMs, due to their size, provides them with a larger surface area and more reactivity. Furthermore, their quantum effect dispenses spectacular catalytic reactivity, mechanical properties, optical absorption, and electrical conductivity, making them better than its bulk use (Pattan and Kaul 2014; Ali et al. 2021). The term 'nanomaterial' was defined by the European Commission (EC) in 2011 as a material that is either natural or manufactured, having a free or aggregated or agglomerated state containing more than 50% of its particles under a size range of 1–100 nm (EU, 2/696/EU, 2011). NMs are classified based on varied shapes, sizes, nature of material, and surface functionalization.

a. **Based on Material and Composition**: NPs can be broadly categories into four form based on material and composition employed
   - **Carbon-Based Nanomaterials (CNMs)**: Due to the unique catenation property of carbon atoms, these CNMs possess spherical, ellipsoidal, or hollow tube-based morphological forms. These forms are graphene (Gr), carbon nanotubes (CNTs), fullerenes (C60), carbon nanofibers, carbon onions, and carbon black. CNMs are considered biocompatible since the major part of the human body is composed of carbon (Yadav and Awasthi 2022).
   - **Organic NMs**: They are wrenched out from living organisms like liposomes, micelles, polymers, and dendrimer NPs (Shukla et al. 2013). These serve as environment-friendly materials due to their biodegradable and non-toxic properties.
   - **Inorganic NMs**: Inorganic NMs include metals and metal oxides in nanoscale form. These materials are commonly synthesized through the chemical reduction of metal salts using suitable reducing agents, resulting in the formation of zero-valent metallic NPs as well as ionic NP species, whereas metal oxides are bismuth(III) oxide (Bi_2O_3), cerium dioxide (CeO_2), chromium dioxide (CrO_2), barium titanate ($BaTiO_3$),

titanium oxide (TiO_2), zinc oxide (ZnO), molybdenum trioxide (MoO_3), etc. (Shukla et al. 2013; Jang et al. 2015).

- **Composite-Based NMs**: They have combinations of other NMs. Being in the multiphase having at least one phase in the nanoscale dimension forms hybrid NMs like hybrid nanofibers. Other examples are Polymer/TiO_2 and Polymer/CNT etc. (Lodahl et al. 2015; Srivastava et al. 2015).

b. **Based on Dimensions**: Another way to look at the varied varieties of NMs is their dimensions. These are classified as 0D, 1D, 2D, 3D (Ashby et al. 2009)

- **Zero-Dimensional NMs (0-D)**: These materials have all three dimensions confined to the nanoscale range, as seen in quantum dots, fullerenes, and NPs (Joudeh and Linke 2022).
- **One-Dimensional NMs (1-D)**: In this category, only two dimensions fall within the nanoscale, while the remaining dimension extends beyond it; typical examples include nanotubes, nanofibers, nanorods, nanowires, and nanohorns.
- **Two-Dimensional NMs (2-D)**: This class is characterized by two of the three dimensions extending beyond the nanoscale, with nanosheets, nanofilms, and nanolayers serving as representative examples.
- **Three-Dimensional NMs (3-D)**: These are called 'bulk nanomaterials' as none of their dimension is under nanoscales. Examples include bulk powders, dispersions of NPs, arrays of nanowires and nanotubes, etc. (Ashby et al. 2009; Maurice and Hochella 2008; Singh 2016).

c. **Biologically Mediated Synthesis**

Biologically facilitated NPs synthesis has gaining significant importance nowadays owing to green, nontoxic, and eco-friendly alternate to conventional methods (Diallo et al. 2017). These strategies harness the inherent properties of biological systems primarily microbes, algae plants, etc. which enable the development of environmentally sustainable synthesis routes (Konishi et al. 2007). Among these, microbial synthesis is most promising characterizes pathway, that comprising bacteria, yeast, microalgae, fungi, etc. driven NP synthesis (Ghosh et al. 2021).

In biological process, metal ions are highly adsorbed on the surface of cell and transported through cell membrane inside cells via enzymatic reduction into their nanoscale form (Murray et al. 2017; Harris et al. 2018). Further, nucleation reactions coupled with natural capping with macro-biomolecules result in the formation of NPs (Kim et al. 2018). Different microbes have different potential based on their structural and functional behaviour that offer great advantages in the synthesis of more effective and stable form of NPs.

- For instance, the growth rate of bacteria, their adaptability to physicochemical conditions due to the nature of cell walls revealed most suitable candidate for NP synthesis (Khosravi-Darani et al. 2019). Similarly, fungi added more advantages on account of their high protein secretion ability, and contribute to yield, amended morphology, and stabilization, which employed in the synthesis of metallic and metal oxide NPs, (Owaid et al. 2017; Tyagi et al. 2019; Shokrzadeh et al. 2020). Further, yeasts, being a unicellular organism, have also gained attention in NP synthesis because of their high-density, scalability and fermentation optimization for variety of products (Ammar et al. 2021). Algae, demonstrates great affinity for heavy metal uptake on account of their intrinsic biochemical, metabolic and photosynthetic mechanisms, thereby translating into an efficient "bio-nano factory" for metallic NPs production by using both fresh and dried biomass (Pandit et al. 2022).

- Another key biosynthetic route i.e., plant-mediated synthesis exemplifies a natural capping and reducing agent to synthesis non-toxic NPs by exploring different plant parts (Küünal et al. 2018). The efficacy of plant-regulated synthesis is associated with varied bioactive compounds such as saponins, flavonoids, polyphenols, tannins, etc. which serve as functionalizing, reducing and stabilizing mediators (Khan 2023). Recently, many techniques demonstrated the application of plant-derived pure metabolites for controlled reproducibility and functional NPs (Ahmad et al. 2022a; Ali et al. 2020). Mainly, flavonoid-catalysed NPs revealed vital promise in bioremediation of toxic metals particularly with silver such as quercetin, unveiling progressive antibacterial efficacy (Aisida et al. 2019).

11.3 NMs FOR IMPROVED CROP GROWTH

Advances in science and technology have contributed significantly to agriculture by offering innovative solutions to critical challenges. The advent of nanotechnology has enabled the development of more efficient and less polluting nanoformulations that support sustainable agricultural practices. These next-generation materials can influence plant physiology once they enter the plant–soil system, allowing their subsequent effects to be readily assessed. Moreover, effective regulation of active compound delivery requires comprehensive knowledge of the potential beneficial as well as adverse impacts of NMs on plants.

Seed germination is the first and most sensitive stage in a plant's life cycle which is affected by some genetic factors, soil microenvironment, and availability of water. Nanotechnology and NM improve the seed germination process, thus increasing crop growth and fruiting rates. Application of silicon dioxide (SiO_2)-NPs improved seed germination and chlorophyll content by accumulating proline in *Solanum lycopodium* (tomato), *Cucurbita pepo* L. (squash plant) under salt stress. NMs are having a major impact on sustainable agriculture, and various studies have shown their positive effects on various crops. For example, the application of SiO_2 and TiO_2 revealed improved soybean germination (Hassanisaadi et al. 2022).

The use of nano boron and boron spraying on the leaves of mango trees shows that nanotechnology is effective in improving overall yields and chemical properties of fruit and leaf content (Genaidy et al. 2020). Similarly, nanofertilizer has a positive effect on the growth of olive trees, increasing fruit weight, size, oil in fruit, and concentration of elements (Hmmam et al. 2025). On mango trees, nano zinc spraying increases fruit weight, quantity, leaf content of chlorophyll, carotene, and concentration of elements (Abdel-Sattar et al. 2024). Nano zinc and nano boron spraying were applied to pomegranate trees before full flowering. It increases the number of fruits, ripeness index, total sugar, and phenols, and also increases fruit yield. Spraying Nano zinc–iron–manganese–boron (ZnFeMnB) on palm trees has a positive effect on leaf area and all chlorophyll, carotenoids, nitrogen (N), phosphorus (P), and potassium (K) in leaf content, reducing precipitation and increasing fruit weight, thereby increasing yield (Raliya et al. 2016). Research studies evidenced aerosol-mediated NP transportation, penetration, and translocation in watermelons and tomatoes. It has been found out that the NPs with less than 100 nm enter the leaf through the stomatal path and reach the root of plants.

11.4 NMs FOR CROP PROTECTION

11.4.1 Nanoherbicides

Nanoherbicides induce a smart delivery system that uses a very small amount of herbicide combined with the active ingredient. Several studies have confirmed the successful application of nano-carrier systems in agriculture. For example, (Kumar et al. 2019) highlighted on liposomal formulation of herbicides that exhibited improved prolonged activity, resulting in augmented herbicidal efficiency while minimizing off-target effects. Their studies underscore the versatility and efficacy of nano-carrier systems in delivering agrochemicals with precision.

11.4.2 Nanofungicides

The most common cause for the damage of crops is due to fungal diseases which cause more than 70% damage in major crops i.e., rice, wheat, barley, groundnut, and cotton and cause huge economical as well as qualitative losses in crop yield (Shukla et al. 2022). The usage of conventional fungicides somehow reduces the losses but along with this, it also targets nonspecific living organisms and leads to disturbance in the equilibrium of biodiversity. The work of (El-Mohamedy et al. 2019) revealed the use of chitosan NPs for controlled release of fungicides, foremost to greater disease management and reduced environmental impact.

The development of NPs serves as the best approach against fungal disease-causing pathogens. Silver nanoparticles (AgNPs) are most commonly used for disinfection purposes due to their antimicrobial properties (Matras et al. 2022).

11.4.3 NMs in Abiotic Stress Tolerance

Abiotic stresses such as salinity, drought, high or cold temperature, organic pollutants, metals, etc. adversely impact the growth and productivity of crops. NMs, because of unique physicochemical features, have integrated as active carriers for agro-nutrients to mitigate wide type of stress factors in plants. Many metals and their oxides like silver, gold, titanium dioxide (TiO_2), ZnO, SiO_2 demonstrated enhanced tolerance against stresses by moderating biosynthetic plant responses. Silva et al. (2022) evidenced that TiO_2 NPs showing high enzymatic, photosynthetic and more water-uptake efficiency in drought stressed *Oryza sativa* plants. Likewise, ZnO NPs enhanced *Triticum aestivum* germination rate, osmolyte accumulation and chlorophyll content when grown in salinity stress (Faizan et al. 2021). AgNPs have been also described as oxidative stress modulator via accelerating reactive oxygen species (ROS) scavenging potential in heavy metal-stressed plants (Alabdallah and Hasan 2021). Beyond metal-based NPs, CNMs, including carbon nanotubes and graphene oxide, have demonstrated improved nutrient uptake, root growth, and stress-responsive gene expression, thereby promoting abiotic stress tolerance in crops.

Therefore, NPs functionally ameliorating wide stresses by regulating different cellular, physiological, biochemicals and molecular pathways via stimulating antioxidant defense enzymes, stress-responsive gene expression, hormonal regulation, uptake and translocation of nutrient under adverse conditions. NPs' evident potential as a carrier for targeted and controlled delivery of herbicides, pesticides, fertilizers, essential nutrients added as a next-generation tool for stress-resilient and smart agriculture (Thabet and Alqudah 2024).

11.5 WATER MANAGEMENT THROUGH NMs

Effective water management is critically important in affected regions, particularly in developing countries where access to safe drinking water remains a major challenge. Recent advances in nanotechnology offer promising opportunities to upgrade water supply systems for next-generation needs. Recent research indicates that many water quality–related challenges can be effectively addressed through the application of NMs. The use of various types of NPs and nanostructured materials enables water purification by facilitating contaminant removal and allowing the reuse of treated water. Some of the commonly employed NMs for water purification are listed below:

11.5.1 Nanosorbents

Sorbents are found to play pivotal roles in removal of organic as well as inorganic contaminants from polluted water. The larger surface area of NPs and their broad affinity for vast array of target compounds make them very vital for nanosorption. Optimum levels of sorption capacities of lead, Copper and Cadmium onto multi-walled carbon nanotubes (MWCNTs) have been found in previous research studies (Rodríguez et al. 2020). In addition to this, it has also been found out that in comparison to powder and granular activated carbon, metal–ion sorption capacities of the MWCNTs were quite high.

11.5.2 Nanocatalysts and Redox Active Nanoparticles

NPs have remarkable ability to purify water because of their redox active media, large surface areas and shape - and size -dependent optical, electronic, and catalytic properties. Newly emerged NPs, TiO_2, have auspicious photocatalytic properties in the process of water purification (Adesina 2024). Moreover, these NPs are useful for both inorganic and organic pollutants because of their oxido-reductive catalytic nature. According to Liu et al. (2024), purification of water contaminated with organic wastes was significantly boosted by adding of TiO_2 NPs along with

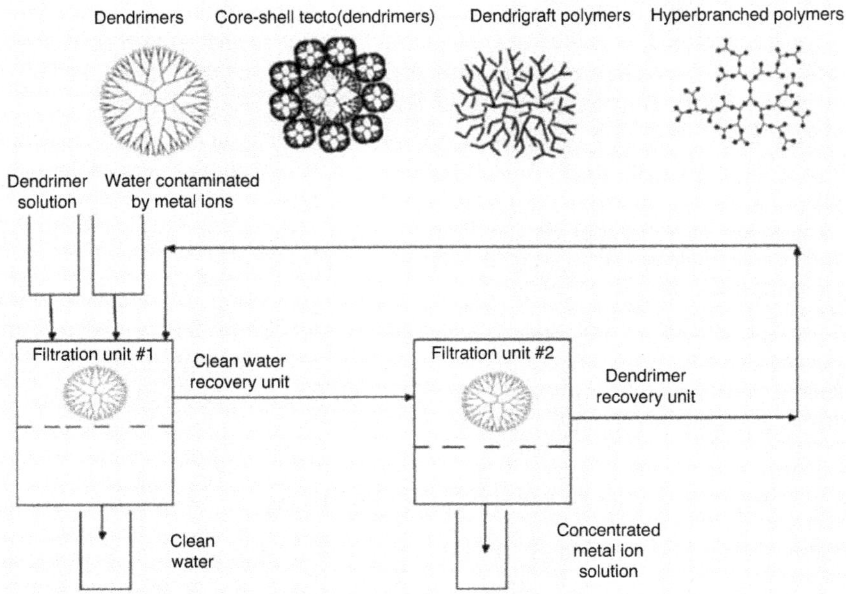

FIGURE 11.1 Dendrimers for water purification systematic diagram.

ultraviolet light. Similar results were found by Mao et al. (2024) under the review of the use of photocatalysts in water purification treatments, contaminated by organic and inorganic impurities. They suggested effective utilization of TiO_2 NPs in degradation of organic compounds and in reducing aqueous solutions of toxic metal ions in presence of UV rays (Kholief et al. 2024).

11.5.3 Polymeric Nano Adsorbents

Dendrimers function as effective adsorbents due to their distinctive architecture, in which the hydrophobic inner cavities facilitate the sorption of organic compounds, while the outer branched groups enable the removal of heavy metals from aqueous systems. This adsorption behaviour depends on factors such as electrostatic interactions, complex formation, hydrogen bonding, and hydrophobic effects. Using a copper(II) ion (Cu^{2+}) to poly(amidoamine) (PAMAM) dendrimer-NH_2 ratio of 0.2, dendrimers were able to remove nearly all Cu^{2+} ions from solutions containing 10 ppm of copper (Pawlaczyk and Schroeder 2019). Following adsorption, metal-loaded dendrimers were separated by ultrafiltration and subsequently regenerated by reducing the solution pH to 4, as shown in Figure 11.1.

11.6 NANO-BIOSENSORS: NMS IN PRECISION AGRICULTURE

In the face of challenges like food security, and climate change, researchers turn to nanotechnology for innovative solutions. Nanotechnology explores materials or structures within the 1–100 nm size ranges, exhibiting unique properties with

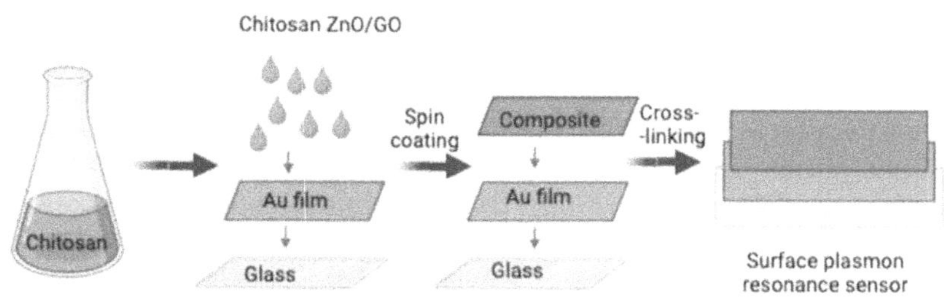

FIGURE 11.2 Scheme for constructing a surface plasmon resonance sensor.

quantum effects. This ranges of NMs, boasting increased surface-to-volume ratios, enable the development of highly sensitive nanosensors. Nanostructured sensory systems that convert nanoscale biological events, such as reactions or bindings, into measurable signals is called a nanobiosensor (Salouti and Khadivi Derakhshan 2020). In agriculture, sustainability and precision are essential, emphasizing practices that are cost-effective, energy-efficient, and environmentally friendly. Such approaches include the use of nanoformulations, biosensors, remote sensing technologies, and global positioning systems. Among these technologies, nanostructured biosensors are becoming more popular. They can assess crop health, detect plant pathogens, and estimate the amount of fertilizer and water needed by crops. While fertilizers and pesticides are important for agriculture, they also have environmental and health risks. That's why we need cutting-edge technology like Nanobiosensors (Figure 11.2).

In precision agriculture, nanotechnology supports genetic improvement, enhancement of crop yield, plant protection, and real-time monitoring of crop health through the use of nanosensors. Owing to their extremely small size, low weight, and high surface-to-volume ratio, nanosensors are well suited for signal amplification and the detection of trace-level analytes, thereby expanding agricultural applications. The regulated use of pesticides, fertilizers, and heavy metals is essential to reduce environmental impacts, while emerging technologies are increasingly applied for soil health assessment and disease surveillance. Advances in nanofabrication and analytical methods have facilitated the development of portable, accurate, and responsive sensing systems for agricultural use. Due to their unique properties, nanosensors hold significant potential for crop management, pathogen identification, and monitoring of soil parameters (Ranjita et al. 2020).

Nanoscale materials enhance biosensor sensitivity and performance, advancing technology in health care, agriculture, and more. Nanotechnology-based biosensors, incorporating materials like metal NPs, CNTs and QDs, represent a frontier in biological detection and material science. Although nanobiosensors show potential in plant pathogen diagnosis, further exploration in field application is needed. Their application extends to detecting fungi, viruses, and phytopathogens, offering rapid environmental pollution analysis (Johnson et al. 2021). Nanotechnology-based devices contribute to infection diagnosis in crops, minimizing pesticide use in the face of low pollutant concentrations and detectable pathogens. Nanotechnology's impact on agriculture is multifaceted. As mentioned earlier, to enhance productivity in a sustainable way, we need to embrace a 'smart' agriculture system. This system enables us to detect the nutritional qualities of the soil, the structure of the soil microbiome community, the presence of pests, and the quality of crops as they grow in the field, all in real time. Unlike traditional methods that only detect these issues after damage has been incurred, real-time detection also relays this data to the farmer's laptop simultaneously for prompt action. Nanobiosensors offer an excellent solution in this field. Typically, the presence of pests is only investigated after symptoms appear, and scientists currently employ time-consuming PCR-based method. However, Nanosensors, such as dextrin-capped gold particles, revolutionize disease detection by offering real-time insights into pathogen deoxyribonucleic acid (DNA) without the need for time-consuming PCR amplification (Karnwal et al. 2024).

Meanwhile, gene delivery using NPs addresses genetic improvement challenges, overcoming barriers like species specificity and enabling advancements such as chloroplast transformation. By bringing together these different platforms, such as indirect detection of biomarkers like hydrogen peroxide (H_2O_2), we can streamline the process of alerting farmers early and assessing crop health in real time. Nanosensors are being effectively utilized to convert a living spinach plant into an auto-sampler, relaying information through infrared technology. Nanosensors detect contaminants, pests, and stress in crops, optimizing resource use for farmers. Nanotechnology and Global Positioning System (GPS)-linked autonomous sensors enable real-time monitoring for precise application of nutrients and protection. In water purification, NMs like carbon nanotubes and nanoporous ceramics offer cost-effective solutions, removing contaminants from irrigation and potable water. Carbon nanotube filters target pathogens and heavy metals, while nanocream filters trap bacteria and viruses, enhancing water quality. Overall, smart sensors in precision farming and NMs in water treatment enhance agriculture productivity and decision-making globally (Figure 11.3). Nanobiosensors swiftly detect contaminants in seafood and vegetables, reducing testing time for microbial and immunoassays. Applications extend to water, raw food, and products. Microorganisms produce volatile compounds, impacting food quality. Bacteria cause food rot, identified by foul odour. Rapid biosensors surpass human senses, practical in preventing poisoning. Global research highlights NPs efficacy in drug delivery, biosensing (Ditta et al. 2015). Hence, nanotechnology's role in agriculture holds transformative potential, offering efficient pollutant removal, real-time monitoring with nanobiosensors, and sustainable waste management. Future prospects rely on continued research, systematic evaluation, and responsible implementation for a more sustainable and technologically advanced agricultural landscape (Srivastava et al. 2018).

11.7 NMs IN AGRICULTURAL BIOTECHNOLOGY

The main aim of NMs in agriculture is to reduce inputs of pesticides and insecticides, minimize nutrient loss by rapid disease diagnosis, enhancing plant nutrients absorption capacity. It further includes utilization of nanofertilizers and nanopesticides that

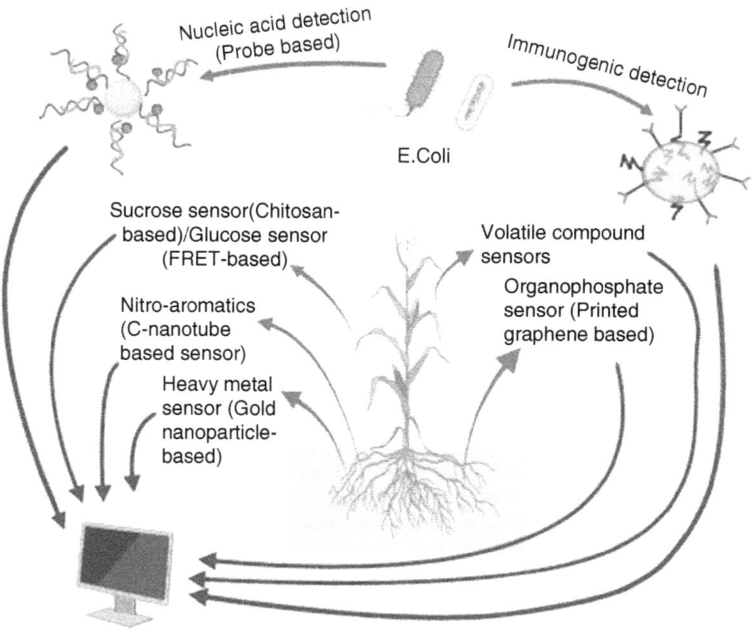

FIGURE 11.3 Real-time monitoring of plant health with nanobiosensors.

decrease soil and water contamination and killing of friendly microbes (Prasad et al. 2017). In contemporary world, there is the need of development of high-tech agricultural system with the use of nanotools to completely revolutionize the current modern agricultural practices (Sekhon 2014; Liu and Lal 2015). The development of nanotechnology alongside biotechnology have led to development of nanobiotechnology, which means combination of biology to nanotechnology (Kah 2015) it resulted in increased utilization of NMs in various areas including agricultural sector (Vishwakarma et al. 2018). NMs are made up of different types of components such as nonmetallic NMs, carbon NMs, metallic NMs, organic polymeric compounds and inorganic NMs (Vishwakarma et al. 2018). Nowadays, a shift is seen towards green nanotechnology that means use of clean technologies that reduce human health and environmental risks.

11.7.1 Carbon Nanotubes

Carbon nanotubes are allotropes of carbon structurally analogous to rolled graphene sheets, and they occur mainly in two forms: single-walled carbon nanotubes (SWCNTs) with diameters ranging from 0.8 to 2.0 nm, and multi-walled carbon nanotubes (MWCNTs) with diameters of 5.0–20.0 nm. The length of these nanotubes can vary from approximately 100 nm–0.5 m. The application of MWCNTs has been reported to significantly enhance seed germination and vegetative biomass in tomato plants, as well as to improve water and nutrient uptake, particularly calcium and iron, thereby promoting seed germination and overall plant growth (Tiwari et al. 2017;

Sigala-Aguilar et al. 2025). Additionally, sulphur- and iron-containing carbon nanotube sponges exhibit high efficiency in removing various water pollutants, including pesticides, fertilizers, oils, and pharmaceuticals, demonstrating up to 3.5-fold greater removal capacity compared to CNT powder (Prasad 2017). Despite these advantages, carbon nanotubes have certain limitations, as their high penetrability and tendency to accumulate within the cytoplasm may pose risks to human cells (Prasad 2017).

11.7.2 Quantum Dots

Quantum dots exhibit molar extinction coefficient, high quantum yield, broad absorption spectra and symmetric fluorescence spectra in the ultraviolet region to near infrared region along with high resistance to photochemical degradation and photobleaching (Prasad 2017). Because of quantum dots exceptional spectral property, they are applied to the new generation fluorophores in bioimaging and biosensing than the traditional organic dyes. Furthermore, due to the presence of photocatalysts they are used for the different light-driven chemical reactions such as part of the process towards solar fuel production (Konstantatos and Sargent 2009). Furthermore, seed germination and seedling row has not negatively affected by quantum dots at low concentration because it has no toxic effect, therefore, it can also be used to verify the various non-physiological process (Szőllősi et al. 2020).

11.7.3 Nano Rods

Nano rods are made up of multifunctional plasmonic materials. Gold nanorods significantly verified the biochemical changes occurring in the watermelon plants and it also confirmed the toxicity towards plants at a particular high concentration (Wan et al. 2023; Prasad 2017). Along with this, it also has significant effect on regulation of tobacco cell culture growth because of the ability to transfer the 2, 4-D auxin plant hormone (Nima et al. 2014).

11.7.4 Nano Emulsions

There is no major distinction between nanoemulsions and microemulsions except in terms of their physical characteristics (Prasad 2017). Compared with conventional emulsions, nano emulsions exhibit significantly higher Laplace pressure and elastic modulus, and they are optically transparent. Nano emulsions have been widely reported for use in the development of pesticides aimed at controlling plant pathogens (Vishwakarma et al. 2018).

11.7.5 Nanofoods

With the use of nanotechnology and NPs or Nanotools in the process from cultivation to final stage, packaging of the foods, then this food is known as nanofood. Similarly, Vishwakarma et al. (2018) reported that beyond its applications in agrifood systems, nanotechnology offers several additional approaches to improve the quality and sustainability of production and processing. They also found the positive effect on soya bean grain yield by the use of nanotechnology in agronomic practices such as use of Fe oxide NPs during nutrients management (Vishwakarma et al. 2018). Moreover, in a study on nutrition rich vegetables crops in which on carbon containing NP fullerol treatment led to an improvement of up to 128% in yield (Kole et al. 2013).

11.8 CHALLENGES AND SAFETY CONCERNS

According to US Environmental Protection Agency, nanotechnology is the science that deals with the understanding and control of the matter, having dimensions of 1–100 nm where novel properties arise due to unique physical properties. In other words, from the viewpoint of agriculture nanotechnology includes the particulate having the size of 10–1,000 nm that are simultaneously colloidal particles (Abobatta 2018). Limited knowledge is available about the use of nano-sized particles and their production on industrial scale which results in unsafe handling procedure. There are always problems with the patent procurement like difficulty in looking for it among many patents so as to use terms like tiny, micro level etc. which results in consumption of much time in logistics and fulfilling the patentability requirements. The special requirement of purification, conditions for specialize operations and complexity of the process makes the production method difficult and requires a greater control over the structure, morphology and particle size of NPs. Reliable work is required to scale up the process of investigation of nanoabsorbant used in the removal of specific pollutants. Cost is being a major challenge and much work is required to lessen the cost of production, regeneration, reuse of nanosized particles and devices. Handling and disposal of nanosized particles should be assessed in such a way so as to cause no damage to environment. The size of NPs should be 1–100 nm and any deviation from size makes it difficult to be produced within a particular product standard. Particles having size <2.5 nm revealed harmful effects. Sustainability is the major concern that restricted the wide use of nanotechnology in agriculture (Usman et al. 2020). Lack of awareness about eco-toxicity, environmental implications and their adverse effects in long term make their use limited in agriculture. Ethical concerns and perceptions regarding nanotechnology are the major and foremost challenge. People use heuristics like trust to win the assurance about safety of nanotechnology particularly in agriculture. Public perception about nanotechnology is greatly influenced by media. In less developed countries, the coverage by media is low which leads to lower level of research in the field in comparison to developed countries (Pandey 2018).

Along with this, non-control release of NPs into the environment have always been a topic of safety concerns, due to their toxic impact on ecosystem and human health. The size of NPs is an important property that affects their properties and applications. Thus, the size regulation is always the basic concern. NMs of size 1–100 nm are comparable to the size of DNA helix, many proteins and thickness of the cell membrane and thus these may easily penetrate the cells. US Food and Drug Administration drafted and published in 2012 a risk guidance document regarding the safety concerns of NP's use in agriculture. Uncontrolled release has adverse effects on basically 3 domains i.e., environment, human and plant health, aquatic ecosystem. NPs have the diameter of 30 nm or less pose a great problem to surrounding. NPs such as carbon, inorganic and organic based NPs readily enter the environment and food chain through use of NPs in various processes from production to processing, preservation, safety, packaging, transportation, storage and delivery. Therefore, this field should always be assessed for nanotoxicity. And some NPs do not come in contact with humans but can actually be consumed with food.

REFERENCES

Abdel-Sattar, M., Makhasha, E., Al-Obeed, R.S. 2024. Conventional and nano-zinc foliar spray strategies to improve the physico-chemical properties and nutritional and antioxidant compounds of timor mango fruits under abiotic stress. *Horticulturae* 10(10), 1096.

Abobatta, W.F. 2018. Nanotechnology application in agriculture. *Acta Scientific Agriculture* 2(6), 99–102.

Adesina, A.A. 2004. Industrial exploitation of photocatalysis: progress, perspectives and prospects. *Catalysis Surveys from Asia* 8(4), 265–273. https://doi.org/10.1007/s10563-004-9127-6

Ahmad, K.S., Yaqoob, S., Gul, M.M. 2022a. Dynamic green synthesis of iron oxide and manganese oxide nanoparticles and their cogent antimicrobial, environmental and electrical applications. *Reviews in Inorganic Chemistry* 42, 239–263.

Aisida, S.O.K., Ugwu, P.A. Akpa, A.C. Nwanya, U. Nwankwo, S.S. Botha, P.M. et al. 2019. Biosynthesis of silver nanoparticles using bitter leaf (*Veronica amygdalina*) for antibacterial activities. *Surfaces and Interfaces* 17, 100359.

Alabdallah, N.M., Hasan, M.M. 2021. Plant-based green synthesis of silver nanoparticles and its effective role in abiotic stress tolerance in crop plants. *Saudi Journal of Biological Sciences* 28, 5631–5639.

Ali, M.A., Ahmed, T., Wu, W., Hossain, A., Hafeez, R., Masum, M.M.I., Wang, Y., An, Q., Sun, G., Li, B. 2020. Advancements in plant- and microbe-based synthesis of metallic nanoparticles and their antimicrobial activity against plant pathogens. *Nanomaterials* 10, 1146. https://doi.org/10.3390/nano10061146

Ali, S.S., Al-Tohamy, R., Koutra, E., Kornaros, M.M., Mustafa, A.M., Mahmoud, Y.A.-G. et al. 2021. Nanobiotechnological advancements in agriculture and food industry: application, nanotoxicity, and future perspective. *Science of the Total Environment* 792, 148359. https://doi.org/10.1016/j.scitotenv.2021.148359

Ammar, H.A., Abd, A.A., Aty, E., El-Awdan, S.A. 2021. Extracellular myco-synthesis of nano-silver using fermentable yeasts and their biomedical applications. *Bioprocess and Biosystems Engineering* 44, 841–854. https://doi.org/10.1007/s00449-020-02494-3

Ashby, M., Ferreira, P., Schodek, D. 2009. *Nanomaterials, Nanotechnologies and Design*. Elsevier, Oxford. https://doi.org/10.1016/B978-0-7506-8149-0.00002-7

Diallo, T.B., Doyle, B.M., Mothudi, E., Manikandan, V., Rajendran, M., Maaza, M. 2017. Magnetic behavior of biosynthesized Co_3O_4 nanoparticles. *Journal of Magnetism and Magnetic Materials* 424, 251–255. https://doi.org/10.1016/j.jmmm.2016.10.063

Ditta, A., Arshad, M., Ibrahim, M. 2015. Nanoparticles in sustainable agricultural crop production: applications and perspectives. In: Siddiqui, M., Al-Whaibi, M., Mohammad, F. (eds), *Nanotechnology and Plant Sciences: Nanoparticles and Their Impact on Plants*, pp. 55–75. Springer, Cham.

El-Mohamedy, R.S.R., Abd El-Aziz, M.E., Kamel, S. 2019. Antifungal activity of chitosan nanoparticles against some plant pathogenic fungi in vitro. *Agricultural Engineering International: CIGR Journal* 21, 201–209.

European Commission. 2011. Commission recommendation of 18 October 2011 on the definition of nanomaterial. *Official Journal of the European Union* 275, 38–40.

Faizan, M., Yu, F., Chen, C., Faraz, A., Hayat, S. 2021. Zinc oxide nanoparticles help to enhance plant growth and alleviate abiotic stress: a review. *Current Protein and Peptide Science* 22, 362–375.

Genaidy, E.A., Abd-Alhamid, N., Hassan, H.S., Hassan, A.M., Hagagg, L.F. 2020. Effect of foliar application of boron trioxide and zinc oxide nanoparticles on leaves chemical composition, yield and fruit quality of *Olea europaea* L. cv. *Picual. Bulletin of the National Research Centre* 44, 106.

Ghosh, S., Ahmad, R., Zeyaullah, M., Khare, S.K. 2021. Microbial nano-factories: synthesis and biomedical applications. *Frontiers in Chemistry* 9, 626834. https://doi.org/10.3389/fchem.2021.626834

Harris, H.W., Sanchez-Andrea, I., McLean, J.S., Salas, E.C., Tran, W., El-Naggar, M.Y. et al. 2018. Redox sensing within the genus *Shewanella*. *Frontiers in Microbiology* 8, 2568. https://doi.org/10.3389/fmicb.2017.02568

Hassanisaadi, M., Barani, M., Rahdar, A., Heidary, M., Thysiadou, A., Kyzas, G.Z. 2022. Role of agrochemical-based nanomaterials in plants: biotic and abiotic stress with germination improvement of seeds. *Plant Growth Regulation* 97, 375–418.

Hmmam, I., Korkar, H.M., Ahmed, I., El-Hamied, A.G.A., Shaban, A. 2025. NPK nano-fertilizers enhance growth, oil quality, and yield regularity in Picual olive trees. *Scientific Reports* 15, 32495. https://doi.org/10.1038/s41598-025-17267-9

Jang, H., Lee, W., Kim, M.-S., Ahn, J.-H. 2015. Inorganic semiconductor nanomaterials for flexible electronics. In: Caironi, M., Noh, Y.-Y. (eds), *Large Area and Flexible Electronics*. Wiley-VCH, Weinheim. https://doi.org/10.1002/9783527679973.ch6

Johnson, M.S., Sajeev, S., Nair, R.S. 2021. Role of nanosensors in agriculture. In: *Proceedings of the International Conference on Computational Intelligence and Knowledge Economy (ICCIKE)*, IEEE, Dubai, pp. 58–63.

Joudeh, N., Linke, D. 2022. Nanoparticle classification, physicochemical properties, characterization, and applications: a comprehensive review for biologists. *Journal of Nanobiotechnology* 20, 262. https://doi.org/10.1186/s12951-022-01477-8

Kah, M. 2015. Nanopesticides and nanofertilizers: emerging contaminants or opportunities for risk mitigation? *Frontiers in Chemistry* 3, 64. https://doi.org/10.3389/fchem.2015.00064

Kaphle, A., Navya, P.N., Umapathi, A., Daima, H.K. 2018. Nanomaterials for agriculture, food and environment: applications, toxicity and regulation. *Environmental Chemistry Letters* 16, 43–58.

Karnwal, A., Sachan, R.S.K., Devgon, I., Devgon, J., Pant, G., Panchpuri, M., Ahmad, A., Alshammari, M.B., Hossain, K., Kumar, G. 2024. Gold nanoparticles in nanobiotechnology: from synthesis to biosensing applications. *ACS Omega* 9, 29966–29982. https://doi.org/10.1021/acsomega.3c10352

Khan, M.K. 2023. Plant-derived metal nanoparticles: synthesis, characterization, and oxidative stress-mediated therapeutic actions. *Future Pharmacology* 3, 252–295. https://doi.org/10.3390/futurepharmacol3010018

Kholief, M.G., Hesham, A.E.L., Hashem, F.S., Mohamed, F.M. 2024. Synthesis and utilization of titanium dioxide nanoparticles for photocatalytic degradation of organics. *Scientific Reports* 14, 11327. https://doi.org/10.1038/s41598-024-53617-9

Khosravi-Darani, K., da Cruz, A., Mozafari, M., Abdi, Z., Ahmadi, N. 2019. Biosynthesis of metal nanoparticles by probiotic bacteria. *Letters in Applied NanoBioScience* 8, 619–626.

Kim, T.-Y., Kim, M.G., Lee, J.- H., Hur, H.- G. 2018. Biosynthesis of nanomaterials by *Shewanella* species for application in lithium ion batteries. *Frontiers in Microbiology* 9, 2817. https://doi.org/10.3389/fmicb.2018.02817

Kole, C., Kole, P., Randunu, K.M., Choudhary, P., Podila, R., Ke, P.C., Marcus, R.K. 2013. Nanobiotechnology can boost crop production and quality. *BMC Biotechnology* 13, 1–10.

Konishi, Y., Ohno, K., Saitoh, N., Nomura, T., Nagamine, S., Hishida, H., Uruga, T. 2007. Bioreductive deposition of platinum nanoparticles on *Shewanella algae*. *Journal of Biotechnology* 128, 648–653.

Konstantatos, G., Sargent, E.H. 2009. Solution-processed quantum dot photodetectors. *Proceedings of the IEEE* 97, 1666–1683.

Kumar, S., Nehra, M., Dilbaghi, N., Marrazza, G., Hassan, A.A., Kim, K.-H. 2019. Nano-based smart pesticide formulations: emerging opportunities for agriculture. *Journal of Controlled Release* 294, 131–153.

Küünal, S., Rähn, M., Kikas, T. 2018. Plant extract mediated synthesis of nanoparticles. In: Barhoum, A., Makhlouf, A.S.H. (eds), *Emerging Applications of Nanoparticles: Architectures, Nanostructures, and Current Prospects*, pp. 161–182. Elsevier, Oxford.

Liu, R., Lal, R. 2015. Potentials of engineered nanoparticles as fertilizers for increasing agronomic productions. *Science of the Total Environment* 514, 131–139. https://doi.org/10.1016/j.scitotenv.2015.01.104

Liu, Y., Dai, X., Li, J., Cheng, S., Zhang, J., Ma, Y. 2024. Recent progress in TiO_2–biochar-based photocatalysts for water contaminants treatment: Strategies to improve photocatalytic performance. *RSC Advances* 14, 478–491. https://doi.org/10.1039/D3RA06910A

Lodahl, P., Mahmoodian, S., Stobbe, S. 2015. Interfacing single photons and single quantum dots with photonic nanostructures. *Reviews of Modern Physics* 87, 347–400. https://doi.org/10.1103/RevModPhys.87.347

Mao, T., Zha, J., Hu, Y., Chen, Q., Zhang, J., Luo, X. 2024. *Research progress of TiO_2 modification and photodegradation of organic pollutants. Inorganics* 12(7), 178. https://doi.org/10.3390/inorganics12070178

Matras, E., Gorczyca, A., Przemieniecki, S.W., Oćwieja, M. 2022. Surface properties-dependent antifungal activity of silver nanoparticles. *Scientific Reports* 12,18046. https://doi.org/10.1038/s41598-022-22659-2

Maurice, P.A., Hochella, M.F. 2008. Nanoscale particles and processes. *Advances in Agronomy* 100, 1–58. https://doi.org/10.3390/plants14050716

Murray, A.J., Zhu, J., Wood, J., Macaskie, L.E. 2017. A novel biorefinery: biorecovery of precious metals from spent automotive catalyst leachates. *Minerals Engineering* 113, 102–108. https://doi.org/10.1016/j.mineng.2017.07.012

Nima, Z.A., Mahmood, M., Xu, Y., Mustafa, T., Watanabe, F., Nedosekin, D.A., Biris, A.S. 2014. Circulating tumor cell identification by functionalized silver–gold nanorods. *Scientific Reports* 4, 4752. https://doi.org/10.1038/srep04752

Okey-Onyesolu, C.F., Hassanisaadi, M., Bilal, M., Barani, M., Rahdar, A., Iqbal, J., Kyzas, G.Z. 2021. Nanomaterials as nanofertilizers and nanopesticides: an overview. *Chemistry Select* 6, 8645–8663. https://doi.org/10.1002/slct.202102379

Owaid, M.N., Al-Saeedi, S.S.S., Abed, I.A. 2017. Biosynthesis of gold nanoparticles using yellow oyster mushroom *Pleurotus cornucopiae*. *Environmental Nanotechnology, Monitoring & Management* 8, 157–162. https://doi.org/10.1016/j.enmm.2017.07.005

Pandey, G. 2018. Challenges and future prospects of agri-nanotechnology for sustainable agriculture in India. *Environmental Technology & Innovation* 11, 299–307. https://doi.org/10.1016/j.eti.2018.06.014

Pandit, C., Roy, A., Ghotekar, S., Khusro, A., Islam, M.N., Emran, T.B., Lam, S.E., Khandaker, M.U., Bradley, D.A. 2022. Biological agents for synthesis of nanoparticles and their applications. *Journal of King Saud University – Science* 34, 101869. https://doi.org/10.1016/j.jksus.2022.101869

Pattan, G., Kaul, G. 2014. Health hazards associated with nanomaterials. *Toxicology and Industrial Health* 30, 499–519. https://doi.org/10.1177/0748233712459900

Pawlaczyk, M., Schroeder, G. 2019. Adsorption studies of Cu(II) ions on dendrimer-grafted silica-based materials. *Journal of Molecular Liquids* 281, 176–185. https://doi.org/10.1016/j.molliq.2019.02.093

Prasad, R., Bhattacharyya, A., Nguyen, Q.D. 2017. Nanotechnology in sustainable agriculture: recent developments, challenges, and perspectives. *Frontiers in Microbiology* 8, 1014. https://doi.org/10.3389/fmicb.2017.01014

Raliya, R., Franke, C., Chavalmane, S., Nair, R., Reed, N., Biswas, P. 2016. Quantitative understanding of nanoparticle uptake in watermelon plants. *Frontiers in Plant Science* 26(7), 1288. doi: 10.3389/fpls.2016.01288

Rodríguez, C., Briano, S., Leiva, E. 2020. Increased Adsorption of Heavy Metal Ions in Multi-Walled Carbon Nanotubes with Improved Dispersion Stability. *Molecules* 8;25(14), 3106. doi: 10.3390/molecules25143106.

Salouti, M., Khadivi Derakhshan, F. 2020. Biosensors and nanobiosensors in environmental applications. In: Ghorbanpour, M., Bhargava, P., Varma, A., Choudhary, D. (eds), *Biogenic Nano-Particles and Their Use in Agro-ecosystems*. Springer, Singapore. https://doi.org/10.1007/978-981-15-2985-6_26

Sekhon, B.S. 2014. Nanotechnology in agri-food production: an overview. *Nanotechnology, Science and Applications* 7, 31–53. https://doi.org/10.2147/NSA.S39406

Shokrzadeh, L., Mohammadi, P., Mahmoudian, M.R., Basirun, W.J., Bahreini, M. 2020. L-glycine-assisted synthesis of SnO_2/Pd nanoparticles. *Materials Chemistry and Physics* 240, 122172. https://doi.org/10.1016/j.matchemphys.2019.122172

Shukla, S., Upadhyay, D., Mishra, A., Jindal, T., Shukla, K. 2022. Challenges faced by farmers due to fungal pathogens. In: Rajpal, V.R., Singh, I., Navi, S.S. (eds), *Fungal Diversity, Ecology and Control Management*, pp. 495–505. Springer Nature, Singapore.

Shukla, S.K., Mishra, A.K., Arotiba, O.A., Mamba, B.B. 2013. Chitosan-based nanomaterials: a state-of-the-art review. *International Journal of Biological Macromolecules* 59, 46–58. https://doi.org/10.1016/j.ijbiomac.2013.04.043

Sigala-Aguilar, N.A., Flores-Rentería, D., Vera-Reyes, I., Moya-Cadena, S., Fernández-Luqueño, F., López, M.G. 2025. Multi-walled carbon nanotubes: effect on the growth-development of tomato plants (*Solanum lycopersicum* L.), biocompounds content in tomato fruits and quality indicators of cultivated soil. *Journal of Soil Science and Plant Nutrition* 25, 6840–6860. https://doi.org/10.1007/s42729-025-02566-x

Silva, S., Dias, M.C., Silva, A.M. 2022. Titanium- and zinc-based nanomaterials in agriculture. *Toxics* 10, 172. https://doi.org/10.3390/toxics10040172

Singh, A.K. 2016. Introduction to nanoparticles and nanotoxicology. In: Singh, A.K. (ed), *Engineered Nanoparticles*, pp. 1–18. Elsevier, Oxford. https://doi.org/10.1016/B978-0-12-801406-6.00001-7

Srivastava, A.K., Dev, A., Karmakar, S. 2018. Nanosensors and nanobiosensors in food and agriculture. *Environmental Chemistry Letters* 16, 161–182. https://doi.org/10.1007/s10311-017-0674-7

Srivastava, M., Sidler, A.V., Allain, D.S., Lembke, A., Kis, A., Imamoglu, A. 2015. Optically active quantum dots in monolayer WSe_2. *Nature Nanotechnology* 10, 491–496. https://doi.org/10.1038/nnano.2015.67

Szőllősi, R., Molnár, Á., Kondak, S., Kolbert, Z. 2020. Dual effect of nanomaterials on germination and seedling growth: Stimulation vs. phytotoxicity. *Plants (Basel)* 9(12), 1745. doi: 10.3390/plants9121745

Thabet, S.G., Alqudah, A.M. 2024. Role of nanoparticles in improving plant resilience under environmental stress. *Plant and Soil* 503, 313–330. https://doi.org/10.1007/s11104-024-06581-2

Tiwari, M., Sharma, N.C., Fleischmann, P., Burbage, J., Venkatachalam, P., Sahi, S.V. 2017. Nanotitania exposure alters physiological and stress responses in tomato. *Frontiers in Plant Science* 8, 633. https://doi.org/10.3389/fpls.2017.00633

Tyagi, S., Kumar, A., Khan, M.S. 2019. Extracellular synthesis of silver nanoparticles using entomopathogenic fungus. *SN Applied Sciences* 1, 1545. https://doi.org/10.1007/s42452-019-1593-y

Usman, M., Farooq, M., Wakeel, A., Nawaz, A., Cheema, S.A., ur Rehman, H., Ashraf, I., Sanaullah, M. 2020. Nanotechnology in agriculture: current status and future opportunities. *Science of the Total Environment* 721, 137778. https://doi.org/10.1016/j.scitotenv.2020.137778

Vishwakarma, K., Upadhyay, N., Kumar, N., Tripathi, D.K., Chauhan, D.K., Sharma, S., Sahi, S.V. 2018. Potential applications of nanotechnology in sustainable agriculture. In: Tripathi, D.K., Ahmad, P., Sharma, S., Chauhan, D.K., Dubey, N.K. (eds), *Nanomaterials in Plants, Algae, and Microorganisms*, pp. 473–500. Academic Press, New York.

Wan, X., Liu, J., Zhang, Y., Tian, D., Liu, Y., Shen, F. 2023. Conversion of biomass into bioethanol and lignin nanoparticles. *Industrial Crops and Products* 191, 115969. https://doi.org/10.1016/j.indcrop.2022.115969

Wang, Q., Liao, Y., Zhao, W., Yi, T., Jiang, Y., Zhu, G., Sun, Y., Huang, L., Chen, F., Zhang, P. 2024. Potassium-based nanomaterials enhance nutrient utilization efficiency. *Environmental Science: Nano* 11, 2906–2922. https://doi.org/10.1039/D4EN00148F

Yadav, T.P., Awasthi, K. 2022. Carbon Nanomaterials: Fullerene to Graphene. *Transactions of the Indian National Academy of Engineering* 7, 715–737.https://doi.org/10.1007/s41403-022-00348-w

Yu, S., Sun, J., Shi, Y., Wang, Q., Wu, J., Liu, J. 2020. Nanocellulose from biomass wastes. *Environmental Science & Ecotechnology* 5, 100077. https://doi.org/10.1016/j.ese.2020.100077

12 Challenges and Future Perspectives of Nanomaterials in Plant Stress Management

Palampreet Singh, Gurvarinder Kaur, and Geetika Sirhindi

12.1 INTRODUCTION

Plant stress factors, such as elevated or low temperatures, waterlogging, droughts, salt, heavy metals (HMs), and ultraviolet (UV) radiation, can have an impact on global agricultural output in the present time (Badawy et al. 2021). Based on the type of crop, the kind of stress, and exposure period, a plant's reaction to stress involves changes in a variety of morphological, physiological, and biochemical processes (Rady et al. 2021). Therefore, the soil properties, water-holding capacity, and other resources necessary for plants can be improved through sustainable agriculture and yield productivity (Desoky et al. 2020). Nanomaterials (NMs) have proved themselves to have great promise in improving plant stress control, especially nanoparticles (NPs). NPs are minuscules with high surface area-to-volume ratio, high adsorption efficacy, and increased connecting and working efficiencies, thereby making them exceptional candidates for enhancing plant resistance to a variety of abiotic stresses, such as drought, heat, salinity, and HM toxicity (Dubchak et al. 2010). NPs can be used as nanosensors, nanofertilizers, nanoherbicides/pesticides, and nanoremediators in the agrifood and agricultural industries (El-Saadony et al. 2021).

Plants are able to produce NPs naturally, which enhance plant morphology without causing harm to the plants (Faraji and Sepehri 2020). These NPs functioning as a carrier for the efficient and targeted release of herbicides, insecticides, fertilizers, and other substances in a desired ratio to target plant cellular organelles, and have the potential to enhance plant morphogenesis. However, some of the potential of NPs remains untapped because there aren't enough established or researched processes. Depending on the use and application, several NP types are created, such as those that contain organic polymeric materials, carbon-based NPs, metallic NPs, and inorganic nonmetallic NPs (Khan et al. 2017).

Because of being rooted to a single spot, plants must withstand a variety of severe environmental stresses, including flooding, HMs, salinity, drought, fluctuating light intensities, UV radiation, and others. Reactive oxygen species (ROS) bursts brought on by the harsh climatic circumstances damage macromolecules and membranes,

DOI: 10.1201/9781032659435-12

initiate cell toxicity, and stunt plant growth. To reduce oxidative stress, antioxidant machinery scavenges ROS using enzymatic and non-enzymatic methods. NPs participate in plant growth and development and then shield plants from diverse abiotic stresses (Mittal et al. 2020). As per the study of Elemike et al. (2019), NPs altered genes and their expressions governing in energy transport, electron transport, and cell biosynthesis and organization during stress responses. Research evidence has demonstrated that NPs are crucial to the development of crop plants; however, early comprehension of the proper mechanism and how NPs interact with plants at various levels is still necessary (Choudhary et al. 2019; Vishwakarma et al. 2018). Besides demonstrating a lot of beneficial roles on the plant growth, NPs also pose some deleterious impacts which will be briefly discussed in this chapter in forthcoming sections. This information is imperative to make use of NPs sustainably so that they can be utilized only for growth promoting aspects.

12.2 DUAL EFFECTS OF NANOPARTICLES IN CROP RESILIENCE AND RISKS

While adding NPs to crops has helped them resist a variety of environmental challenges, not all NPs offer sufficient protection since species-to-species variation exists in their efficiency (Figure 12.1). Numerous studies suggest that the NPs cause phytotoxicity, which triggers the production of ROS and results in oxidative damage (Chichiriccò and Poma 2015; Ebbs et al. 2016). Gottschalk et al. (2013) claim that while NPs can be hazardous when applied in high levels, they can also be beneficial when used in moderate doses to strengthen the antioxidant defense system and prevent oxidative damage brought on by abiotic stress (Solaiman et al. 2015). As per the study of Shen et al. (2010), NPs cause genotoxicity and oxidative stress in plants. They also interfere with physicochemical metabolic processes by preventing agricultural crops from absorbing minerals (Yadav et al. 2014). Peralta-Videa et al.

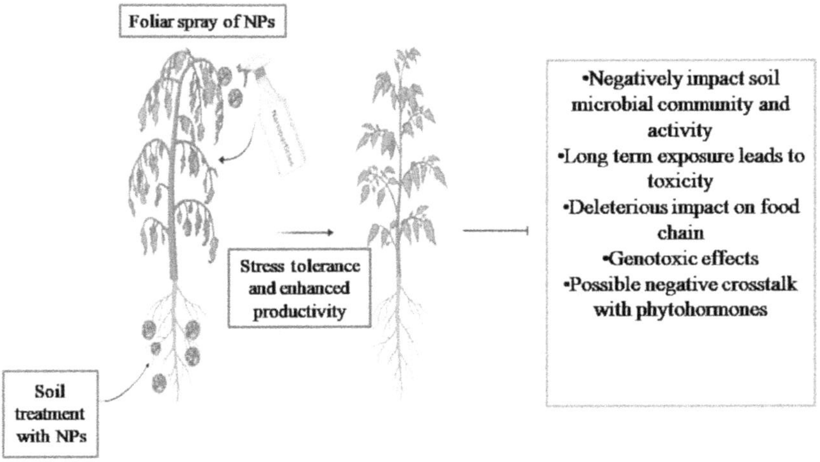

FIGURE 12.1 Dual role of NPs in agriculture.

(2011) also reported negative impacts of NPs on plants. In addition to the dosage administered, the administration method, as well as the NPs' size and form, all affect how harmful they are (Chichiriccò and Poma 2015; Syu et al. 2014). Manke et al. (2013) claim that the conformational change in the NPs' size and shape can impact biological metabolism and cause ROS generation. Additionally, they showed that NPs' phytotoxicity causes genotoxicity, cell signaling, and inflammation, all of which led to serious physiological decline. According to Ebbs et al. (2016), the surface area and composition of plants also affect the degree of NPs' toxicity with regard to uptake, transportation, and accumulation. Similarly, Huang et al. (2010) showed that metal/metalloid-category NPs cause Fenton reactions, which release free radicals and ultimately cause ROS to develop in plants. Certain variables lead to an imbalance in the redox status of NPs, which in turn downregulates the antioxidant activities and increases the production of free radicals (Huang et al. 2010). Nevertheless, in order to prevent ecotoxicological risks for both humans and plants, the factors viz. shape, composition, size, surface area, application procedures and dose, redox state, etc. should be thoroughly investigated prior to applying NPs in agricultural fields.

12.3 ECOTOXICOLOGICAL IMPACTS OF NANOPARTICLES IN SOIL AND PLANT SYSTEMS

Zinc oxide Engineered NMs (ZnO ENMs) production was projected to reach 30,000 metric tons annually in 2010 (Keller et al. 2013). Further, 8%–28% of the zinc (Zn) containing NPs are disposed of in the soil, indicating the substantial impact these NMs have on soil. Moreover, a variety of soil physiochemical characteristics, such as pH, humic acid, organic matter, salinity, and clay, are crucial for the movement and retention of metals. Consequently, it is anticipated that these characteristics will influence how ZnO NPs behave in the soil. According to Cornelis et al. (2014), ZnO NPs in soil have a tendency to cluster or interact with other soil particles, which may cause a decrease in the rate of diffusion and reactivity, toxicity, and bioavailability to plants and microbes (Milani et al. 2015). ZnO NPs are gaining a lot of attention since they may be used in a variety of fields and are also being used as plant growth regulators (PGRs) in agriculture (Wang et al. 2023). However, over time, their time span in plants elevates them to bioaccumulate in plants, which adversely affects the food chain (Rajput et al. 2020). As a result, numerous investigations looked into the potential reasons underlying diverse NP toxicity in diverse plant species (Azhar et al. 2023; Khan et al. 2023b). Higher concentrations of ZnO NPs have been reported to have negative impacts on plants. ZnO NPs poisoned some plant species at increasing exposure amounts. The recent study indicates that ZnO NPs cause stomatal and ultrastructural damage to rice seedlings by impacting biomass, chlorophyll molecule content, and stress markers level (Khan et al. 2021). Additionally, some NPs have been reported to bioaccumulate in plant tissues, even comprising in edible plant parts, and adversely downregulate the expression of genes allied with photosynthetic efficiency, growth and yield of crops (Rajput et al. 2018a, 2018b; Chen et al. 2018). Moreover, their findings demonstrated that ZnO stimulates antioxidative defense enzymes by upregulating the expression of antioxidative defense-responsive genes in rice seedlings. Similarly, Kim et al. (2012) reported that ZnO NPs at higher concentration inhibit germination rate, morphology of plants, photosynthetic content, triggered ROS level, etc. At the molecular level, these NPs show genotoxicity by cause damage to chromosome, deoxyribonucleic acid (DNA), cell cycle arrest, etc. and cytotoxicity through damaging the cell membrane in plants (Sun et al. 2021; Plaksenkova et al. 2020).

Earthworms are widely recognized as reliable indicators of soil ecological quality, and comprehensive ecotoxicological assessments covering their life-cycle traits were first reported by Heckmann et al. (2011). In that study, the effects of several inorganic NPs, including nickel (Ni), copper (Cu), silver (Ag), aluminum oxide (Al_2O_3), zirconium dioxide (ZrO_2), silicon dioxide (SiO_2), and titanium dioxide (TiO_2), were evaluated in earthworms. The findings demonstrated pronounced toxic responses to Ag, Cu, and TiO_2 NPs, with Ag exposure resulting in a complete inhibition of reproductive activity in treated earthworms (Heckmann et al. 2011). The ability of earthworms to detect and steer clear of soils containing Ag NPs suggests that the behavior of non-target animals may be affected in an unforeseen way. Exposure of earthworms to silver nitrate ($AgNO_3$) (94.21 mg/kg) or Ag NPs caused a considerable reduction in their reproduction. In contrast to the same quantities of Ag NPs, tissues treated to $AgNO_3$ were found to accumulate Ag more heavily (Shoults-Wilson et al. 2011). Unrine et al. (2010) examined how particle size affected the oxidation, bioavailability, and unfavorable impacts of produced Cu NPs in soils. They found that only at relatively high concentrations (N65 mg Cu/kg soil) did oxidized Cu NPs have unfavorable effects in earthworms. Toxicity assessments conducted over different exposure durations demonstrated that atrazine exerts concentration-dependent effects on earthworms (*Nsukkadrilus mbae*), as reflected by substantial variation in LC_{50} values. Histopathological investigations further revealed cellular abnormalities, including nuclear pyknosis, pronounced vacuolation, and damage to epithelial tissues as well as the chloragogenous layer (Oluah et al. 2010). In comparison with conventional herbicide formulations, poly(ε-caprolactone) nanocapsule-encapsulated herbicides exhibited lower toxic effects on *Pseudokirchneriella subcapitata* and *Prochilodus lineatus*, whereas increased sensitivity was observed in *Daphnia similis* (Andrade et al. 2019). These findings underscore the necessity for comprehensive NP risk-assessment frameworks to identify safer and more effective nanopesticide formulations at cellular and molecular scales, while emphasizing the need for further investigation into NP-induced toxicity (Pandey et al. 2018; Tiwari et al. 2020).

12.4 LIMITATIONS OF NANOMATERIALS AS BIOREMEDIATION AGENTS

Bioremediation is widely regarded as a site-specific, environmentally benign, cost-effective, and adaptable strategy for removing organic and inorganic pollutants from contaminated environments. Despite these advantages, its practical application is often constrained by prolonged remediation durations, limited contaminant bioavailability, reduced efficiency in heavily polluted soils due to toxic effects of pollutants on biological agents (such as bacteria, fungi, and plants), and the potential generation of secondary toxic metabolites (Cecchin et al. 2017). The incorporation of NMs into bioremediation strategies has emerged as a promising approach to overcome several of these limitations. Although NMs have been employed for nearly two decades in chemical remediation processes, their integration with biological remediation systems remains a relatively nascent research area. Nano-bio integrated approaches offer the potential to remediate contaminated soils either concurrently or through stepwise treatment processes.

NMs have been shown to alleviate metal-induced toxicity in plants by modulating gene networks associated with metal stress responses, oxidative balance, water regulation, cell wall development, photosynthetic activity, and cell cycle progression (Nair and Chung 2014). Previous studies have indicated that NMs can alter cell wall permeability, regulate the expression of metal transporter genes, and facilitate the co-movement of HMs with NPs, potentially enhancing metal accumulation within plant tissues (Hu et al. 2015). Conversely, there is also evidence suggesting that NMs may impair phytoremediation efficiency under certain conditions. For example, Tang et al. (2015) reported that the combined exposure to graphene oxide (GO) and cadmium intensified Cd toxicity in *Microcystis aeruginosa* through elevated ROS production.

Due to varying experimental settings, soil features, and ENM dosages utilized in the studies, recent research has also shown inconsistent impacts of ENMs on soil organic matter (SOM) dynamics (Rahmatpour et al. 2017; Shi et al. 2018). At environmentally relevant concentrations, Ag NPs do not appear to significantly alter SOM turnover (Rahmatpour et al. 2017). Likewise, metal oxide NPs such as copper(II) oxide (CuO) and magnetite (Fe_3O_4) exert minimal influence on total SOM content, apart from inducing shifts in soil microbial community structure. In contrast, certain metal oxides, particularly TiO_2 NPs, have been shown to enhance SOM stabilization by promoting photo-oxidative cross-linking of humic constituents via covalent interactions (Nuzzo et al. 2016). Although TiO_2 NPs generally do not markedly affect microbially derived SOM formation, they can suppress SOM mineralization by increasing organic matter stability through interactions with humic substances (Simonin and Richaume 2015).

Conversely, reduced microbial activity following ZnO NP application has been associated with a substantial decline in the decomposition of litter-derived organic carbon (Rashid et al. 2017b). Similar outcomes were reported for ferric oxide (Fe_2O_3) NPs, which significantly lowered soil carbon dioxide (CO_2) efflux, indicating decreased SOM breakdown (Rashid et al. 2017a). Collectively, these findings suggest that certain NMs may contribute to climate change mitigation by limiting soil carbon losses and enhancing carbon sequestration. However, contrasting evidence indicates that elevated concentrations of CuO NPs in paddy soils can lead to reductions in SOM levels (Shi et al. 2018).

12.5 IMPACT OF NANOMATERIALS ON SOIL MICROORGANISMS

12.5.1 Nanoparticle Effects on Soil Microbial Communities

Given their direct influence on soil quality, particularly nutrient cycling and SOM dynamics, soil microorganisms are excellent markers of soil quality.

Understanding how these NMs applied to soil can impact the diversity, structure, and activity of microorganisms is crucial. According to reports, the toxicity of newly added NMs might directly damage soil microorganisms, or it can indirectly affect them by increasing the bioavailability of other hazardous chemicals that are already present in the soil. Depending on the type of NP, the impact of these particles on microbial activities and functioning can vary greatly. For example, fullerenes and carbon nanotubes are less toxic than inorganic NPs, such as metal and metal oxide (Simonin and Richaume 2015; Rajput et al. 2018b). Because of their distinct cell wall architectures, Gram-ve bacteria are typically less responsive to ENMs than $Gram^{+ve}$ bacteria (McKee and Filser 2016). However, there is disagreement in the literature in this regard because Ag NPs are more harmful to $Gram^{-ve}$ bacteria (Ingle et al. 2014). The functional genes and pathways of soil microbial communities (Archaea, Bacteria, and Eukarya) engaged in the carbon (C) and nitrogen (N) cycles were severely impacted by the carbon-based NPs; however, the sulfur (S) and phosphorus (P) cycles were less susceptible (Wu et al. 2020). Microbial biomass carbon (MBC), particularly in Gram-ve bacteria, was reduced by nano-sized metal oxides like ZnO and TiO_2 (Rajput et al. 2018a; Rashid et al. 2017b).

12.5.2 Influence of Nanoparticles on Soil Microbial Activity

ZnO and Fe_2O_3 NPs were shown to markedly decrease MBC as well as the abundance of culturable heterotrophic bacteria and fungi, whereas C_{60} NMs of different particle sizes exerted only minor effects on microbial biomass and associated enzyme activities (Rashid et al. 2017a, 2017b). According to Ben-Moshe et al. (2010), bacterial/fungal biomass ratios revealed a considerable alteration in microbial biodiversity, whereas other metal oxide NPs (cerium dioxide (CeO_2), Fe_3O_4, and tin dioxide (SnO_2)) did not exhibit any effect on MBC. In contrast, through increasing the activity of the enzymes urease and invertase, Fe-oxide NPs promoted the proliferation of bacteria, particularly Actinobacteria. Bacterial community architecture changed significantly, despite the fact that bacterial quantity remained unchanged. A recent two-year long-term study also demonstrated the beneficial effects of administering *n*ZnO and *n*CuO at low concentrations (10 mg/kg) on the microbial population and enzymatic activity (Jośko et al. 2019). Zhao et al. (2020) found that denitrification was strongly slowed in another investigation, with an 11-fold rise in NO_3 concentration. However, no significant inhibitory effect was observed when lower quantities of *n*CuO (10 and 100 mg/kg) were used. Studies have shown that Ag NPs adversely affect soil microbial processes only when applied at elevated concentrations, indicating that NP-induced changes in microbial biomass and activity are largely governed by application dose (He et al. 2016; Rahmatpour et al. 2017). In general, several NPs, including Ag, Al_2O_3, TiO_2, CuO, and ZnO, have been reported to negatively influence soil microbial communities, whereas minimal or negligible effects have been observed for NPs composed of Si, Fe, Au, Pd, and Ag_2S (Simonin and Richaume 2015; Rajput et al. 2020). Beyond chemical composition, NP toxicity is further modulated by multiple environmental and physicochemical factors, such as particle type and concentration, soil characteristics, and moisture conditions, which collectively determine the extent of microbial stress (Peng et al. 2020).

12.5.3 Influence of Nanoparticles on Soil Microbial Biomass and Enzyme-Related Biological Functions

Elevated concentrations of NPs are known to substantially alter soil microbial biomass as well as the composition and functioning of microbial communities (Peng et al. 2020). In response to concerns regarding nanotoxicity in soil ecosystems, increasing attention has been directed toward biogenic NPs, as several studies indicate that biologically synthesized NPs exert comparatively lower adverse effects on soil microorganisms than chemically produced counterparts (Mishra et al. 2020; Ottoni et al. 2020). Despite these promising observations, the application of biogenic NMs in soil systems remains insufficiently explored, highlighting the need for broader and more systematic investigations.

Although NMs have frequently been reported to exert beneficial effects on plant growth and soil processes, their application can also induce phytotoxic responses under certain conditions (Aslani et al. 2014; Mazaheri-Tirani and Dayani 2020). The direction and magnitude of these effects are strongly influenced by NP characteristics, including concentration, chemical composition, particle size, exposure duration, and environmental context (Mazaheri-Tirani and Dayani 2020; Noori et al. 2020). For example, Yusefi-Tanha et al. (2020) demonstrated that dissolved Cu^{2+} ions and smaller CuO NPs (25 nm) caused greater toxicity in soybean plants than larger CuO particles (50 and 250 nm). Their findings further revealed that smaller NPs exhibited a linear dose-response relationship, whereas larger particles followed a nonlinear response pattern.

NP-induced toxicity can influence multiple plant physiological traits, including seed germination, biomass accumulation, leaf development, and root elongation. Previous studies have shown that NP exposure may suppress germination, inhibit plant growth, or, in extreme cases, lead to plant mortality (Yang et al. 2017). Additional adverse effects reported in the literature include reduced photosynthetic efficiency (Barhoumi et al. 2015), oxidative damage to cellular membranes (Noori et al. 2020), growth retardation (Colman et al. 2013), disturbances in subcellular metabolism, chromosomal irregularities (Raskar and Laware 2014), impairment of plant water relations (Martínez-Fernández et al. 2016), and alterations in phytohormone levels accompanied by changes in gene expression profiles (García-Sánchez et al. 2015).

At the molecular level, several studies have documented genotoxic and transcriptomic alterations following NP exposure. Moreno-Olivas et al. (2014) reported DNA damage in plants treated with TiO_2 NPs, while transcriptome-based analyses have revealed disruptions in gene regulatory networks governing stress responses and metabolism (Landa et al. 2015). In maize, exposure to single-walled carbon nanotubes resulted in enhanced expression of *SLR1* and *RTCS* genes, alongside down-regulation of *RTH1* and *RTH3* genes (Yan et al. 2013). Similarly, fullerene NPs were shown to interfere with transcriptional regulation of genes involved in energy production and electron transport (Hossain et al. 2016). Conversely, NP exposure has also been associated with the up-regulation of stress-responsive and aquaporin-related genes, indicating complex regulatory effects (Tripathi et al. 2017). The expression of genes such as *NNtPIP1*, *NtLRX1*, and *CycB* associated with water transport, cell wall biosynthesis, and cell division was significantly enhanced following treatment with multi-walled carbon nanotubes (Khodakovskaya et al. 2012).

Nevertheless, excessive NP concentrations frequently result in pronounced toxicity. For instance, exposure to high levels of CeO_2 NPs (2,000–4,000 mg/L) caused structural damage to soybean DNA (López-Moreno et al. 2010). Similarly, TiO_2 NPs applied at 2 mM induced DNA damage in tobacco, while a concentration of 10 mM produced comparable effects in maize (Ghosh et al. 2010; Ruffini Castiglione et al. 2011). These findings collectively emphasize the necessity of species-specific toxicity assessments and rigorous pre-evaluation of NMs prior to their large-scale field application and commercial deployment.

12.6 GENOTOXIC EFFECTS OF NANOPARTICLES ON PLANT SPECIES

Exposure to NPs has been associated with a range of cellular and molecular disturbances in plants, including impaired cell cycle progression, genomic instability, chromosomal irregularities, and altered microRNA regulation. Tobacco, buckwheat, onion, and radish plant species were among the many plant species that experienced genotoxic effects after being exposed to different NPs, such as ZnO, CeO_2, TiO_2, and CuO (Pakrashi et al. 2014; Chichiriccò and Poma 2015). Wheat and barley showed cytologic alterations after being exposed to Ag NPs. Ag NPs caused distortion of chromosomal fragments, chromosomal aneuploidy, micronuclei production, chromosomal deletion, chromosomal abnormalities, binucleate cells, and sticky chromosomes (Abou-Zeid and Baraka 2014). These NPs decreased the root tips of faba beans' mitotic index. Patlolla et al. (2012) CuO NPs also caused changes in the rice's root-growth-related genes. They have also been discovered to be responsible for Onion DNA damage (Wang et al. 2015). Furthermore, CuO NPs increased the abnormality index and lowered the mitotic index in onion roots as reported by Nagaonkar et al (2015). In a comparative transcriptomic study, Tumburu et al. (2015) demonstrated that exposure to TiO_2 and CeO_2 NPs modified the expression of 204 and 142 genes in *Arabidopsis thaliana*, respectively, highlighting the distinct molecular impacts of different metal oxide nanoparticles.

12.7 HARMFUL EFFECTS OF NPs ON PHYTOHORMONES

Plant growth hormones have a major role in determining crop output and growth under many environmental conditions, including stress. Important toxicity markers in plants are the quantity and activity of phytohormones (Gangwar et al. 2014). NPs significantly impacted the synthesis of phytohormones. It was shown that the concentration of phytohormone decreased in rice seeds exposed to carbon nanotubes (Hao and colleagues 2016). According to research by Bleecker and Kende (2000), the presence of silver ions decreased ethylene synthesis, which in turn impaired the bond between ethylene and indole-3-acetic acid (IAA).

The root hormone concentrations in both the normal cotton and the *Bt*-transgenic cotton were negatively impacted by varying quantities of exogenously administered Fe_2O_3 NPs, which hindered the plant's development. The levels of gibberellic acid

(GA) and abscisic acid (ABA) in leaves of traditional cotton were considerably lower after exposure to Fe_2O_3 NPs than in leaves of transgenic cotton, which resulted in an increase in GA content without influencing ABA content. Exposure of both transgenic and non-transgenic rice plants to γ-Fe_2O_3 NPs resulted in elevated concentrations of IAA and ABA in root tissues (Gui et al. 2015). In contrast, Van Hai et al. (2015) observed that treatment with CeO_2 NPs did not significantly alter the levels of GA, IAA, or ABA in the leaves of either conventional or Bt-transgenic cotton relative to untreated controls. However, when conventional cotton plants were exposed to a higher concentration (500 mg/L) of CeO_2 NPs, a marked reduction in trans-zeatin riboside (t-ZR) content was recorded, with levels declining by approximately 25% compared to control plants.

Similarly, Vankova et al. (2017) demonstrated that increasing doses of ZnO NPs suppressed the biosynthesis of key phytohormones, including auxins and cytokinins, in the shoot apices of *Arabidopsis*. In contrast, the accumulation of cis-zeatin, a cytokinin commonly associated with stress responses, was significantly enhanced in root tissues, showing increases of 28% and 59% at NP concentrations of 20 and 100 mg/L, respectively.

12.8 FUTURE PROSPECTS

There are difficulties involved with using nanotechnology in agriculture. Despite these obstacles, there are a number of exciting opportunities for NMs in agriculture in the future. Green synthesis process advancements have great promise. By using biological processes to create NPs, these techniques can minimize damage to the environment and perhaps save expenditures. These environmentally benign methods support sustainable practices and increase public acceptability of nanotechnologies (Balusamy et al. 2023).

Global crop production has faced a number of climatic and stress-related difficulties. Nanotechnology has emerged as a crucial element for sustainable development in order to address these issues. By enhancing the plant's antioxidant defense mechanism, NMs can counteract the negative impacts of abiotic stressors. They have more effective adsorption and targeted distribution because of their wide surface area and ability to penetrate plants. This may be in charge of controlling photosynthetic efficiency and water absorption, eliminating ROS, and improving crop output, growth, and seed germination. Better productivity in agriculture can be achieved by using different NMs in a sustainable way by carefully analyzing their dosage. Risk assessment and knowledge of the destiny of NMs in plants and soil, as well as their interactions with the environment, are still necessary, nevertheless (Muzammil et al. 2023).

The development of safer and more ecologically friendly NMs will probably be the main focus of future research. This includes NMs that are less potentially hazardous and biodegradable as well as ones that do not build up in the environment.

12.9 CONCLUSION

Increasing crop resilience and agricultural productivity can be achieved through the application of nanotechnology, especially using NPs. Because of the distinct physicochemical characteristics of NPs, abiotic stressors such as salt, drought, and HM toxicity can be better managed in agricultural settings. They can strengthen antioxidant defense, optimize nutrient uptake efficiency, and promote plant development mechanisms. But there are also serious drawbacks to using NPs, most notably the possibility of phytotoxic effects that could negatively impact soil health and plant growth. The chapter emphasized how crucial it is to take a balanced approach to NMs in agriculture, taking into account both the possible advantages and ecological hazards. The development of biodegradable, ecologically benign NMs with low health concerns should be the main goal of future research. Furthermore, in order to guarantee the safe and efficient application of nanotechnology in agriculture and open the door for environmentally sound farming methods that can satisfy the world's food demand while maintaining environmental integrity, it is imperative to set strict risk assessment protocols.

REFERENCES

Abou-Zeid, L., Baraka, H.N. 2014. Combating oxidative stress as a hallmark of cancer and aging: computational modeling and synthesis of phenylene diamine analogs as potential antioxidant. *Saudi Pharmaceutical Journal* 22, 264–272.

Andrade, L.L.D., Pereira, A.E.S., Fernandes Fraceto, L., Bueno dos Reis Martinez, C. 2019. Can atrazine loaded nanocapsules reduce the toxic effects of this herbicide on the fish *Prochilodus lineatus*? A multibiomarker approach. *Science of the Total Environment* 663, 548–559.

Aslani, F., Bagheri, S., Muhd Julkapli, N., Juraimi, A.S., Hashemi, F.S.G. 2014. Effects of engineered nanomaterials on plants growth: an overview. *The Scientific World Journal* 2014, 641759.

Azhar, W., Khan, A.R., Salam, A., Ulhassan, Z., Qi, J., Shah, G., Liu, Y., Chunyan, Y., Yang, S., Gan, Y. 2023. Ethylene accelerates copper oxide nanoparticle-induced toxicity at physiological, biochemical, and ultrastructural levels in rice seedlings. *Environmental Science and Pollution Research* 30, 26137–26149.

Badawy, A.A., Abdelfattah, N.A.H., Salem, S.S., Awad, M.F., Fouda, A. 2021. Efficacy assessment of biosynthesized copper oxide nanoparticles (CuO-NPs) on stored grain insects and their impacts on morphological and physiological traits of wheat (*Triticum aestivum* L.). *Biology* 10, 233.

Balusamy, S.R., Joshi, A.S., Perumalsamy, H., Mijakovic, I., Singh, P. 2023. Advancing sustainable agriculture: a critical review of smart and eco-friendly nanomaterial applications. *Journal of Nanobiotechnology* 21, 372.

Barhoumi, L., Oukarroum, A., Taher, L.B., Smiri, L.S., Abdelmelek, H. 2015. Effects of superparamagnetic iron oxide nanoparticles on photosynthesis and growth of the aquatic plant *Lemna gibba*. *Archives of Environmental Contamination and Toxicology* 68, 510–520.

Ben-Moshe, T., Dror, I., Berkowitz, B. 2010. Transport of metal oxide nanoparticles in saturated porous media. *Chemosphere* 81, 387–393.

Bleecker, A.B., Kende, H. 2000. Ethylene: a gaseous signal molecule in plants. *Annual Review of Cell and Developmental Biology* 16, 1–18.

Cecchin, I., Reddy, K.R., Thomé, A., Tessaro, E.F., Schnaid, F. 2017. Nanobioremediation: integration of nanoparticles and bioremediation for sustainable remediation of chlorinated organic contaminants in soils. *International Biodeterioration & Biodegradation* 119, 419–428.

Chen, J., Dou, R., Yang, Z., You, T., Gao, X., Wang, L. 2018. Phytotoxicity and bioaccumulation of zinc oxide nanoparticles in rice (*Oryza sativa* L.). *Plant Physiology and Biochemistry* 130, 604–612.

Chichiriccò, G., Poma, A. 2015. Penetration and toxicity of nanomaterials in higher plants. *Nanomaterials* 5, 851–873. https://doi.org/10.3390/nano5020851

Choudhary, R.C., Kumaraswamy, R.V., Kumari, S., Sharma, S.S., Pal, A., Raliya, R., Biswas, P., Saharan, V. 2019. Zinc encapsulated chitosan nanoparticle to promote maize crop yield. *International Journal of Biological Macromolecules* 127, 126–135. https://doi.org/10.1016/j.ijbiomac.2019.01.046

Colman, B.P., Arnaout, C.L., Anciaux, S., Gunsch, C.K., Hochella Jr., M.F., Kim, B., Lowry, G.V., McGill, B.M., Reinsch, B.C., Richardson, C.J. 2013. Low concentrations of silver nanoparticles in biosolids cause adverse ecosystem responses under realistic field scenario. *PLoS One* 8, e57189.

Cornelis, G., Hund-Rinke, K., Kuhlbusch, T., Van den Brink, N., Nickel, C. 2014. Fate and bioavailability of engineered nanoparticles in soils: a review. *Critical Reviews in Environmental Science and Technology* 44, 2720–2764.

Desoky, E.-S.M., Merwad, A.-R.M., Semida, W.M., Ibrahim, S.A., El-Saadony, M.T., Rady, M.M. 2020. Heavy metals-resistant bacteria (HM-RB): potential bioremediators of heavy metals-stressed *Spinacia oleracea* plant. *Ecotoxicology and Environmental Safety* 198, 110685.

Dubchak, S., Ogar, A., Mietelski, J.W., Turnau, K. 2010. Influence of silver and titanium nanoparticles on arbuscular mycorrhiza colonization and accumulation of radiocaesium in *Helianthus annuus*. *Spanish Journal of Agricultural Research* 8, 103–108.

Ebbs, S.D., Bradfield, S.J., Kumar, P., White, J.C., Musante, C., Ma, X. 2016. Accumulation of zinc, copper, or cerium in carrot (*Daucus carota*) exposed to metal oxide nanoparticles and metal ions. *Environmental Science: Nano* 3, 114–126. https://doi.org/10.1039/C5EN00161G

Elemike, E.E., Uzoh, I.M., Onwudiwe, D.C., Babalola, O.O. 2019. The role of nanotechnology in the fortification of plant nutrients and improvement of crop production. *Applied Sciences* 9, 499. https://doi.org/10.3390/app9030499

El-Saadony, M.T., Saad, A.M., Taha, T.F., Najjar, A.A., Zabermawi, N.M., Nader, M.M., AbuQamar, S.F., El-Tarabily, K.A., Salama, A. 2021. Selenium nanoparticles from *Lactobacillus paracasei* HM1 capable of antagonizing animal pathogenic fungi as a new source from human breast milk. *Saudi Journal of Biological Sciences* 28, 6782–6794.

Faraji, J., Sepehri, A. 2020. Exogenous nitric oxide improves the protective effects of TiO_2 nanoparticles on growth, antioxidant system, and photosynthetic performance of wheat seedlings under drought stress. *Journal of Soil Science and Plant Nutrition* 20, 703–714. https://doi.org/10.1007/s42729-019-00158-0

Gangwar, S., Singh, V.P., Tripathi, D.K., Chauhan, D.K., Prasad, S.M., Maurya, J.N. 2014. Plant responses to metal stress: the emerging role of plant growth hormones in toxicity alleviation. In: Sitko, K. (eds), *Emerging Technologies and Management of Crop Stress Tolerance, Volume 2: A Sustainable Approach*, pp. 215–248. Academic Press, Amsterdam, Netherlands.

García-Sánchez, S., Bernales, I., Cristobal, S. 2015. Early response to nanoparticles in the *Arabidopsis* transcriptome compromises plant defence and root-hair development through salicylic acid signalling. *BMC Genomics* 16, 341.

Ghosh, M., Bandyopadhyay, M., Mukherjee, A. 2010. Genotoxicity of titanium dioxide (TiO_2) nanoparticles at two trophic levels: plant and human lymphocytes. *Chemosphere* 81, 1253–1262. https://doi.org/10.1016/j.chemosphere.2010.09.033

Gottschalk, F., Sun, T., Nowak, B. 2013. Environmental concentrations of engineered nanomaterials: review of modeling and analytical studies. *Environmental Pollution* 181, 287–300. https://doi.org/10.1016/j.envpol.2013.06.003

Gui, X., Deng, Y., Rui, Y. 2015. Response difference of transgenic and conventional rice (*Oryza sativa*) to nanoparticles (γFe_2O_3). *Environmental Science and Pollution Research International* 22, 17716–17723.

Hao, Y., Yu, F., Lv, R. 2016. Carbon nanotubes filled with different ferromagnetic alloys affect the growth and development of rice seedlings by changing the C:N ratio and plant hormones concentrations. *PLoS One* 11, e0157264.

He, S., Feng, Y., Ni, J., Sun, Y., Xue, L., Feng, Y., Yu, Y., Lin, X., Yang, L. 2016. Different responses of soil microbial metabolic activity to silver and iron oxide nanoparticles. *Chemosphere* 147, 195–202.

Heckmann, L.-H., Hovgaard, M.B., Sutherland, D.S., Autrup, H., Besenbacher, F., Scott Fordsmand, J.J. 2011. Limit-test toxicity screening of selected inorganic nanoparticles to the earthworm *Eisenia fetida*. *Ecotoxicology* 20, 226–233.

Hossain, Z., Mustafa, G., Sakata, K., Komatsu, S. 2016. Insights into the proteomic response of soybean towards Al_2O_3, ZnO, and Ag nanoparticles stress. *Journal of Hazardous Materials* 304, 291–305.

Hu, X., Kang, J., Lu, K., Zhou, R., Mu, L., Zhou, Q. 2015. Graphene oxide amplifies the phytotoxicity of arsenic in wheat. *Scientific Reports* 4, 6122.

Huang, Y., Wu, C., Aronstam, R. 2010. Toxicity of transition metal oxide nanoparticles: recent insights from in vitro studies. *Materials* 3, 4842–4859. https://doi.org/10.3390/ma3104842

Ingle, A.P., Duran, N., Rai, M. 2014. Bioactivity, mechanism of action, and cytotoxicity of copper-based nanoparticles: a review. *Applied Microbiology and Biotechnology* 98, 1001–1009.

Jośko, I., Oleszczuk, P., Dobrzyńska, J., Futa, B., Joniec, J., Dobrowolski, R. 2019. Long-term effect of ZnO and CuO nanoparticles on soil microbial community in different types of soil. *Geoderma* 352, 204–212.

Keller, A.A., McFerran, S., Lazareva, A., Suh, S. 2013. Global life cycle releases of engineered nanomaterials. *Journal of Nanoparticle Research* 15, 1–17.

Khan, A.R., Azhar, W., Wu, J., Ulhassan, Z., Salam, A., Zaidi, S.H.R., Yang, S., Song, G., Gan, Y. 2021. Ethylene participates in zinc oxide nanoparticles induced biochemical, molecular and ultrastructural changes in rice seedlings. *Ecotoxicology and Environmental Safety* 226, 112844.

Khan, A.R., Fan, X., Salam, A., Azhar, W., Ulhassan, Z., Qi, J., Liaquat, F., Yang, S., Gan, Y. 2023a. Melatonin-mediated resistance to copper oxide nanoparticles-induced toxicity by regulating the photosynthetic apparatus, cellular damages and antioxidant defense system in maize seedlings. *Environmental Pollution* 316, 120639.

Khan, A.R., Ulhassan, Z., Yang, S., Hamid, A.R., Muhammad, S., Azhar, W., Salam, A., Sheteiwy, M.S., Aftab, T., Wei, J.A. 2023b. Potential toxic effects of metal or metallic nanoparticles in plants and their detoxification mechanisms. In: Aftab, T. (eds), *Emerging Contaminants and Plants: Interactions, Adaptations and Remediation Technologies*, pp. 67–85. Springer, Cham.

Khan, M.N., Mobin, M., Abbas, Z.K., AlMutairi, K.A., Siddiqui, Z.H. 2017. Role of nanomaterials in plants under challenging environments. *Plant Physiology and Biochemistry* 110, 194–209. https://doi.org/10.1016/j.plaphy.2016.05.038

Khodakovskaya, M.V., Kim, B.S., Kim, J.N., Alimohammadi, M., Dervishi, E., Mustafa, T., Cernigla, C.E. 2013. Carbon nanotubes as plant growth regulators: effects on tomato growth, reproductive system, and soil microbial community. *Small* 9, 115–123.

Kim, S., Lee, S., Lee, I. 2012. Alteration of phytotoxicity and oxidant stress potential by metal oxide nanoparticles in *Cucumis sativus*. *Water, Air, & Soil Pollution* 223, 2799–2806.

Landa, P., Prerostova, S., Petrova, S., Knirsch, V., Vankova, R., Vanek, T. 2015. The transcriptomic response of *Arabidopsis thaliana* to zinc oxide: a comparison of the impact of nanoparticle, bulk, and ionic zinc. *Environmental Science & Technology* 49, 14537–14545.

López-Moreno, M.L., de la Rosa, G., Hernández-Viezcas, J.A., Castillo-Michel, H., Botez, C.E., Peralta-Videa, J.R., Gardea-Torresdey, J.L. 2010. Evidence of the differential biotransformation and genotoxicity of CeO_2 and ZnO nanoparticles on soybean (*Glycine max*) plants. *Environmental Science & Technology* 44, 7315–7320.

Manke, A., Wang, L., Rojanasakul, Y. 2013. Mechanisms of nanoparticle-induced oxidative stress and toxicity. *BioMed Research International* 2013, 942916. https://doi.org/10.1155/2013/942916

Martínez-Fernández, D., Barroso, D., Komárek, M. 2016. Root water transport of *Helianthus annuus* L. under iron oxide nanoparticle exposure. *Environmental Science and Pollution Research* 23, 1732–1741.

Mazaheri-Tirani, M., Dayani, S. 2020. In vitro effect of zinc oxide nanoparticles on *Nicotiana tabacum* callus compared to ZnO microparticles and zinc sulfate ($ZnSO_4$). *Plant Cell, Tissue and Organ Culture* 140, 279–289.

McKee, M.S., Filser, J. 2016. Impacts of metal-based engineered nanomaterials on soil communities. *Environmental Science: Nano* 3, 506–533.

Milani, N., Hettiarachchi, G.M., Kirby, J.K., Beak, D.G., Stacey, S.P., McLaughlin, M.J. 2015. Fate of zinc oxide nanoparticles coated onto macronutrient fertilizers in an alkaline calcareous soil. *PLoS One* 10, e0126275.

Mishra, S., Yang, X., Singh, H.B. 2020. Evidence for positive response of soil bacterial community structure and functions to biosynthesized silver nanoparticles: an approach to conquer nanotoxicity? *Journal of Environmental Management* 253, 109584.

Mittal, D., Kaur, G., Singh, P., Yadav, K., Ali, S.A. 2020. Nanoparticle-based sustainable agriculture and food science: recent advances and future outlook. *Frontiers in Nanotechnology* 2, 10. https://doi.org/10.3389/fnano.2020.579954

Moreno-Olivas, F., Gant, V.U., Johnson, K.L., Peralta-Videa, J.R., Gardea-Torresdey, J.L. 2014. Random amplified polymorphic DNA reveals that TiO_2 nanoparticles are genotoxic to *Cucurbita pepo*. *Journal of Zhejiang University Science A* 15, 618–623.

Muzammil, S., Ashraf, A., Siddique, M.H., Aslam, B., Rasul, I., Abbas, R., Afzal, M., Faisal, M., Hayat, S. 2023. A review on toxicity of nanomaterials in agriculture: current scenario and future prospects. *Science Progress* 106, 00368504231221672. https://doi.org/10.1177/00368504231221672

Nagaonkar, D., Shende, S., Rai, M. 2015. Biosynthesis of copper nanoparticles and its effect on actively dividing cells of mitosis in *Allium cepa*. *Biotechnology Progress*, 31(2), 557–565. https://doi.org/10.1002/btpr.2040

Nair, P.M.G., Chung, I.-M. 2014. Cell cycle and mismatch repair genes as potential biomarkers in *Arabidopsis thaliana* seedlings exposed to silver nanoparticles. *Bulletin of Environmental Contamination and Toxicology* 92, 719–725.

Noori, A., Donnelly, T., Colbert, J., Cai, W., Newman, L.A., White, J.C. 2020. Exposure of tomato (*Lycopersicon esculentum*) to silver nanoparticles and silver nitrate: physiological and molecular response. *International Journal of Phytoremediation* 22, 40.

Nuzzo, A., Madonna, E., Mazzei, P., Spaccini, R., Piccolo, A. 2016. In situ photopolymerization of soil organic matter by heterogeneous nano-TiO_2 and biomimetic metal-porphyrin catalysts. *Biology and Fertility of Soils* 52, 585–593.

Oluah, M., Obiezue, R.N.N., Ochulor, A.J., Onuoha, E. 2010. Toxicity and histopathological effect of atrazine (herbicide) on the earthworm *Nsukkadrilus mbae* under laboratory conditions. *Animal Research International* 7, 1287–1293.

Ottoni, C.A., Lima Neto, M.C., Léo, P., Ortolan, B.D., Barbieri, E., De Souza, A.O. 2020. Environmental impact of biogenic silver nanoparticles in soil and aquatic organisms. *Chemosphere* 239, 124698.

Pakrashi, S., Jain, N., Dalai, S., Jayakumar, J., Chandrasekaran, P.T., Raichur, A.M., Chandrasekaran, N., Mukherjee, A. 2014. In vivo genotoxicity assessment of titanium dioxide nanoparticles by *Allium cepa* root tip assay at high exposure concentrations. *PLoS One* 9, e87789.

Pandey, S., Giri, K., Kumar, R., Mishra, G., Rishi, R.R. 2018. Nanopesticides: opportunities in crop protection and associated environmental risks. *Proceedings of the National Academy of Sciences, India, Section B* 88, 1287–1308.

Patlolla, A.K., Berry, A., May, L., Tchounwou, P.B. 2012. Genotoxicity of silver nanoparticles in *Vicia faba*: a pilot study on the environmental monitoring of nanoparticles. *International Journal of Environmental Research and Public Health* 9, 1649–1662.

Peng, Z., Liu, X., Zhang, W., Zeng, Z., Liu, Z., Zhang, C. et al. 2020. Advances in the application, toxicity and degradation of carbon nanomaterials in environment: a review. *Environmental International* 134, 105298.

Peralta-Videa, J.R., Zhao, L., Lopez-Moreno, M.L., de la Rosa, G., Hong, J., Gardea-Torresdey, J.L. 2011. Nanomaterials and the environment: a review for the biennium 2008–2010. *Journal of Hazardous Materials* 186, 1–15. https://doi.org/10.1016/j.jhazmat.2010.11.020

Plaksenkova, I., Kokina, I., Petrova, A., Jermaļonoka, M., Gerbreders, V., Krasovska, M. 2020. The impact of zinc oxide nanoparticles on cytotoxicity, genotoxicity, and miRNA expression in barley (*Hordeum vulgare* L.) seedlings. *The Scientific World Journal* 2020, 2.

Rady, M.M., Boriek, S.H.K., Abd El-Mageed, T.A., Seif El-Yazal, M.A., Ali, E.F., Hassan, F.A.S., Abdelkhalik, A. 2021. Exogenous gibberellic acid or dilute bee honey boosts drought stress tolerance in *Vicia faba* by rebalancing osmoprotectants, antioxidants, nutrients, and phytohormones. *Plants* 10, 748.

Rahmatpour, S., Shirvani, M., Mosaddeghi, M.R., Nourbakhsh, F., Bazarganipour, M. 2017. Dose–response effects of silver nanoparticles and silver nitrate on microbial and enzyme activities in calcareous soils. *Geoderma* 285, 313–322.

Rajput, V., Minkina, T., Sushkova, S., Behal, A., Maksimov, A., Blicharska, E., Ghazaryan, K., Movsesyan, H., Barsova, N. 2020. ZnO and CuO nanoparticles: a threat to soil organisms, plants, and human health. *Environmental Geochemistry and Health* 42, 147–158.

Rajput, V., Minkina, T., Sushkova, S., Tsitsuashvili, V., Mandzhieva, S., Gorovtsov, A., Nevidomskyaya, D., Gromakova, N. 2018a. Effect of nanoparticles on crops and soil microbial communities. *Journal of Soils and Sediments* 18, 2179–2187.

Rajput, V.D., Minkina, T.M., Behal, A., Sushkova, S.N., Mandzhieva, S., Singh, R., Gorovtsov, A., Tsitsuashvili, V.S., Purvis, W.O., Ghazaryan, K.A. 2018b. Effects of zinc-oxide nanoparticles on soil, plants, animals and soil organisms: a review. *Environmental Nanotechnology, Monitoring & Management* 9, 76–84.

Rashid, M.I., Shahzad, T., Shahid, M., Ismail, I.M.I., Dhavamani, J., Basahi, J.M., Almeelbi, T. 2017a. Toxicity of iron oxide nanoparticles to grass litter decomposition in a sandy soil. *Scientific Reports* 7, 41965.

Rashid, M.I., Shahzad, T., Shahid, M., Ismail, I.M.I., Shah, G.M., Almeelbi, T. 2017b. Zinc oxide nanoparticles affect carbon and nitrogen mineralization of *Phoenix dactylifera* leaf litter in a sandy soil. *Journal of Hazardous Materials* 324, 298–305.

Raskar, S., Laware, S. 2014. Effect of zinc oxide nanoparticles on cytology and seed germination in onion. *International Journal of Current Microbiology and Applied Sciences* 3, 467–473.

Ruffini Castiglione, M., Giorgetti, L., Geri, C., Cremonini, R. 2011. The effects of nano-TiO_2 on seed germination, development and mitosis of root tip cells of *Vicia narbonensis* and *Zea mays*. *Journal of Nanoparticle Research* 13, 2443–2449.

Shen, C.X., Zhang, Q.F., Li, J., Bi, F.C., Yao, N. 2010. Induction of programmed cell death in *Arabidopsis* and rice by single-wall carbon nanotubes. *American Journal of Botany* 97, 1602–1609. https://doi.org/10.3732/ajb.1000048

Shi, J., Ye, J., Fang, H., Zhang, S., Xu, C. 2018. Effects of copper oxide nanoparticles on paddy soil properties and components. *Nanomaterials* 8, 839.

Shoults-Wilson, W.A., Reinsch, B.C., Tsyusko, O.V., Bertsch, P.M., Lowry, G.V., Unrine, J.M. 2011. Effect of silver nanoparticle surface coating on bioaccumulation and reproductive toxicity in earthworms (*Eisenia fetida*). *Nanotoxicology* 5, 432–444.

Simonin, M., Richaume, A. 2015. Impact of engineered nanoparticles on the activity, abundance, and diversity of soil microbial communities: a review. *Environmental Science and Pollution Research* 22, 13710–13723.

Solaiman, A.S., El-Feky, S.A., Darwish, E. 2015. Alleviation of salt stress on *Moringa peregrina* using foliar application of nanofertilizers. *Journal of Horticulture and Forestry* 7, 36–47.

Sun, L., Song, F., Zhu, X., Liu, S., Liu, F., Wang, Y., Li, X. 2021. Nano ZnO alleviates drought stress via modulating the plant water use and carbohydrate metabolism in maize. *Archives of Agronomy and Soil Science* 67, 245–259.

Syu, Y.Y., Hung, J.H., Chen, J.C., Chuang, H.W. 2014. Impact of size and shape of silver nanoparticles on *Arabidopsis* plant growth and gene expression. *Plant Physiology and Biochemistry* 83, 57–64. https://doi.org/10.1016/j.plaphy.2014.07.003

Tang, Y., Tian, J., Li, S., Xue, C., Xue, Z., Yin, D., Yu, S. 2015. Combined effects of graphene oxide and Cd on the photosynthetic capacity and survival of *Microcystis aeruginosa*. *Science of the Total Environment* 532, 154–161.

Tiwari, E., Mondal, M., Singh, N., Khandelwal, N., Monikh, F.A., Darbha, G.K. 2020. Effect of the irrigation water type and other environmental parameters on CeO_2 nanopesticide–clay colloid interactions. *Environmental Science: Processes & Impacts* 22, 84–94.

Tripathi, D.K., Singh, S., Singh, S., Srivastava, P.K., Singh, V.P., Singh, S., Prasad, S.M., Dubey, N.K., Chauhan, D.K. 2017. Nitric oxide alleviates silver nanoparticles (AgNPs)-induced phytotoxicity in *Pisum sativum* seedlings. *Plant Physiology and Biochemistry* 110, 167–177.

Tumburu, L., Andersen, C.P., Rygiewicz, P.T., Reichman, J.R. 2015. Phenotypic and genomic responses to titanium dioxide and cerium oxide nanoparticles in *Arabidopsis thaliana* germinants. *Environmental Toxicology and Chemistry* 34, 70–83. https://doi.org/10.1002/etc.2756

Unrine, J.M., Tsyusko, O.V., Hunyadi, S.E., Judy, J.D., Bertsch, P.M. 2010. Effects of particle size on chemical speciation and bioavailability of copper to earthworms (*Eisenia fetida*) exposed to copper nanoparticles. *Journal of Environmental Quality* 39, 1942–1953.

Van Hai, L., Son, H.N., Seo, Y.B. 2015. Physical and bio-composite properties of nanocrystalline cellulose from wood, cotton linters, cattail, and red algae. *Cellulose* 22, 1789–1798.

Vankova, R., Landa, P., Bellinvia, P., Špíšek, R., Tylová, E., Vágner, M., Vaněk, T. 2017. ZnO nanoparticle effects on hormonal pools in *Arabidopsis thaliana*. *Science of the Total Environment* 593, 535–542. https://doi.org/10.1016/j.scitotenv.2017.03.160

Vishwakarma, K., Upadhyay, N., Kumar, N., Tripathi, D.K., Chauhan, D.K., Sharma, S., Sahi, S. 2018. Potential applications and avenues of nanotechnology in sustainable agriculture. In: Tripathi, D.K., Ahmad, P., Sharma, S., Chauhan, D.K., Dubey, N.K. (eds), *Nanomaterials in Plants, Algae, and Microorganisms*, pp. 473–500. Academic Press, Cambridge, MA.

Wang, J., Zhang, X., Chen, Y., Sommerfeld, M., Hu, Q. 2015. Influences of zinc oxide nanoparticles on *Allium cepa* root cells and the primary cause of phytotoxicity. *Environmental Science and Pollution Research* 22, 19227–19236.

Wang, R., Sun, L., Zhang, P., Wan, J., Wang, Y., Xu, J. 2023. Zinc oxide nanoparticles alleviate cadmium stress by modulating plant metabolism and decreasing cadmium accumulation in *Perilla frutescens*. *Plant Growth Regulation* 100, 85–96.

Wu, F., You, Y., Werner, D., Jiao, S., Hu, J., Zhang, X., Wan, Y., Liu, J., Wang, B., Wang, X. 2020. Carbon nanomaterials affect carbon cycle-related functions of the soil microbial community and the coupling of nutrient cycles. *Journal of Hazardous Materials* 390, 122144.

Yadav, T., Mungray, A.A., Mungray, A.K. 2014. Fabricated nanoparticles: current status and potential phytotoxic threats. In: Whitacre, D.M. (ed), *Reviews of Environmental Contamination and Toxicology*, Vol. 230, pp. 83–110. Springer International Publishing, New York.

Yan, S., Zhao, L., Li, H., Zhang, Q., Tan, J., Huang, M., He, S. 2013. Single-walled carbon nanotubes selectively influence maize root tissue development accompanied by the change in the related gene expression. *Journal of Hazardous Materials* 246–247, 110–118.

Yang, J., Cao, W., Rui, Y. 2017. Interactions between nanoparticles and plants: phytotoxicity and defense mechanisms. *Journal of Plant Interactions* 12, 158–169.

Yusefi-Tanha, E., Fallah, S., Rostamnejadi, A., Pokhrel, L.R. 2020. Particle size and concentration dependent toxicity of copper oxide nanoparticles (CuONPs) on seed yield and antioxidant defense system in soil grown soybean (*Glycine max* cv. Kowsar). *Science of The Total Environment* 715, 136994. https://doi.org/10.1007/978-3-319-03777-6_3

Zhao, S., Su, X., Wang, Y., Yang, X., Bi, M., He, Q., Chen, Y. 2020. Copper oxide nanoparticles inhibited denitrifying enzymes and electron transport system activities to influence soil denitrification and N_2O emission. *Chemosphere* 245, 125394.

Index

Note: **Bold** page numbers refer to tables and *italic* page numbers refer to figures.

For Product Safety Concerns and Information please contact our EU representative GPSR@taylorandfrancis.com
Taylor & Francis Verlag GmbH, Kaufingerstraße 24, 80331 München, Germany

www.ingramcontent.com/pod-product-compliance
Ingram Content Group UK Ltd.
Pitfield, Milton Keynes, MK11 3LW, UK
UKHW020202200926
475702UK00004B/571
9781032659404